A Modern Introduction to
Classical
Number Theory

Revised Edition of The Book of Numbers

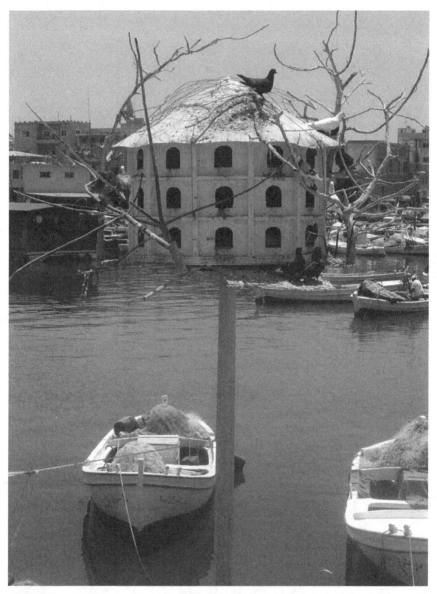

The bird-cage of Tyre, photo taken by the author in Lebanon. It was a Phoenician port, the hometown of Pythagoras and the birthplace of number theory.

The cover photograph was taken by the author at the former residence of the Pan family in Suzhou, renowned in China and around the world for its gardens. Professor Chengdong Pan, the author's doctoral supervisor, was a member of this famous lineage.

A Modern Introduction to
Classical Number Theory

Revised Edition of The Book of Numbers

Cai Tianxin

Zhejiang University, China

World Scientific

NEW JERSEY · LONDON · SINGAPORE · BEIJING · SHANGHAI · HONG KONG · TAIPEI · CHENNAI · TOKYO

Published by

World Scientific Publishing Co. Pte. Ltd.

5 Toh Tuck Link, Singapore 596224

USA office: 27 Warren Street, Suite 401-402, Hackensack, NJ 07601

UK office: 57 Shelton Street, Covent Garden, London WC2H 9HE

Library of Congress Cataloging-in-Publication Data
Names: Cai, Tianxin, 1963– author.
Title: A modern introduction to classical number theory / Tianxin Cai, Zhejiang University, China.
Other titles: Book of numbers
Description: Revised edition. | New Jersey : World Scientific, [2021] |
 Revised edition of: The books of numbers. 2017. | Includes bibliographical references.
Identifiers: LCCN 2021015352 | ISBN 9789811218293 (hardcover) |
 ISBN 9789811218309 (ebook for institutions) | ISBN 9789811218316 (ebook for individuals)
Subjects: LCSH: Number theory--History. | Numeration--History. | Numbers, Natural--History.
Classification: LCC QA141.2 .C35 2021 | DDC 512.7--dc23
LC record available at https://lccn.loc.gov/2021015352

British Library Cataloguing-in-Publication Data
A catalogue record for this book is available from the British Library.

For photographs/figures not taken by the author, they were obtained from the wikipedia site. While every effect has been made to contact the owners of the photographs/figures prior to publication, we have not been successful in some cases. Proper credit will be accorded to these copyright owners in future edition of this work.

Preface translated by Professor Jing Du.

For any available supplementary material, please visit
https://www.worldscientific.com/worldscibooks/10.1142/11771#t=suppl

Desk Editor: Liu Yumeng

Typeset by Stallion Press
Email: enquiries@stallionpress.com

Printed in Singapore

Au fond de l'inconnu pour trouver du nouveau

— Charles Baudelaire

Preface

The most classical is also the most modern. — Azra Pound
True laws of nature cannot be linear. — Albert Einstein

Needless to say, this is a book about integers (sometimes also involving rational numbers), the content of which everyone can understand wholly or partly with a secondary or higher education. One day in the early spring of 2013, when I was preparing for my Elementary Number Theory class, I accidentally realized that our human body contains mysterious prime numbers. In fact, under normal circumstances, in our hands, each thumb has two joints, the

Left: The cylindrical seal of Sumerian 4000 years ago. Photo taken by the author in Baghdad. Right: Clay tablet no. 7289, 18 BC–16 BC, Yale University. $\sqrt{2}$ represented by Babylonian in sexagesimal, accurate to 6 decimal places

other finger has three joints, and every hand has five fingers. That is, each of our hands is related to the minimal three primes (the number on the top and bottom tiles and the number on the front cover also happen to be these three prime numbers). This discovery made me so excited that for some time, I mentioned the mystery to everybody I met. I also thought of the words of the Russian painter Kandinsky: "Numbers are the final abstract expression of all kinds of art."

Natural numbers are the oldest among the creations and inventions that are essential to human beings, and number theory on the properties of natural numbers or integers is also the oldest in the branches of science that are still active so far. In the pre-agricultural society, still burrowing in the cave, people had livestock as their possessions, and cattle and sheep grazed out every day must be counted after returning. Therefore, counting was necessary for human survival; it was above all a means of livelihood. Later, only with the economic and social development, it gradually grew to be a science or an art — number theory. This book retains the basic content of elementary number theory, which not only allows us to get a glimpse of its growth process, but also helps to accommodate some of the new ideas, and furthermore, it makes beginners and enthusiasts easier to enter the subject; at the same time, we constantly refine and reveal the mystery of natural numbers.

Application of the Euclidean algorithm in the 3rd century BC to knot theory in the late 20th century

Nearly a century ago, the American-born British mathematician Mordell wrote in an essay, "Number theory is unique because of integers and a variety of conclusions, and because of the beauty and richness of the argument. Higher arithmetic (number theory) appears to contain most of the romance of mathematics. As (prince of mathematics) Gauss wrote to Sophie Germain (a French businessman's wife who loved primes) in a letter, 'Such pure research will show the most charming magic only to those who have the courage to explore her.'" Perhaps one day, the world's gold and diamond mining will be exhausted, but number theory is an inexhaustible treasure.

In a sense, number theory is the cause of young people. In 1801, the 24-year-old German youth Gauss published *Disquisitiones Arithmeticae* as his debut, beginning a new era of number theory. This great book had been posted to the French Academy of Sciences and unnoticed, but a friend of Gauss funded for its publication. At the end of the century, a mathematics historian Cantor commented: "'Arithmetic Research' is the Charter of number theory … Gauss' publication is a Code higher than any other human invention, because no mistake has ever been found anywhere in it any time." Gauss himself also praised, "Mathematics is the Queen of science; number theory is the Queen of mathematics."

In his book's opening paragraph, Gauss defined congruence that is so easy to understand and has a stake in our daily life: any two integers a and b are considered to be congruent modulo n, if their difference $a - b$ is divisible by n. Gauss first introduced the notation for congruence by using the symbol "\equiv" to indicate congruence.

The sum of numbers in each line is equal to the Fibonacci number on the left

Thus, the above definition can be expressed as

$$a \equiv b \pmod{n}.$$

With this convenient congruence notation, a number theory textbook appears to be concise and beautiful. Today, basic number theory textbooks mostly start with the introduction of division or divisibility. Divisibility and congruence constitute the two major contents of this book. In fact, the division with a remainder

$$a = bq + r, \quad 0 \le r < b$$

is equivalent to the congruence $a \equiv r \pmod{b}$ (the resulting Euclidean algorithm has its own original stories and names in different civilizations of China, India and Greece, etc.)

Next, linear Diophantine equations, exponents, primitive roots, or indices are all related to congruence, let alone the first, second, and nth-order remainder. Moreover, most famous theorems in elementary number theory are also related to congruence, except for the fundamental theorem of arithmetic. Their examples are the Euler–Fermat theorem, the Wilson–Gauss theorem, the Lagrange theorem, the Wolstenholme theorem, the Lucas theorem, the Kummer theorem, the von Staudt–Clausen theorem, and the Chinese remainder theorem, which should be called the Qin Jiushao-Sun Zi theorem, or the Qin Jiushao theorem. This book names the "Qin Jiushao theorem" for the first time not only to restore the truth, but also in line with the academic practice. We also take the opportunity to point out that the public key cryptography system relies on not only Euler's theorem, but also a technique closely related to the Dayan method from Qin Jiushao. We also expect that Qin Jiushao's scientific achievements, legend, and controversial life can attract such kind of directors as Ang Lee to shoot his biographical film.

In this book we described Gauss' quadratic reciprocity law that he called an "arithmetic gem," which he was most proud of and spent a lot of effort to demonstrate repeatedly (8 times). Let p and q be two different odd prime numbers. Then

$$\left(\frac{p}{q}\right)\left(\frac{q}{p}\right) = (-1)^{\frac{(p-1)(q-1)}{4}}.$$

Here $\left(\frac{a}{p}\right)$ is the Legendre symbol; it takes either 1 or -1 to indicate that the quadratic congruence $x^2 \equiv a \pmod{p}$ has a solution or no solution, respectively. The result is perfect, which will be proved with geometric and algebraic methods respectively. In Section 14, we introduced a newly discovered congruence that has the same beautiful symmetry. Let p and q be two different odd prime numbers. Then

$$\binom{pq-1}{(pq-1)/2} \equiv \binom{p-1}{(p-1)/2}\binom{q-1}{(q-1)/2} \pmod{pq},$$

where $\binom{m}{n}$ are binomial coefficients for all integers $m \geq n$. Such coefficients can be arranged as the Pascal triangle, and in the East,

Snooker pool invented by English soldiers in India in the 19th century. It consists of semi-circle and triangle number 21

such as China, both of the Jia Xian triangle or the Yang Xian triangle occurred earlier. The name of binomial coefficients is undoubtedly related to the expanded binomial coefficients from $(x + y)^m$, a job done by Newton, which is an important contribution that he made to number theory.

Later on, Gauss gave two proofs on the existence of primitive roots and discussed them, one of which is a constructive proof. Although this theory was rather complete, it still needs a supplement (such as the congruence of the product and the sum of primitive roots), and the current book will have innovative contributions to this aspect. Speaking of the existence of primitive roots, one needs the congruence module prime powers. In this book we not only explain the secret relation between Gauss' theorem (the generalization of Wilson's theorem) and the existence of primitive roots, but also give a lot of new results on the congruence module prime powers or even integer powers, including the generalization of Lehmer's congruence, Morley's theorem and Jacobsthal's theorem, the first of which was the main tool for the study of Fermat's last theorem until Wiles' proof came out. As two colleagues from Canada and Ireland have pointed out, this generalization (referring to from module prime powers to module integer powers) is the first time attempt since 1906. As another example, let n be an arbitrary odd integer, we have proved that

$$(-1)^{\phi(n)/2} \prod_{d|n} \binom{d-1}{(d-1)/2}^{\mu\left(\frac{n}{d}\right)} \equiv 4^{\phi(n)} \begin{cases} (\bmod n^3) & \text{if } 3 \nmid n, \\ \left(\bmod \dfrac{n^3}{3}\right) & \text{if } 3 \mid n, \end{cases}$$

where $\phi(n)$ and $\mu(n)$ are the Euler function and the Möbius function, respectively. When n is a prime number p, that is just the famous Morley theorem:

$$(-1)^{\frac{p-1}{2}} \binom{p-1}{(p-1)/2} \equiv 4^{p-1} \pmod{p^3}, \quad p \geq 5.$$

The quadratic form is the main topic of Gauss' works, especially the problem of representation of integers. In 1770, the French mathematician Lagrange proved that every natural number can be

expressed as a sum of four integer squares. In the same year, the British mathematician Waring extended it to the problem of representing integers as sums of integer powers, which is the famous Waring problem, with regard to which this book also has unique descriptions and discoveries. Let k and s be positive integers and consider the Diophantine equation

$$n = x_1 + x_2 + \cdots + x_s,$$

where

$$x_1 x_2 \ldots x_s = x^k.$$

By Hilbert's demonstration in 1909, there must exist $s = s'(k)$ such that any positive integer n can be expressed as the sum of no more than s positive integers, the product of which is the kth power of some integer. Denote by $g'(k)(G'(k))$ the smallest positive integer s such that the above equation holds for any (sufficiently large) positive integer n. We have given the exact value of $g'(k)$ and an estimate of $G'(k)$ with a conjecture that $G'(3) = 3, G'(4) = 4$. A more sophisticated conjecture is that, except for $2, 5,$ and 11, each prime number can be the sum of three positive integers whose product is a cube. These results and conjectures are more concise and beautiful than the Waring question, but it is equally difficulty to prove these results and conjectures.

The reason why we have been able to raise such questions is that we can consider the addition and multiplication of integers together, which is inspired by the format of the abc conjectures that can easily lead to Fermat's last theorem and a series of well-known conjectures and theorems, the influence of which in number theory has quickly replaced Fermat's Last Theorem proved already. In fact, the problems of perfect numbers and amicable numbers are a combination of these two basic operations; they have a long-lasting charm. It is exactly from here that our imagination is lifted, gradually getting out of the idea of congruence.

In addition to the Waring problem, we have also generalized the well-known Fermat's Last Theorem (Section 32), that is, we

considered the positive integer solutions of the Diophantine equation

$$\begin{cases} a + b = c, \\ abc = x^n. \end{cases}$$

When $n = 2$, the Pythagorean theory of integer arrays implies that there are infinitely many solutions. Assume that $n \geq 3$ and $d = \gcd(a, b, c)$. Clearly, if $d = 1$, then the above equation is equivalent to the Fermat equation $x^n + y^n = z^n$. When $d > 1$, there are all possibilities for the solvability of the equation. For example, if $d = 2$ and $n = 4$, then there is a solution $(a, b, c; x) = (2, 2, 4; 2)$. If $d = 7$ and $n = 4$, then there is a solution $(a, b, c; x) = (17, 272, 289; 34)$. We were reminded that there might be counterexamples when $n = 9$, but so far we haven't found one.

After some exploration, we can obtain several results and also put forward a new class of problems. One conjecture among them is that, if d and n are odd primes, then the above equation has no solution.

THE COLLATZ CONJECTURE STATES THAT IF YOU PICK A NUMBER, AND IF IT'S EVEN DIVIDE IT BY TWO AND IF IT'S ODD MULTIPLY IT BY THREE AND ADD ONE, AND YOU REPEAT THIS PROCEDURE LONG ENOUGH, EVENTUALLY YOUR FRIENDS WILL STOP CALLING TO SEE IF YOU WANT TO HANG OUT.

The 21st century mathematical puzzle: the $3x + 1$ problem

This is entirely a generalization of Fermat's Last Theorem, which cannot be deduced by the abc conjecture. Taking this opportunity, we have also shown the relationship between the two well-known generalizations — the Fermat–Catalan conjecture and Bill's conjecture (offered a reward of one million US dollars) and the abc conjecture. In addition, we have also studied the solvability of the new Fermat's equation in the quadratic field $\mathbb{Q}(\sqrt{t})$. For example, the original Fermat's equation has no solution in $\mathbb{Q}(\sqrt{-3})$.

In 1796, the 19-year-old Gauss discovered the Euclidean graphing theory for the regular 17-gon (only with a compass and no scale ruler), which had been an open problem of mathematics for over 20 centuries. In the same year, Gauss also proved that each natural number can be expressed as the sum of three triangular numbers. This was the first affirmative answer to another problem of Fermat's, who wrote the 18th note in the margin of the Diophantine *Arithmetic* of the Latin version:

Each natural number can be expressed as the sum of n n-gonal numbers.

The so-called n-gonal number is the number represented as dots arranged in the shape of a regular polygon, as defined by the Pythagorean school. In particular, the triangular numbers are $0, 1, 3, 6, 10, 15, 21, \ldots$ that are all binomial coefficients of the form $\binom{x}{2}$, in which 10 and 21 are, respectively, the numbers of bowling pins and snooker's object balls.

We have noted that the binomial coefficients are of special and wonderful nature, which are the simplest integers next to the powers, so they are worthy of particular attention from number theorists. Binomial coefficients and polynomial coefficients will appear several times in this book, each time being of stunning presence. For example, to test if a positive integer n is a prime, we need only to check if

$$\binom{n-1}{s} \equiv (-1)^s \pmod{n}$$

holds for all s with $0 \leq s \leq \sqrt{n}$. It is sometimes more convenient than testing by using Wilson's theorem. As another example, supposing

p is a prime number, we have found that the Lucas-type congruence

$$\binom{p-1}{s} \equiv (-1)^s \pmod{p^2}$$

has a solution $1 < s < p - 1$ if and only if

$$p \mid s! \left(1 + \frac{1}{2} + \cdots + \frac{1}{s}\right).$$

It is not difficult to find that the collection of prime numbers no more than 100 that satisfy the above condition is $\{11, 29, 37, 43, 53, 61, 97\}$. The ratio of number of such prime numbers over the number of all the prime numbers among $1, 2, \ldots, x$ may approach a constant (approximately $0.4\cdots$) as x goes to infinity, but we are still unable to determine whether there are infinitely many such primes. Anyway, it is meaningful to classify prime numbers under different situations.

As a further example, Gauss once proved that, when $p = 2$ or a prime that is congruent to 3 modulo 4, the congruence

$$n \equiv x^2 + y^2 \pmod{m}$$

has a solution for all integers n if and only if m is not divisible by the square of such p. We have found that for any positive integer m, the congruence

$$n \equiv \binom{x}{2} + \binom{y}{2} \pmod{m}$$

has a solution for all integers n.

As another example, let's consider the positive integer solutions of the Diophantine equation

$$\begin{cases} n = a + tb, \\ ab = \binom{c}{2}. \end{cases}$$

This is a variant of the quadratic form problem. We have concluded that, when $t = 1$, the equation has a solution if and only if $2n^2 + 1$ is a composite number; if $t = 2$, then the equation has a solution if and only if $n^2 + 1$ is a composite number. When t is some odd prime

number p, the situation will be more subtle. This not only touches on the problem of representing integers with quadratic forms, but also provides new clues for us to judge the authenticity of primes. In summary, after introducing binomial coefficients and addition–multiplication equations, we have added new research topics into the theory of quadratic forms and many other number theory problems.

As Gauss noted, "numbers are just creation of our mind", we have given new concepts to natural numbers. For example, we have defined figurate primes $\binom{p^i}{j}$, in which p is a prime, and i and j are positive integers. These are a special kind of binomial coefficients, possessing the characteristics of primes and figurate numbers, and their number and the number of prime numbers are equivalent in the sense of infinity. After some exploration, we conjectured (verified up to 10^7; if limited to positive integers for the figurate primes, then the conjecture is true for all integers greater than 1):

> *Each natural number can be represented as the sum*
> *of two figurate primes.*

The above conjecture is no doubt inspired by the Goldbach conjecture, which states that each odd (even) number greater than or equal to 9(6) can be expressed as the sum of 3(2) odd prime numbers. In our view, the statement of the Goldbach conjecture lacks consistency, and while primes constitute the basic unit integers for multiplication, they may not be the best choice for addition. On the other hand, as the number increases, the number of representations in the Goldbach problem is also growing, quite a waste. Further, if non-prime figurate primes are defined as true figurate primes, then we have the following stronger conjecture (if figurate primes are limited to be positive, then the conjecture is true for all integers greater than 5):

> *Any integer greater than 5 can be expressed as the sum*
> *of a prime number and a true figurate prime.*

At the same time, we also had the following conjecture which might be more difficult to prove than the twin prime conjecture:

> *There exist infinitely many pairs of neighboring figurate primes.*

However, we are not able to prove the most fundamental claim:

All the figurate primes are distinct.

It is worth mentioning that, regardless of the Goldbach conjecture or the twin prime conjecture, by far the best results were achieved by Chinese mathematicians (Jingrun Chen and Yitang Zhang).

Taking this opportunity, we look back at the last paragraph of Hilbert's 8th problem. He said: "After a thorough discussion of the Riemann prime number formula, perhaps we are able to solve the Goldbach question rigorously, namely whether every even number can be written as the sum of two prime numbers, can further address whether there are infinitely many prime numbers of difference 2, and even solve the more general problem, namely whether there are always prime solutions (x, y) of the linear Diophantine equation $ax + by + c = 0$, $(a, b) = 1$."

After introducing the concept of figurate primes, we try to make linear Diophantine equation, which Hilbert cared, meaningful, while combining the Goldbach conjecture and the twin prime conjecture together. We conjecture (its latter part can be deduced from the properties of Diophantine equations and the Schinzel conjecture): If a and b are two given positive integers that are relatively prime, then for every integer $n \geq (a-1)(b-1)$, the equation

$$ax + by = n$$

has figurate prime solutions (x, y); and as long as $n \equiv a + b \pmod 2$, the equation

$$ax - by = n$$

has infinitely many prime number solutions (x, y).

This book was born out of my textbook *Number Theory, from the Viewpoint of Congruence* and *the Book of Numbers*, after polishing and revisions, and can be viewed as a commentary or a revised book, the new title obviously comes from *A Classical Introduction to Modern Number Theory* by Kenneth Ireland and Michael Rosen. The author has been inspired by Fermat's comments of *Arithmetic* by Diophantus. Almost every chapter of the book has a new discovery, and each section contains new material. The first five parts

constitute the main content of basic number theory, and the material of Chapter 7 is almost brand new. A major feature of this book is the supplementary reading after each section, which is the author's attempt with the hope of extending readers' knowledge, lifting their imagination, and increasing their interest in and love of number theory. These supplementary readings are at least of two functions.

First, we presented the preliminary knowledge of other problems in number theory and research, such as Euler numbers and Euler prime numbers, the Hadamard matrix and Egypt fractions, the Pell equation and the Diophantine m-tuple, the Artin conjecture and special exponential sums, elliptic curves and the congruent number problem, the Goldbach conjecture and the twin prime problem, the abc conjecture and the BSD conjecture, automorphic forms and modular forms, and so on. Among them, the theory of elliptic curves is an important tool for us (especially in Chapter 7). Second, we introduced new questions and conjectures related to each other in basic number theory. In addition to the aforementioned, we included Graham's conjecture, the $3x + 1$ problem, the generalized Euler function, covering congruence systems, prime number chains and composite number chains, Catalan's conjecture, non-power polynomial coefficients, and so on.

The so-called Euler (prime) numbers are a kind of generalization of Fermat (prime) numbers. There are many mysteries about them, which may offer clues for finding the 6th Fermat prime. Another interesting generalization is the *GM* number named by the later Colombian writer Garcia Marquez (his masterpiece, *Hundred Years of Solitude*, is reminiscent of those unanswered number theory problems of centuries). We define a positive integer

$$s = 2^{\alpha} + t$$

a *GM* number, where t is the sum of the proper factors of s and α is a non-negative integer. Obviously, odd primes in the *GM* numbers are the same as Fermat primes. In addition, we have found two odd *GM* numbers that are not prime; they are 19649 and 22073325, which are the only such numbers up to 2×10^9, and the second one is much greater than the 5th Fermat prime. As for the even GM number, we believe that there are infinitely many.

Elliptic curves, the symbol of modern number theory

Let's look at the $3x+1$ problem, which is easy to verify by a child who knows elementary arithmetic but also a well-known world-level difficult problem. Erdös even believed that it could not be attacked with existing mathematical methods. Moreover, it was also claimed that this problem cannot be extended, and the corresponding problems $qx+1$ or $3x-1$ do not exist. We have found that the latter is equivalent to

$$
g_c(x) = \begin{cases} \dfrac{x}{2} & \text{if } x \equiv 0 \ (\mathrm{mod}\, 2), \\[2ex] \left[\dfrac{c+2}{c+1} x \right] + 1 & \text{if } x \equiv 1 \ (\mathrm{mod}\, 2), \end{cases}
$$

when $c = 1$. Here $c \geq 1$, $[x]$ represents the largest integer not exceeding x. When c is even, $c + 2$ provides a counterexample. There is also a counterexample $x = 37$ when $c = 3$, that is, there is the cycle $\{37, 47, 59, 74\}$. When c is an odd number greater than 3, the iteration of $g_c(x)$ can make any integer x back to 1. We have verified it for all even numbers $c \leq 50000$ and $x \leq 10^7$.

This book also studied the oldest mathematical problem — the perfect number problem, which was possibly initiated by the Pythagorean school in ancient Greek. In 2000, the perfect number problem was put by Italian mathematician Odifredi as the first of four unsolved mathematical problems. Euclid gave a detailed description

in his Elements, the perfect numbers are those natural numbers that
are equal to the sum of their proper factors, that is

$$\sum_{d|n,d<n} d = n.$$

For example, $6 = 1+2+3$ and $28 = 1+2+4+7+14$. So 6 and 28 are
perfect numbers. After the efforts of Plato's student Archytas, who
probably invented the kite, and Euler (2200 years later), a necessary
and sufficient condition was found for an even number n to be a
perfect number:

$$n = 2^{p-1}(2^p - 1),$$

where p and $2^p - 1$ are primes. Such primes $2^p - 1$ are also called
Mersenne primes. With the help of the most advanced computer
networking, i.e., GIMPA, people have found a total of 51 Mersenne
primes. In other words, we have known 51 even perfect numbers,
but still do not know whether there are infinitely many of them, or
whether there exists an odd perfect number.

At the same time, mathematicians, including Fibonacci,
Descartes, Fermat, and Euler, considered a natural generalization
to the equality $\sum_{d|n,d<n} d = kn(k > 1)$ (called a kth-order perfect
number), but only got fragmented results. In this book, we consider
the situation of square-sum perfect numbers, namely

$$\sum_{d|n,d<n} d^2 = 3n. \tag{F}$$

We have found that all the solutions of (F) are (no matter whether
n is odd or even) as follows:

$$n = F_{2k-1}F_{2k+1} \quad (k \geq 1),$$

where F_{2k-1} and F_{2k+1} are Fibonacci twin primes (their subscripts
are naturally also twin primes). This number sequence satisfies $F_0 =
0, F_1 = 1, F_n = F_{n-2} + F_{n-1}(n > 1)$.

By employing our currently available computing technology, we
have found five such perfect numbers, the first two of which are 10
and 65, but the latter two are astronomical. We call classic perfect
numbers as M perfect numbers, and square-sum perfect numbers as

F perfect numbers. It must be noted that 10 is the only even F perfect number. And if we can improve the performance of computers and computing skills, it should be possible to discover more and bigger perfect numbers. It is worth mentioning that the page numbers of the seven illustrations in the preface are precisely arranged according to the Fibonacci sequence. That is, their page numbers are v, v, vi, vii, ix, xii, and xvii (which are pages $1, 1, 2, 3, 5, 8,$ and 13, respectively of the preface).

Square-sum perfect numbers are of nonlinear fashion, as Eisenstein pointed out in his *Autobiographical Notes* that true laws of nature cannot be linear. Recently, we have found more profound and wonderful laws. Let s be any positive integer and consider the equation

$$\sum_{d|n,d<n} d^2 = L_{2s}n - F_{2s}^2 + 1. \qquad \text{(LF)}$$

We have found all of its solutions, which are $F_{2k+1}F_{2k+2s+1}(k \geq 1)$ except for finitely many computable solutions. Here L_{2s} are the Lucas numbers, $L_0 = 2$, $L_1 = 1$, $L_n = L_{n-2} + L_{n-1}$ $(n > 1)$, F_{2k+1} and $F_{2k+2s+1}$ the Fibonacci primes. When $s = 1$, (LF) is (F).

In addition, we let the twin prime conjecture be equivalent to a problem of square-sum perfect numbers, that is, the twin prime conjecture is true if and only if the following equation

$$\sum_{d|n,d<n} d^2 = 2n + 5$$

has infinitely many solutions.

Chapter VII of this book is undoubtedly the most modern and most creative, in which we have made intensive observations and extensions to the five big problems (the Waring problem, Fermat's Last Theorem, Euler's conjecture, and the congruent number problem in addition to the perfect number problem), and got a lot of new results and conjectures. Since there are many problems and conjectures (sometimes more daring) that are incorporated with the personal research experience, especially ideas out of deep thinking in the recent years (some unpublished), there must be inevitable errors

(not being the words of politeness), so the readers are expected to discover and correct them.

In the last section (Section 36) of the book, we proposed the *abcd* equation that is original. Let n be a non-negative integer, and consider the equation

$$n = (a + b)(c + d),$$

where

$$abcd = 1.$$

Here, a, b, c, d are positive rational numbers. For what n, does the above equation have a solution? When solutions do exist, how many? This is both a wonderful and complex issue, and might be comparable to the congruence number problem. We have found an infinite number of n for which the equation is solvable, but also ruled out more unsolvable cases. Studying this problem not only needs skills from elementary number theory, but is also closely related to the theory of elliptic curves and the BSD conjecture.

Finally, we raise the following new Egyptian fraction problem: For any positive integer n, consider the Diophantine equation

$$\begin{cases} n = x + y + z, \\ xyz = \dfrac{1}{A}. \end{cases}$$

What is the least positive integer $A = A(n)$ such that the above equation has positive rational number solutions? In particular, for $n = 4$ or multiple of 4, it is difficult to determine the value of $A(n)$.

This book is also a summary of my teaching "Elementary Number Theory" and directing graduate students at Zhejiang University in the past 30 years. Its writing and translation into English were assisted by almost ten of my graduate students, undergraduates, and collaborators, and some of their research results are included in the book as well as those related to some domestic and international researchers. Unfortunately, due to our limited vision and knowledge, we may miss many treasures that have been found by number theorists, even though we have been quite aggressive in learning. It is worth mentioning that a lot of work in progress and predictions being

made were helped by computers, which is our advantage over the ancient researchers. We believe that the importance of computers to number theorists is like telescopes to astronomers.

On the occasion of publishing the book, I thank late Professor Alan Baker of Cambridge University, a Fields medalist, and Professor Preda Mihailescu of the University of Gottingen, who proved the Catalan conjecture, for their warm words in communications. Professor Baker called the new Waring problem "a truly original contribution to the subject." Professor Mihailescu encouraged the author "you have found your own territory in today's complex world of mathematics." In a survey article on the abc conjecture entitled Around ABC, which is contained in the *Newsletter of the European Mathematical Society*, he discussed the additive and multiplicative equations defined by the author in the special section Diophantine Variations, which he called the "yin-yang equations."

Number theory is lucky because she attracted many great mathematicians in the history, such as Pythagoras, Euclid, Qin Jiushao, Fibonacci, Descartes, Fermat, Leibniz, Euler, Lagrange, Gauss, and Hilbert. Among them, three were hermits, who wrote down their sophisticated discoveries quietly in a notebook. There were some other people who explored tirelessly the mysteries of natural numbers: Diophantus, Legendre, Waring, Dirichlet, Kummer, Lucas, Hardy, Mordell, Ramanujan, Erdos, Wiles; and some amateurs: Mersenne, Goldbach, Sophie Germain, Wilson, Paganini, Pollock, Paganini, and de Polignac. A few of them left their name in the history of mathematics by just doing a little bit.

Going over some of the problems raised in the book, especially Chapter 7, there is still a lot of fascinating work needing to be done. This should be the most valuable part of the book, and we expect it to blossom beautiful flowers and bear fruits. Since Fermat's Last Theorem was solved in 1995, there have been many pieces of exciting news in number theory: first the Catalan conjecture was proved in 2002, then the abc conjecture (2012) and the odd Goldbach problem (2013) were declared resolved (the former still has gaps to fill); there was also a major breakthrough towards the twin prime conjecture (2013) . On the other hand, this is a double-edged sword; it sounds

an alarm to the number theory community: the chickens that lay golden eggs will become fewer and fewer. Hopefully, the publication of the book would be a breath of fresh blood, and could give birth to one or two chicks.

Finally, I would like to quote Gauss to end the Preface. In the Introduction of the collected works of his late student Eisenstein, he said: "Number theory provides us with an inexhaustible treasure trove, full of interesting stored truths. These truths are not isolated, but most closely linked to each other. With every success of this science, we continue to find new and sometimes totally unexpected start. The special charm of arithmetic theory mostly comes from important propositions by easy induction. These propositions have simple expressions, but their proofs are buried deeply. After numerous futile efforts we were able to dig them out; even if we succeed through lengthy artificial means, more fresh and natural proofs are still hidden."

Tianxin Cai
West Brook, Hangzhou, China, Autumn 2020

Contents

Preface vii

1. The Division Algorithm 1

 1. The Origin of Natural Numbers 1

 Perfect numbers and amicable numbers 7

 2. The Mystery of Natural Numbers 11

 Mosaic geometry and Euler's characteristic 17

 3. The Division Algorithm 19

 Mersenne primes and Fermat primes 26

 4. The Greatest Common Divisor 33

 Graham's conjecture 38

 5. The Fundamental Theorem of Arithmetic 40

 Hilbert's 8th problem 47

 Exercises 1 . 55

2. The Concept of Congruence 57

 6. The Concept of Congruence 57

 Gauss's Disquisitiones Arithmeticae 61

 7. Residue Classes and Residue Systems 64

 Function $[x]$ and the $3x + 1$ problem 69

 8. The Fermat–Euler Theorem 75

 The Euler number and the Euler prime 83

 9. The Fraction Expressed as Repeating Decimal 84

 Möbius' function . 88

10. Application in Cryptology 93
 The generalized Euler function 97

3. Congruences 103

 11. Qin Jiushao's Theorem 103
 Fibonacci's rabbits 109
 12. Wilson's Theorem 114
 A theorem that Gauss did not prove 123
 13. Diophantine Equation 126
 The Pythagorean triple 132
 14. Lucas' Congruence 135
 Covering system 142
 15. The Truth of Primes 145
 The chain of primes or composite numbers 151
 Exercises 3 . 154

4. Quadratic Residue 157

 16. Quadratic Congruences 157
 Integers in the Gaussian ring 161
 17. The Legendre Symbol 165
 Representing integers as the sum of squares 171
 18. The Law of Quadratic Reciprocity 179
 N-gonal number and Fermat 181
 19. The Jacobi symbol 184
 The Hadamard matrix and Hadamard conjecture . . 189
 20. Congruences Modulo a Composite 191
 Constructibility of the regular 17-gon 194
 Exercises 4 . 197

5. Nth Power Residues 199

 21. Definition of Order 199
 Egyptian fraction 202
 22. The Existence of Primitive Roots 205
 Artin's conjecture 208
 23. The nth Power Residues 210
 Pell's equation 219

24. The Case of Composite Modulus 224
 Diophantine arrays 227
25. Dirichlet Character 230
 Three special exponent sums 235
Exercises 5 . 240

6. Congruence Modulo Integer Powers 243

26. Bernoulli Number and Bernoulli Polynomial 243
 Kummer's congruence 248
27. Wolstenholme Theorem 252
 Elliptic curve 257
28. Lehmer's Congruence 263
 The abc conjecture 273
29. Morley's Theorem and Jacobstahl's Theorem 278
 Automorphic form and modular form 288
30. Congruence on Harmonic Sum 292
 Multinomial coefficient of non-power 297

7. Additive and Multiplicative Equations 303

31. New Waring's Problem 303
32. New Fermat's Last Theorem 313
33. Euler's Conjecture 324
34. F-Perfect Number Problem 332
35. New Congruent Number Problem 342
36. The ABCD Equation 352

Solutions to Exercises 365

Appendix: The List of Prime Numbers Less Than 10000 389

Bibliography 393

Author Index 395

Chapter 1

The Division Algorithm

1. The Origin of Natural Numbers

Mathematics and poetry are no doubt the oldest invention by human beings. Ever there was a history of human beings, there were these two things. If we say that poetry originated from praying for the harvest by farmers, then mathematics came from the counting of livestock by herders. In view of this, they were both born for the survival needs. With the passage of time, human beings gradually had the definite concept of the integers: $1, 2, 3, \ldots$. These positive integers are known as natural numbers (later people also take 0 as a natural number). Apparently, they have the meaning of naturalness or from the natural world. However, the creation was not going smoothly.

At the beginning, because the property of livestock was so important, people used different quantifiers in the expression of different objects with the same amount. For example, in old English, a team of horses was used for two horses pulling a cart or a plow together, yoke of oxen for two conjugated cows, a span of mules for two mules, a brace of dogs for two dogs, and a pair of shoes for two shoes, etc. Slowly, only one phrase, a pair of, is used more commonly. While in Chinese, there are more varieties for quantifiers, and many of them are still in use even today.

After a long time, from numerous life experiences and social practices, human beings abstracted such numbers like 2 as a common property, which meant the birth of natural numbers. In fact, it is far

1

beyond this, as the British philosopher Bertrand Russell (1872–1970) pointed out,

> *When people found the common thing (number 2) between a pair of chicks and two days, then the mathematics was born.*

But in my opinion, mathematics might be born a little later, i.e., when people abstracted "2 + 3 = 5" from "3 eggs plus 2 eggs equals to 5 eggs, 3 pieces of arrows plus 2 pieces of arrows equals 5 pieces of arrows, etc." That is to say, when we have the operation of addition and subtraction for natural numbers, then the birth time of mathematics is perhaps thousands of years later than what was defined by Russell.

When people needed more extensive and profound communication about numbers, they had to systemize the counting method. Different nations around the world simultaneously adopted the following methods: choosing several consecutive natural numbers starting from 1 as a base and using their combinations to represent numbers bigger than them. In other words, they used systems. In different nations

Six apples in a triangular shape

Bamboo slips of Tsinghua around the 4th century, the earliest multiplication table in decimal in the world

and original tribes, the bases which they used and were well documented are 2, 3, 4, 5, 8, 9, 10, 12, 16, 20 and 60.

As the all-around Greek philosopher Aristotle (384–322 BC) said, that 10 was eventually adopted widely as a base of number system is just "because of the anatomic fact that the vast majority of people have 10 fingers." Of course, the hexadecimal number system invented by the ancient Babylonian people is still retained, as a kind of well-known time unit. Some people guess that the counting method of Babylon people came in this way, namely the use of 5 fingers of one hand with 12 joints of the other hand. As for the binary system, there is evidence that both Queensland natives in Australia and African dwarves used it, while in the ancient and profound Chinese book

I Ching, which appeared more than three thousand years ago, had hidden this mystery in the 64 hexagrams.

Next came the emerging and symbol of 0. The *Bakhshali manuscript* excavated in a village near Peshawar, India (now Pakistan) in 1881, recorded the mathematics of Jainism several centuries before and after Jesus. There was the complete decimal system, including the use of zero written as a solid point. No later than the 9th century, the Indians used the circle 0 instead of the solid point. After that, 0 together with other 9 digits were transferred to the West, through a modification by the Arabs, and another modification by the westerners, then spread throughout the world, and is called incorrectly as Arabia digital.

In the 17th century, the German mathematician and philosopher Gottfried Wilhelm Leibniz (1646–1716) established a strict binary system, before he invented the pinwheel calculator which could calculate the multiplication and the division. He presented the vacancy by 0, and real by 1. Then each natural number can be expressed by these two numbers. For example, $1 = 1$, $2 = 10$, $3 = 11$, $4 = 100$, $5 = 101, \ldots$. Unfortunately, Leibniz failed to connect the two things. Only in the middle of the 20th century, the Hungarian born American mathematician John von Neumann (1903–1957) made a series of important reforms in the computer program, that he designed by used the binary system instead of the decimal one, and thus he was known as "the father of modern computers."

For any integer $b > 1$, the number system with the base b was established long time ago. However, for a prime number p (defined in the third section), the p-adic number was a new concept just a century ago. In 1897, the German mathematician Kurt Hensel (1861–1941) extended the arithmetic of rational numbers in a way different from the method of the real or complex number systems, through the re-interpretation of the absolute value. For example, if $x = \frac{12}{5} = 2^2 \cdot 3 \cdot 5^{-1}$, then $|x|_2 = \frac{1}{4}$, $|x|_3 = \frac{1}{3}$, $|x|_5 = 5$, while for the other primes p, $|x|_p = 1$. Therefore, in the number system with the base 5, the following series is convergent:

$$S = 1 + 5 + 5^2 + 5^3 + \cdots.$$

This is because $|5^i|_5 = \frac{1}{5^i}$. Moreover, we can also find the value of S.

Multiple both sides by 5, we have

$$5S = 5 + 5^2 + 5^3 + \cdots.$$

Make the subtraction, we get $-4S = 1$, namely, $S = -\frac{1}{4}$.

However in this book, the integers we want to discuss is in the decimal system. In some oriental countries, there were examples finding the rules of numbers extracted from the nature, such as Babylonian and Chinese, they both found triples of squares, what our ancestors called the Gougu numbers or the ancient Greeks called the Pythagorean numbers. In China, the possible earliest mathematical book *Zhou Bi Suan Jing* (to 1st century BC), recorded a dialog on the measurement between the Duke of Zhou (11th century BC), the brother of the first emperor of the Western Zhou Dynasty, and Minister Shang Gao:

勾广三，股修四，径隅五。

It means that *if the short sides of a right triangle are 3 and 4, then the longer side is 5.*

This should be the first record of the minimum Pythagorean triple $(3, 4, 5)$. Although the Babylon clay tablet Plimpton 322, collected by Columbia University in New York, has a longer history, the smallest Pythagorean triple there is $(45, 60, 75)$. The Duke of Zhou is the one whom Confucius admired mostly in his lifetime and the author of *Zhou Li*, in which he first talked about "six arts" with numbers one of them. The *Zhou Bi Suan Jing* also described a conversation on measurement through the sun shadow, between Rong Fang, the descendant of the Duke of Zhou, and the scholar Chen Zi (about 6th or 7th century BC):

以日下为勾，日高为股，勾股各自乘，并而开方除之，得斜至日。

This is just the Gougu theorem, using modern mathematical language to express:

The sum of two right angle side squares of each right triangle is equal to the square of the hypotenuse.

However, the theorem is called Pythagorean theorem in the West by convention, as in the ancient time, the exchanges of culture, science and technology between the Eastern and Western were almost isolated, and Pythagoras (580–500 BC) was the first to give the proof. He was the first great ancient Greek mathematician, lived in the time a little later than Rong Fang and Chen Zi. It is worth mentioning that Pythagoras used a poetic language to describe the first theorem he found,

> *The square of the hypotenuse,*
> *If I'm not mistaken,*
> *Is equal to the sum of squares*
> *Of the other two sides.*

Pythagoras was born on an island of Samos (now Greece), which is located in the eastern end of the Mediterranean. He had traveled

The monument of Pythagoras in Samos

Site of Pythagora's school, photo taken by the author in Croton

to Egypt and Babylon, and later to Croton (now southern Italy), setting up a secret society there. He absorbed many of his disciples, forming the so-called Pythagoreans. Its effect lasted for two thousand years behind him. Pythagoras is the first to realize the importance of numbers in the objective world and in music, he believed that everything was related to mathematics and that numbers were the ultimate reality, or "everything is a number". This is his philosophical proposition. In all sorts of discoveries he and his disciples about the natural numbers, the most interesting might be perfect numbers and amicable numbers.

Perfect numbers and amicable numbers

Many civilizations and peoples have a religious faith or fantasy of natural numbers, perfect number is one of them. The so-called perfect number is the natural number n such that it is equal to the sum

of its real factors, i.e.,

$$\sum_{d|n,d<n} d = n.$$

Here $d|n$ denotes d divide n (cf. Definition 1.1). The least two perfect numbers are 6 and 28, since

$$6 = 1 + 2 + 3,$$

$$28 = 1 + 2 + 4 + 7 + 14.$$

Perhaps Pythagoras was the first studying the perfect numbers. No later than the 1st century, the Greeks knew the third perfect number 496 and the fourth perfect number 8128. Both Pythagoras and ancient Rome thinker Saint Augustin (354–430 AD) were identified 6 as perfect. Even in the first volume *Genesis of Old Testament,* it is mentioned that God created the world in 6 days (the seventh days is a day of rest). The ancient Greeks who believed the geocentric theory thought that the moon revolves around the earth every 28 days each time.

We will know in the third section that, an even number n is a perfect number if and only if it is in the following form:

$$n = 2^{p-1}(2^p - 1),$$

where both p and $2^p - 1$ are prime numbers. The latter is the famous Mersenne prime. That is to say, there are even perfect numbers as many as Mersenne primes, and vice versa. So far, people have found 48 Mersenne prime numbers, which is also the amount of even perfect numbers. On the other hand, no one could find one odd perfect number (if exists, it must be greater than 10^{1500}), or deny its existence. These problems make perfect numbers very attractive, we will give the proof of a sufficient and necessary condition in Section 3, and after the definition of Mersenne primes for composite numbers, we even raise a new form for this condition. Meanwhile, in Section 33 we will make a more in-depth study, which might be one of the most valuable parts in the book.

The so-called amicable numbers refer to a pair of numbers, each one of them equals the sum of the real factors of other one. Obviously, any perfect number is amicable to itself, so people usually only

consider a pair of different amicable numbers. Pythagoreans found the smallest amicable numbers, i.e., 220 and 284, because

$$220 = 1 + 2 + 4 + 71 + 142,$$

$$284 = 1 + 2 + 4 + 5 + 10 + 11 + 20 + 22 + 44 + 55 + 110.$$

In *Genesis*, it is also mentioned that Jacob gave 220 sheep to his twin brother Esau to show love. Later people added these numbers with mystery, which has been used in magic and astrology. During the Middle Ages in Europe, Syria (now Turkey) mathematician, the translator of Archimedes Sinan Ibn Thābit (826–901 AD) attributed to Pythagoras and his school the employment of amicable and perfect numbers in the illustration of their philosophy, he was the first to give a method to determine the amicable numbers. Thābit pointed out that, for any integer, $n > 1$, if

$$p = 3 \times 2^{n-1} - 1, \quad q = 3 \times 2^n - 1, \quad r = 9 \times 2^{2n-1} - 1$$

are all primes, then $(2^n pq, 2^n r)$ must be amicable numbers. When $n = 2$, it corresponds to the pair which Pythagoras discovered. It's a pity that the Arabs did not use this method to find new amicable numbers. In fact, there are two pairs which are as easy to get as just lifting a finger.

Only in 1636, the second pair of amicable numbers (17296, 18416) was found by "the king of amateur mathematicians", the French mathematician Pierre de Fermat (1601–1665). In the same year, his fellow mathematician and philosopher René Descartes (1596–1650) found the third one (9363584, 9437056). Before that, the Iranian Muhammad Baqir Yazdi, the last main mathematician in the Islamic world in the 16th century, found the latter. The two pairs are respectively correspond to the Thābit's array $n = 4$ and $n = 7$. In 1747, the Swiss mathematician Leonhard Euler (1707–1783) suddenly found more than 30 pairs of amicable numbers, by using his own method. He found totally more than 60 pairs, and one of them, (10744, 10856), is even smaller than that found by Fermat. However, the second smallest amicable numbers (1184, 1210) is found by a 16-year-old Italian boy Nicolò I. Paganini in 1866.

Statue of Thābit, Syrian mathematician

Euler also generalized the Thābit's formula for amicable numbers as follows. For any integers $n > m > 0$, if

$$p = (2^{n-m} + 1)2^m - 1, \quad q = (2^{n-m} + 1)2^n - 1,$$
$$r = (2^{n-m} + 1)^2 2^{m+n} - 1$$

are all primes, then $(2^n pq, 2^n r)$ must be a pair of amicable numbers. When $m = n - 1$, this is Thābit's formula, but Euler's formula only provides two new pairs of amicable numbers, i.e., $(m, n) = (1, 8)$ and $(29, 40)$.

So far, we have found amicable numbers many more than perfect numbers, about 12 millions. However, we do not know whether there are infinitely many amicable or perfect numbers. In 1955, the Hungarian mathematician Paul Erdős (1913–1996) showed that the density of amicable numbers is zero among natural numbers. And we would like to ask whether the ratio of the two numbers of each pair tends to 1 if there are infinitely many pairs of amicable numbers.

2. The Mystery of Natural Numbers

There are endless mystery among natural numbers, which makes number theory — the mathematical branch which takes care of the properties of the integers — full of charming, it attracts numerous mathematical genius of all ages and amateurs. As "Prince of mathematics", the German mathematician Karl Friedrich Gauss (1777–1855) said: "any one who spent a bit of time learning number theory, will feel particular passion and enthusiasm." Even the Russian painter Wassily Kandinsky (1866–1944) pointed out: "the numbers are the final abstract expression of all kinds of arts."

Andrey Kolmogorov (1903–1987) was the most influential mathematician of the former Soviet Union in the 20th century. His research interests are mainly in the theory of probability, harmonic analysis and dynamic system, but he was very sensitive on natural numbers and enjoyed the fun of discovering the relationship of numbers. As is well known, integers are divided into odd and even ones. At the age of 5, Kolmogorov found himself:

$$1 = 1^2,$$
$$1 + 3 = 2^2,$$
$$1 + 3 + 5 = 3^2,$$
$$1 + 3 + 5 + 7 = 4^2,$$
$$\vdots$$

That is to say, for any positive integer n, the sum of the first n odd numbers are squares, he then felt the magic of numbers. This conclusion is not difficult to get, but comparing with the legend that at the age of 9 Gauss quickly calculated the sum of arithmetic progression from 1 to 100, i.e.,

$$1 + 2 + \cdots + 99 + 100 = 5050,$$

the story is of more enlightening significance to teachers and parents.

Of course, a common (or even child prodigy) may not be able to discover intrinsic connections or rules between natural numbers, such as the twin prime conjecture or Goldbach's conjecture. This is why we must learn number theory, from which we can get a glimpse of the

magic of the natural numbers. It is worth mentioning that many of those simple yet profound theorems or formulas in this book, came from great minds.

When he talked about the turning of the master from the theory of algebraic Invariants to number theory, The biographer of the "last know-all of modern mathematics," the German mathematician David Hilbert (1862–1943), pointed out: that "no other branch in mathematics could attract the essence of mathematicians as number theory, with its beauty — an irresistible force." On the other hand, there are also some amateurs and little people who devote their talents and therefore leave their names in the history of mathematics. One example is the Italian boy Paganini who found the second smallest amicable numbers as we mentioned in the previous section. In the 12th section, we shall introduce Wilson's theorem, the casual discovery of which is by an undergraduate student at the University of Cambridge, he was later not engaged in mathematical research any more.

Now, we need to give the Peano axioms of natural numbers. This is proposed by the Italy mathematician Giuseppe Peano (1858–1932) in 1898: let \mathbb{N} (or \mathbb{Z}_+) be the set of all natural numbers, then (1) 1 is a natural number; (2) each natural number has a successor; (3) 1 is not the successor of any natural number; (4) the successors of different natural numbers are different; (5) if a subset S of \mathbb{N} contains 1 and the successor of any element belonging to S, then $S = \mathbb{N}$. The last axiom is the principle of the mathematical induction.

Since the natural numbers came from nature, there must be the rules of integers in the boundless universes. Pythagoras observed that when the blacksmith struck his anvil, different notes were produced according to the weight of the hammer. The number (in this case *amount of weight*) seemed to govern musical tone. By playing the harp, he tried to find the relationship between the pitch and the proportions of the two string lengths clicked and separated by fingers, then he realized that the integral proportion might produce wonderful harmony:

If the length of the strings is in the relation of 3:4 to each other, then the difference in pitch is called a fourth; if the length is in the

relation of 2:3, the difference is called a fifth: and if the length is in the relation of 1:2, the difference is called a eighth.

That is to say, the number (*amount of weight or space*) seemed to govern musical tone. As another example, no matter how to make the tune, no one can play the high eighth tone and the fifth tone accurately in the same piano at the same time. This is because that the frequency of two notes with the eighth difference in pitch is 2:1, and the frequency of two notes with the fifth difference in pitch is 3:1, while the following equation:

$$\left(\frac{3}{2}\right)^x = \left(\frac{2}{1}\right)^y$$

has no positive integer solution. In other words, the two ratios are incommensurable.

Between the area of some geometric figures (irrational or transcendental number), there are also interesting relationships with integer proportion. For example, in Sicily of the Mediterranean Sea, the ancient Greek mathematician Archimedes (287–212 BC) discovered that the ratio of the area of a parabolic (the image cut by a straight line parallel to the alignment) and the area of an isosceles triangle is 4 to 3 (see Figure 1).

He was also the first to point out that the surface area of an arbitrary sphere is as four times as the area of any great circle (with the center at the center of the sphere). Archimedes was not satisfied with the measurement of the known geometric figure. He was also interested in the complicated and non-intuitive curve which he found in the study of angle trisection, one of the three ancient Greek geometry problems. People of later generations called it the Archimedes spiral,

Fig. 1 The area of the parabola shape is 4/3 times that of its inscribed triangle

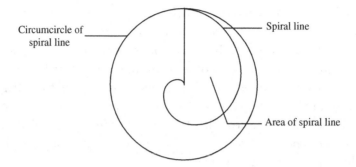

Circumcircle of
spiral line

Spiral line

Area of spiral line

Fig. 2 The area of spiral line is 1/3 of its circumcircle

whose polar coordinate equation is $r = a\theta$. As shown in Figure 2, the ratio of the area of the spiral and the graph enclosed by the vertical line through the starting point and the area of its circumcircle is 1 to 3.

The British man Isaac Newton (1642–1726) might be the greatest scientists in history, and also one of the greatest mathematicians. In addition to the invention of calculus, he also established the Newton binomial theorem. He generalized the binomial power from a positive integer to an arbitrary real number. The binomial coefficients are known as the Pascal triangle in the West (1653), but the earliest record might be from India. Chinese and Persians also found it in the early time. We call it Jia Xian triangle (c. 1050) or Yang Hui triangle (1261). Newton himself didn't make contribution to number theory, but our book will show that there hide many mysteries of natural numbers and primes in binomial coefficients.

The first four satellites of Jupiter — the largest planet in the solar system — were found by the Italian scientist Galileo Galilei (1564–1642) in 1610, which are also known as the Galileo satellites. Among them, the third is the largest one in our solar system, and its revolution period is about 7.15455296 days. Although the French mathematician and astronomer Pierre-Simon Laplace (1749–1827) made no major contribution to number theory as Archimedes, but he found some secret integer ratios between celestial bodies, i.e.,

$$T_3 = 2T_2 = 4T_1,$$

The geometrical figure on the wall of Palermo Cathedral, Sicily. Photo taken by the author

where T_1, T_2, T_3 are the revolution periods of the three satellites. This is known as the Laplace resonance. In his opinion, this problem is the perfect reflect of partition theory (a branch of number theory). Laplace also found that, if the average angular speeds are ω_1, ω_2, and ω_3, respectively, then

$$\omega_1 + 2\omega_2 = 3\omega_3.$$

We also want to mention that the 20 regions of Paris — the capital of France — are not only named in Arabia numbers (the first district including Musée du Louvre is located on the right bank of the Seine River), but also arranged according to the Archimedes spiral. I noticed that at the closing ceremony of the 2004 Athens Olympic Games, thousands of actors for the group calisthenics performance entered in along a spiral of Archimedes (see Figure 3).

In short, since Pythagoras concluded that everything is a number, more and more people observed that, the number of the world is closely combined with the real world, from the vast universe to

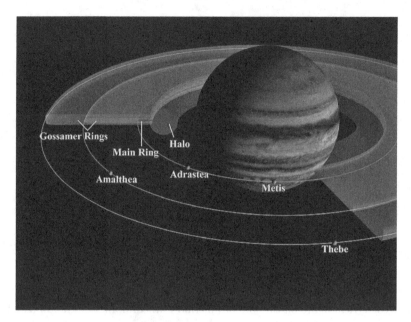

The Jupiter and its satellites

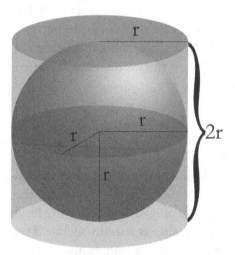

Fig. 3 Cylinder and its inscribed ball

tiny particles, the secrets of the natural numbers might hide everywhere. We have following two examples, they belong to geometry and topology, respectively.

Mosaic geometry and Euler's characteristic

In geometry, there is a very interesting problem of mosaic, i.e., filling a plane with congruent regular polygons. Let n be the number of edges of each polygon, then the internal angle of each vertex is $\theta_n = \frac{n-2}{n}\pi$. For example, $\theta_3 = \frac{\pi}{3} = 60°$, $\theta_4 = \frac{\pi}{2} = 90°$. If we want vertices of each regular polygon to meet only vertices of other regular polygons with the same amount of edges, then the number of polygons at each vertex must be

$$\frac{2\pi}{\theta_n} = \frac{2n}{n-2} = 2 + \frac{4}{n-2}.$$

If we want this number to be an integer, then n must be 3, 4 or 6. That is to say, we can only use the congruent triangles, squares, or hexagons to fill the plane. It is interesting that, in the 28th section of this book, when we discuss Lehmer's congruence which could help to show whether the first case of Fermat's last theorem is true, these three integers will appear again. And if we put the vertices of a polygon in the edges of polygons (such as midpoints), then the number of polygons assembled at each vertex will be reduced by half, that is, $1 + \frac{2}{n-2}$. Hence, n must be 3 or 4.

It is worth mentioning that, when the Holland graphic artist M.C. Escher (1898–1972) traveled to Granada in Spain in 1936, he was fascinated by the geometric patterns of mosaic tiles in Alhambra, and he said "This is the richest source of inspiration I have ever seen." Later he used these basic pattern in his paintings. Also he used the reflection, the transformation, and the rotation of geometry to get more patterns. He carefully made these patterns distorted into the shapes of various animals. They are amazing and beautiful.

In topology there is so-called Euler's characteristic (also known as Euler–Descartes' characteristic), describing the topology relation among the numbers of vertices, edges, and faces of geometric figure, which is defined as $V - E + F$. The conclusion is, for a simple

polygon in a plane ($F = 1$), the characteristic is 1; while the simple polyhedron in the three-dimensional space, the characteristic is 2, i.e.,

$$V - E + F = 2.$$

This formula appeared in a letter Euler wrote to the German mathematician Christian Goldbach (1690–1764) in 1750. A century later, people found that, as early as in 1635, Descartes wrote down this in his notebook.

The painting of Escher: The Limit of Circle (1959)

We can give many examples, such as the Pyramid (tetrahedron) in Cairo, Egypt and Kaaba (cube) in Mecca, Saudi Arabia. The numbers of the vertices, edges and faces are $(4, 6, 4)$ and $(8, 12, 6)$, respectively. With this formula one can also prove that there are only five kinds of regular polyhedrons, namely the so-called Platonic solid, the numbers of their faces were 4, 6, 8, 12, and 20.

Suppose that each face of a regular polyhedron has m sides, and each vertex connects n edges. Note that every two faces share one edge and each edge corresponds to two vertices, hence $mF = nV = 2E$.

Then $F = \frac{2E}{m}$ and $V = \frac{2E}{n}$. By Euler's formula, after eliminating E, we have $\frac{1}{m} + \frac{1}{n} = \frac{1}{2} + \frac{1}{E} > \frac{1}{2}$, so m and n cannot be both greater than 3. On the other hand, m and n are obviously no less than 3, so m or n must be 3. If $m(n) = 3$, then $n(m) = 3, 4$ or 5, which correspond to $E = 6, 12$ (twice), or 30 (twice). These five cases correspond exactly to the Platonic solids.

It is worth mentioning that, in Alexander (now Egypt), the ancient Greek mathematician Euclid (activity around 300 BC) pointed out in his masterpiece *Elements* that, "The name Platonic solids is wrong," since three of them, tetrahedron, cube, and dodecahedron came from the Pythagorean School, while octahedron and icosahedron came from Theaetetus (c. 417–369 BC). Theaetetus was a student of the ancient Greek philosopher Plato (427–347 BC), and was the inventor of solid geometry and the dialogist of Plato's two dialogues — *Theaetetus* and *Sophist*.

3. The Division Algorithm

Generally speaking, in number theory we study not only the natural numbers, but also zero and negative integer (sometimes even rational numbers). The sum, difference and product of two integers are still integers. However, when a non-zero integer divides another integer, the quotient is not always an integer. Therefore, we must introduce the concept of divisibility and multiples.

Definition 1.1. Let $a, b(\neq 0)$ be integers, if there exists an integer q such that the equation

$$a = bq \tag{1.1}$$

holds, then we say that b divides a or a is divisible by b. In this case, we call a multiple of b, and b is a factor of a, denoted as $b|a$. If there is no such q for (1.1), then we say that b does not divide a, or a is not divisible by b, denoted as $b \nmid a$.

Like natural numbers, the concept of divisibility is so easy to understand that it had been widely accepted before earliest well-known mathematicians appeared. However, probably it took a long

time for mathematicians to give the definition of congruence and find that the division is a special kind of congruence, i.e., $b|a$ means that a is congruent to 0 modulo b. We will give the definition of congruence at the beginning of Section 6. Definition 1.1 is very useful, from which we can deduce many properties about divisibility.

Theorem 1.1. *If a is a multiple of b, b is a multiple of c, then a is a multiple of c.*

Proof. We need only to use the definition of divisibility three times. In fact, there exist integers q_1 and q_2 such that $a = bq_1$, $b = cq_2$, hence $a = c(q_1 q_2)$. ▫

Theorem 1.2. *If a and b are both multiples of c, then $a \pm b$ is also a multiple of c.*

Proof. Like Theorem 1.1, we need to use the definition of divisibility three times. In fact, there exist integers q_1 and q_2 such that $a = cq_1$, $b = cq_2$, hence $a \pm b = c(q_1 \pm q_2)$. ▫

Example 1.1. The product of any two consecutive numbers is even, and the product of any three consecutive numbers is multiple of 6.

We have done a preliminary discussion on the properties of division; however, for any two integers a and b, they generally do not have relation of divisibility. Therefore, we need to introduce the following division algorithm.

Theorem 1.3. *(Division algorithm).* *For any integers a and b, with $b > 0$, there exist unique integers q and r such that*

$$a = bq + r, \quad 0 \leq r < b. \tag{1.2}$$

Proof. First we show the existence. Considering the following sequence of integers:

$$\ldots, -3b, -2b, -b, 0, b, 2b, 3b, \ldots .$$

We note that there exists an integer q such that

$$qb \leq a < (q+1)b.$$

Let $a - qb = r$, then $a = bq + r$ with $0 \leq r < b$.

Statue of Euclid

Now we verify uniqueness. Suppose that integers q_1 and r_1 also satisfy equation (1.2), i.e.,

$$a = bq_1 + r_1, \quad 0 \le r_1 < b.$$

Then $bq_1 + r_1 = bq + r$, $b(q - q_1) = r_1 - r$, thus

$$b|q - q_1| = |r - r_1|.$$

Since both r and r_1 are positive integers less than b, so the left-hand side above is also less than b.

Therefore $q = q_1$ and $r = r_1$. □

Note that q and r in Theorem 1.3 are called the incomplete quotient and remainder, respectively.

Example 1.2. Suppose a and b are integers not both zero, $ax_0 + by_0$ is the least integer in the form of $ax + by$, then

$$ax_0 + by_0 | ax + by,$$

where x and y are any integers.

Example 1.3. For any integer $n > 1$, the number $1 + \frac{1}{2} + \frac{1}{3} + \cdots + \frac{1}{n}$ cannot be an integer.

Suppose $2^l \le n < 2^{l+1}$ ($l \ge 1$). Then in the above sum there is exactly one term with its denominator multiple of 2^l, and for the rest terms the denominator is at most divisible by 2^{l-1}. Hence

$$\sum_{i=1}^{n} \frac{1}{i} = \frac{q}{2^{l-1}(2k+1)} + \frac{1}{2^l} = \frac{2q + 2k + 1}{2^l(2k+1)},$$

with the numerator odd and the denominator even, so the sum cannot be an integer.

It is worth mentioning that Erdős and Niven once studied elementary symmetric functions of $\{1, \frac{1}{2}, \ldots, \frac{1}{n}\}$. Denote the kth item by $S(k,n)$. In 2012, Yong-Gao Chen and Min Tang (*Amer. Math. Monthly* **119**, 862–867) showed that $S(1,1) = 1$ and $S(2,3) = \frac{1}{2} + \frac{1}{3} + \frac{1}{6}$ are the only integers.

Example 1.4. The drawer principle, also known as the pigeonhole principle or Dirichlet principle, states that if n items are put into m drawers, with $n > m$, then at least one drawer must contain more than one item. By using drawer principle, one could show that, for $1 \le a_1 < a_2 < \cdots < a_{n+1} \le 2n$, there exist $1 \le i < j \le n+1$ such that $a_i | a_j$.

As a matter of fact, for any $1 \le i \le n$, let $s_i = \{1 \le (2i-1)2^l \le 2n, l \ge 0\}$. Then there must exist 2 of the $n+1$ numbers a_i's that belong to the same set s_i. If we only choose n positive integers, then the situation will be different, i.e., for any positive integer n, there exist n positive integers $1 \le a_1 < a_2 < \cdots < a_n \le 2n$ such that $a_i \nmid a_j (1 \le i < j \le n)$. It is called a primitive sequence. One can deduce that the "largest" primitive sequence is $\{n+1, n+2, \ldots, 2n\}$ and the "smallest" primitive sequence is $\{2^i(2x+1) | \frac{2n}{3^{i+1}} < 2x +$

$1 < \frac{2n}{3^i}$, $x \geq 1, i \geq 0$}. For example, if $n = 12$, then the smallest primitive sequence is $\{4, 6, 9, 10, 11, 13, 14, 15, 17, 19, 21, 23\}$. In 2018, we (Shane Chern, Tianxin Cai and Hao Zhong, *Chinese Adv. Math.* **47**(1), 150–154) discussed the properties of primitive sequences. In particular, making use of the following sum estimate (C is the Euler constant):

$$\sum_{n \leq x} \frac{1}{n} = \log x + C + O\left(\frac{1}{x}\right),$$

we deduced that when n tends to infinity, the harmonic sum of the "smallest" sequence tends to $\ln 3$. More exactly, we have

$$\max \sum_{1 \leq i \leq n} \frac{1}{a_i} = \log 3 + O\left(\frac{1}{n^{\log_3 2}}\right).$$

Now we introduce an application and generalization of the division algorithm, which will be very useful for calculating the greatest common factor in the next section. This method is called Euclidean algorithm in the west, which first appeared in *Elements*. However, in India it is called pulverizer, and in China it is called 辗转相除法 Zhan Zhuan Xiang Chu Fa (method of successive division). They were found by Aryabhata (c. 476–550) in the Gupta dynasty and Qin Jiushao (1202?–1261) in the Southern Song Dynasty, independently.

Let a and b be any positive integers. By the division algorithm,

$$\begin{cases} a = bq_1 + r_1, & 0 < r_1 < b, \\ b = r_1 q_2 + r_2, & 0 < r_2 < r_1, \\ \vdots \\ r_{n-2} = r_{n-1} q_n + r_n, & 0 < r_n < r_{n-1}, \\ r_{n-1} = r_n q_{n+1}. \end{cases} \tag{1.3}$$

We can see that each time we take the division algorithm, the remainder becomes smaller. Hence taking at most b times, the remainder will be zero.

Finally, we will introduce the concept of prime numbers, which is the core problem in number theory. It first appeared in Chinese as the "number root" in a book published in Qing Dynasty.

Female math teacher, an illustration of *Elements* in Latin version (1310)

In no later than the 3rd century BC, the ancient Greek mathematicians were aware of the existence of infinitely many primes. They divided natural numbers into 1, prime numbers and composite numbers. A prime number means an integer greater than 1, which has 1 and itself as its only positive factors, such as $2, 3, 5, 7, 11, \ldots$. Otherwise it is called a composite number, such as $4, 6, 8, 9, 10, \ldots$. The 1 (or unit) is excluded out of the primes or composite numbers because it can't be decomposed into the product of two smaller positive integers, so it could not be a composite number. If 1 is a prime, then the uniqueness of the arithmetic fundamental theorem in the fifth sections will be not true.

In fall of 2012, the author found that human body contains prime numbers. Look at our hand, the thumb has 2 joints, each of the rest fingers has 3 joints, and each hand has 5 fingers. That is to say, each hand contains the first three prime numbers. Although compared with composite numbers, there are fewer prime numbers; for some

polynomial, its value goes through primes when the independent variable goes through consecutive positive integers. For example, Euler found that

$$x^2 - x + 41$$

is always prime when $x = -39, \ldots, -1, 0, 1, \ldots, 40$.

In 1798, the French mathematician Adrien-Marie Legendre (1752–1833) observed that

$$2x^2 + 29$$

is also always prime when $x = -28, \ldots, -1, 0, 1, \ldots, 28$.

The well-known twin prime conjecture and Goldbach conjecture are both about primes. What the former says is that there are infinitely many pairs of primes with difference 2, e.g., $(3, 5), (5, 7)$, $(11, 13), (17, 19), \ldots$. Until 2016, the largest known twin primes are

$$2996863034895 \times 2^{1290000} \pm 1.$$

They have 388342 digits. However, they are very small comparing to the largest known Mersenne prime (2018)

$$2^{82589933} - 1,$$

which is the 51th one and has 24862048 digits.

Portrait of French priest and mathematician Mersenne

Mersenne primes and Fermat primes

A prime in the form of $M_p = 2^p - 1$ is called Mersenne prime, named after the French mathematician Marin Mersenne (1588–1648). It is easy to deduce that if M_p is a prime number, then p must be a prime number. Because of the importance of prime numbers, many mathematicians are interested in finding a mathematical formula or series, so that each value it takes is a prime. This is a bit like the ancient Chinese emperor keen on ever-young medicine, and it is always difficult to succeed.

Mersenne was probably the first person to do this exploration, he was a priest, a philosophy, with the name "father of optics", was very fond of mathematics, especially interested in the problems related with prime numbers in number theory. In addition, he also traveled between countries in the west Europe, as a contact person among mathematicians and scientists from different countries. He went to Italy 15 times. The French Academy of Sciences was founded in Paris, taking his salon as prototype. In 1644, Mersenne observed that $M_2 = 3$, $M_3 = 7$, $M_5 = 31$, $M_7 = 127$, so he conjectured that all of M_p are prime numbers. However, soon it was found that M_{11} is not a prime. In fact,

$$M_{11} = 2^{11} - 1 = 2047 = 23 \times 89.$$

In 1876, after 19 years of efforts, the French mathematician Édouard Lucas (1842–1891) figured out manually that M_{127} is prime (77 digits). During the next 75 years, it had been the largest prime number known to man, until the computer era. Today, with the help of computer network, the largest prime number we can find is still Mersenne prime.

It is worth mentioning that, people usually only remember the person who discovered Mersenne prime. It is Lucas who showed that M_{67} is not prime, that was in 1867. The decomposition of M_{67} was done by the American mathematician Frank Nelson Cole (1861–1926) in 1903, he was the secretary of the American Mathematical Society (AMC). In order to commemorate him, AMS set up the Cole prize,

which is regarded as the highest award in the field of number theory and algebra.

Whether there exist infinitely many Mersenne primes or infinitely many Mersenne composites is still unknown. But if there are infinitely many Sophie Germain primes in the form of $4m + 3$, then there must be infinitely many Mersenne composites. The so-called Sophie Germain prime is a prime p, with $2p + 1$ also a prime. It is named after the French female mathematician and physicist Sophie Germain (1776–1831). The proof needs the properties of quadratic remainders, which will be given in the 17th section.

Sophie was born in a rich family in Paris, but at that time, there was discrimination against women. Sophie studied number theory only through herself and correspondence. She had an important contribution to Fermat's Last Theorem, and communicated with Legendre, Lagrange and Gauss, using a name of man. Later Gauss suggested Gottingen awarded her the honorary doctor degree, but the university did it six years after her death.

On the other hand, as early as the 4th century, the ancient Greek mathematician Archytas (c. 400–350 BC), who might have invented the kite, knew that, if $2^p - 1$ is prime number, then $2^{p-1}(2^p - 1)$ must be a perfect number. Since each factor of 2^{p-1} must be $2^i, 0 \le i \le p - 1$, and since $(2^{p-1}, 2^p - 1) = 1$, the sum $\sigma(n)$ of all the factors of $n = 2^{p-1}(2^p - 1)$ is

$$(1 + 2 + \cdots + 2^{p-1})(2^p - 1 + 1) = 2(2^{p-1})(2^p - 1).$$

This proof appeared in Euclid's Elements. Only in 1747, Euler proved in Berlin that all even perfect numbers are in the above form. In 1772, Euler had returned to Petersburg and was blind, but he found the 8th perfect number, it has 19 digits ($p = 31$).

Euler's Proof. Suppose $n = 2^{r-1}s$ such that $r \ge 2$ and s is odd. Then $\sigma(n) = \sigma(2^{r-1}s) = 2^r s$. Since 2^{r-1} and s have no common factor, the sum of the factors of $2^{r-1}s$ equals to the product of the sum of the factors of s and $(2^r - 1)/(2 - 1) = 2^r - 1$, hence $\sigma(n) = (2^r - 1)\sigma(s)$. Taking $\sigma(s) = s + t$, where t is the sum of real factors

of s, we see that $2^r s = (2^r - 1)(s + t)$, i.e., $s = (2^r - 1)t$, $s > t$. That is to say, t is a real factor of s and also the sum of real factors of s, therefore $t = 1$, and $s = 2^r - 1$ is a prime.

Let p be a prime. Then any prime factor q of $2^p - 1$ is greater than p. This is because that, if $2^p \equiv 1 \pmod q$, then the element 2 has order p in the multiplicative group $Z_p \backslash \{0\}$ that has $q - 1$ elements, so by Lagrange theorem in abstract algebra, $p | q - 1$, hence $p < q$. From this we could also deduce that there are infinitely many prime numbers.

Let α be a positive integer and β a non-negative integer. Then the equation

$$\sum_{d | n \, d < n} d - n = 2^\alpha (2^\beta - 1)$$

has a solution $n = 2^{\alpha + \beta - 1}(2^\alpha - 1)$, here $2^\alpha - 1$ is a Mersenne prime; while the equation

$$\sum_{d | n \, d < n} d - n = 2^\alpha (1 - 2^\beta)$$

has a solution $n = 2^{\alpha - 1}(2^{\alpha + \beta} - 1)$, here $2^{\alpha + \beta} - 1$ is a Mersenne prime.

In particular, when $\beta = 0$, this is Euler's criterion for perfect number $n = 2^\alpha (2^\alpha - 1)$; when $\beta = 1$, all the solution for $\sum_{d | n \, d < n} d - n$ to be a power of 2 are $n = 2^{\alpha - 2}(2^\alpha - 1)$, here $2^\alpha - 1$ is Mersenne prime.

In 1640, four years before his friend Mersenne, Fermat also gave a formula, i.e.,

$$F_n = 2^{2^n} + 1.$$

People called it a Fermat number and a Fermat prime if it is prime. Fermat himself showed that when $n = 0, 1, 2, 3, 4$, the number F_n are all prime, they are 3, 5, 17, 257, and 65537. He then conjectured that F_n is prime for any n.

This conjecture had existed for nearly a century until in 1732, Euler proved that F_5 is not a prime number. He was 25 then as a guest in St Petersburg. In fact, what Euler proved is $641 | F_5$. Here is his proof

Table of perfect numbers

Rank	p	Perfect number	Digits	Year	Discover
1	2	<u>6</u>	1	ancient Greek	Euclid or before
2	3	<u>28</u>	2	ancient Greek	Euclid or before
3	5	<u>496</u>	3	ancient Greek	Euclid or before
4	7	<u>8128</u>	4	ancient Greek	Euclid or before
5	13	33550336	8	1456	Medieval manuscript
6	17	8589869056	10	1588	Cataldi
7	19	137438691328	12	1588	Cataldi
8	31	2305843008139952128	19	1772	Euler
9	61	265845599⋯953842176	37	1883	Pervushin
10	89	191561942⋯548169216	54	1911	Powers
11	107	131640364⋯783728128	65	1914	Powers
12	127	144740111⋯199152128	77	1876	Lucas
13	521	235627234⋯555646976	314	1952	Robinson
14	607	141053783⋯537328128	366	1952	Robinson
15	1279	541625262⋯984291328	770	1952	Robinson
16	2203	108925835⋯453782528	1327	1952	Robinson
17	2281	994970543⋯139915776	1373	1952	Robinson
18	3217	335708321⋯628525056	1937	1957	Riesel
19	4253	182017490⋯133377536	2561	1961	Hurwitz
20	4423	407672717⋯912534528	2663	1961	Hurwitz
21	9689	114347317⋯429577216	5834	1963	Gillies
22	9941	598885496⋯073496576	5985	1963	Gillies
23	11213	395961321⋯691086336	6751	1963	Gillies
24	19937	931144559⋯271942656	12003	1971	Tuckerman
25	21701	100656497⋯141605376	13066	1978	Noll & Nickel
26	23209	811537765⋯941666816	13973	1979	Noll
27	44497	365093519⋯031827456	26790	1979	Nelson & Slowinski
28	86243	144145836⋯360406528	51924	1982	Slowinski
29	110503	136204582⋯603862528	66530	1988	Colquitt & Welsh
30	132049	131451295⋯774550016	79502	1983	Slowinski
31	216091	278327459⋯840880128	130100	1985	Slowinski
32	756839	151616570⋯565731328	455663	1992	Slowinski & Gage
33	859433	838488226⋯416167936	517430	1994	Slowinski & Gage
34	1257787	849732889⋯118704128	757263	1996	Slowinski & Gage
35	1398269	331882354⋯723375616	841842	1996	Armengaude *et al.*
36	2976221	194276425⋯174462976	1791864	1997	Spence *et al.*
37	3021377	811686848⋯022457856	1819050	1998	Clarkson *et al.*
38	6972593	955176030⋯123572736	4197919	1999	Hajratwala *et al.*
39	13466917	427764159⋯863021056	8107892	2001	Cameron *et al.*
40	20996011	793508909⋯206896128	12640858	2003	Shafer *et al.*
41	24036583	448233026⋯572950528	14471465	2004	Findley *et al.*
42	25964951	746209841⋯791088128	15632458	2005	Nowak *et al.*
43	30402457	497437765⋯164704256	18304103	2005	Cooper *et al.*
44	32582657	775946855⋯577120256	19616714	2006	Cooper *et al.*
45	37156667	204534225⋯074480128	22370543	2008	Elvenich *et al.*
46	42643801	144285057⋯377253376	25674127	2009	Strindmo *et al.*
47	43112609	500767156⋯145378816	25956377	2008	Smith *et al.*
48	57885161	169296395⋯270130176	34850340	2013	Cooper *et al.*
49	74207281	451129962⋯930315776	44677236	2016	Cooper *et al.*
50	77232917	109200152⋯016301056	46498850	2017	Pace *et al.*
51	82589933	110847779⋯191207936	49724095	2018	Laroche *et al.*

Euler's proof. Take $a = 2^7$, $b = 5$. Then $a - b^3 = 3$, $1 + ab - b^4 = 1 + 3b = 2^4$, and

$$F_5 = (2a)^4 + 1 = (1 + ab - b^4)a^4 + 1 = (1 + ab)a^4 + 1 - a^4 b^4$$
$$= (1 + ab)\{a^4 + (1 - ab)(1 + a^2 b^2)\},$$

where $1 + ab = 641$, and $F_5 = 641 \times 6700417$.

A simpler proof is, since $641 = 2^7 \times 5 + 1 = 2^4 + 5^4$, then $2^7 \times 5 \equiv -1 \pmod{641}$, and $5^4 \equiv -2^4 \pmod{641}$, hence $2^{28} \times 5^4 \equiv -2^{32} \equiv 1 \pmod{641}$, so $641 | F_5$.

Since then, mathematicians have tested for more than 40 integers n, including $5 \le n \le 32$, and found none of the F_ns are primes, but no one knows the prime factors of F_{20} and F_{24}. That is to say, no more Fermat primes have been found, while Mersenne primes do appear uninterrupted. It is worth mentioning that, $F_n = F_0 F_1 \cdots F_{n-1} + 2$, hence any two Fermat numbers are coprime. This property was discovered by David Goldbach. An interesting fact is $2^{2^n} + 15$ $(0 \le n \le 5)$ are all primes. That is to say, there are consecutive 6 primes, $17, 19, 31, 271, 65551$, and 4294967391.

There are still many mysteries on Fermat numbers. For example, when $n > 4$, is F_n always composite? Are there infinitely many Fermat composites? Or are there infinitely many Fermat primes? The last question was raised by the German mathematician Ferdinand Gotthold Eisenstein (1823–1852) in 1844.

After Fermat's Last Theorem was proved in 1995, the Fermat prime problem became the last one Fermat left.

The so-called generalized Fermat numbers are defined as $a^{2^n} + 1$, where a is an even number. Similarly, no one knows if there exist infinitely many generalized Fermat primes. In 2014, we define a positive integer s as a GM number, if $s = 2^\alpha + t$, where t is the sum of real factors of s and α is a non-negative integer. When the author thought about this problem, the Columbian writer Gabriel García Márquez (1927–2014) died, and his Magnum opus "One Hundred Years of Solitude" reminded me of those mathematical problems which remain unsolved for centuries.

Obviously, an odd prime in GM numbers is a Fermat primes. In addition, we were lucky to have found two odd composites in GM

numbers, which are (there are only two less than 2×10^{10})

$$19649 = 7^2 \times 401 = 2^{14} + 3265,$$

$$22075325 = 5^2 \times 883013 = 2^{24} + 5298109.$$

As for even GM number, the situation is different. For any positive integer α, β, if $p = 2^\alpha + 2^\beta - 1$ is a prime number, then $2^{\beta-1}p$ is a GM number. When $\beta = 1$, the number p is Fermat prime, and if $\beta \geq 2$, then p is an even GM number. We have two conjectures.

Conjecture 1. *There are infinitely many prime numbers in the form of $2^\alpha + 2^\beta - 1$.*

Conjecture 2. *There are infinitely many even GM numbers.*

We note that, from the viewpoint of the binary system, each odd prime could be expressed uniquely as

$$1 + 2^{n_1} + \cdots + 2^{n_k} (1 \leq n_1 < \cdots < n_k).$$

Our problem is, for any fixed positive integer k, how many primes could be expressed in the above way? This is the generalization of Eisenstein's problem ($k = 1$). For any prime number, we could classify it as of k-class if it could be expressed in the above way. Then 1-class primes are Fermat primes, $7, 11, 13, 19, 41, \ldots$ are 2-class primes, and $23, 29, 43, 53, \ldots$ are 3-class primes, etc.

On the other hand, let $t = \sum_{i=1}^{k} n_i$. We can classify odd prime numbers among them according to t so that each class has infinitely many elements. For example, classes 1 to 3 have only one element each, $3, 5, 7$ respectively, each of classes 4–5 has two elements, 11, 13 and 17, 19, respectively, class 6 is an empty set; each of classes 7–8 has 3 elements, 23, 37, 67 and 41, 131, 257, respectively, and so on. Our problem is, in addition to class 6, is any other class empty?

In addition to the above problems and conjectures, we also have some definite results. For example, for any odd number $n > 1$, the number $1 + 2 + 2^{2^n}$ is always composite; for any integer $i > 1$, there exists a unique positive integer k such that $i \equiv 2^{k-1} + 1 \pmod{2^k}$, and for any $n \geq k$, the number $1 + 2^i + 2^{2^n}$ is always composite. In particular, there are infinitely many pairs m, n such that $1 + 2^{2^m} + 2^{2^n}$ is composite.

In fact, if n is odd, $1 + 2 + 2^{2^n}$ is a multiple of 7; and if $i = 2^{k-1} + 1 + 2^k s$, then

$$1 + 2^i + 2^{2^n} = 1 + 2(2^{2^{k-1}})^{2s+1} + (2^{2^{k-1}})^{2^{m-k+1}}$$

$$\equiv 1 - 2 + 1 = 0 \ (\mathrm{mod}\, F_{k-1}).$$

Let's take a look at primes in the form of $\frac{3^{2^n}+1}{2}$. When $n = 0, 1, 2, 4$, we get primes $2, 5, 41, 21523361$, respectively; when $n = 3$, we get a composite $3281 = 17 \times 193$. There are more primes in the form of $3^{2^n} + 2$.

Now we consider a variety of Mersenne primes. For example, in the form of $\frac{3^p-1}{2}$ or, $\frac{5^p-1}{4}$, we have 13 $(p = 3)$, 1093 $(p = 7)$, 797161 $(p = 13)$ and 31 $(p = 3)$, 19531 $(p = 7)$, 12207031 $(p = 11)$, respectively, etc. Like Mersenne primes, their prime factors must be in the form of $2px + 1$, but might not be congruent to ± 1 modulo 8, and one of them is Wieferich prime (cf. the definition in Section 8) 1093.

Next, we want to mention the founder of set theory, the Russian born Danish German mathematician Georg Cantor (1845–1918) who once conjectured

> Let $p_0 = 2, p_{n+1} = 2^{p_n} - 1$ for $n \geq 0$. Then this sequence is always prime.

The sequence is called the Cantor number. It is known that $p_0 = 2, p_1 = 3, p_2 = 7, p_3 = 127, p_4 = 2^{127} - 1$ are all primes, but p_5 has 5×10^{37} digits. Even one century later after his death, it is still larger than the known largest prime. This means that, we are still unable to determine whether it is prime. On the other hand, if p_m is composite, then, for any $n > m$, the number p_n is always composite.

As for Mersenne primes and perfect numbers, although we currently cannot determine whether their numbers are infinity or not, we can raise the following conjectures.

Conjecture 3. *Let n be a positive integer. If both $2^n - 1$ and $2^{2n-1} - 1$ are primes, then n must be $2, 3, 7, 31$.*

Conjecture 4. *There are infinitely many perfect numbers, with the ratio of the number of those ending with 8 to the number of those ending with 6 goes to the golden ratio, i.e., $0.618\ldots$.*

So far among the 51 known perfect numbers, there are 19 ending with 8 and 32 ending with 6, their corresponding (Mersenne) primes are $p \equiv 3 \pmod 4$ and $p \equiv 1 \pmod 4$, respectively, the ratio of the numbers is $\frac{19}{31} = 0.613 \ldots$.

4. The Greatest Common Divisor

With the division algorithm and Euclidean algorithm we could introduce and discuss the greatest common divisor and least common multiple, which are the two basic concepts of number theory. First we give the definition of the greatest common factor.

$(24, 60) = 12$, which means that a 24×60 rectangle can be covered with ten square tiles

Definition 1.2. Let a, b be two integers. If $d|a, d|b$, then d is called a common factor of a and b. The largest common factor of a and b is called the greatest common divisor of a and b, denoted by (a, b).

When $(a, b) = 1$, we say that a and b are coprime. At this time, we also call the fraction $\frac{a}{b}$ the irreducible fraction. Obviously, we can generalize the definition of the common divisor and the greatest common divisor of integers from two to any number. If the greatest common factor of more than 2 integers is equal to 1, we say that they are coprime, but it does not mean that they are pairwise coprime.

By the definition, we can see that the two integers a, b have the same common factors as $|a|$, $|b|$, thus $(a, b) = (|a|, |b|)$. Therefore, we only need to discuss the greatest common factor of positive integers.

The following results can be obtained directly from Definition 1.2: if a is a positive integer, then the common factors of 0 and a are factors of a; $(0, a) = a$.

Theorem 1.4. *Let a, b, c be any integers not all zero, and let*

$$a = bq + c,$$

where q is an integer, then a, b and b, c have same common factors, and $(a, b) = (b, c)$.

It is easy to deduce Theorem 1.4 from Theorem 1.2. By applying Theorem 1.4 to Euclidean algorithm repeatedly, we have $(a, b) = r_n$, where r_n is the last non-zero remainder in (3). Furthermore, we can get the following theorem.

Theorem 1.5. *The common factors of a and b are the same as the factors of r_n.*

By Definition 1.2 and Euclidean algorithm, we have the following theorem.

Theorem 1.6. *If a, b are any two integers not both zero, then there exist two integers s, t such that*

$$as + bt = (a, b).$$

Here s, t are in fact the x_0, y_0 of Example 1.2. This is because on the one hand $(a, b) | ax_0 + by_0$, and on the other hand, by assumption from Example 1.2, $ax_0 + by_0 | a$, $ax_0 + by_0 | b$. It follows from Theorem 1.4 that $ax_0 + by_0 | (a, b)$. Meanwhile, the results can be generalized to the case of n integers.

Example 1.5. Let n be an integer greater than 1. If $2^n + 1$ is prime, then n must be a power of 2. This is what we called the Fermat number F_n in the previous section.

Example 1.6. Let a, b, c be positive integers. If $(a,c) = 1$ or $(b,c) = 1$, and $\frac{1}{a} + \frac{1}{b} = \frac{1}{c}$, then $a + b$ is a square number.

Now we give some properties of the greatest common factor.

Theorem 1.7. *Let a, b be any two integers not both zero. If m is a positive integer, then*

$$(am, bm) = (a, b)m.$$

If d is a common factor of a and b, then

$$\left(\frac{a}{d}, \frac{b}{d} \right) = \frac{(a, b)}{d}.$$

Moreover, we have

$$\left(\frac{a}{(a,b)}, \frac{b}{(a,b)} \right) = 1.$$

If both sides in the Euclidean algorithm are multiplied by m, it is not difficult to show the first part of Theorem 1.7 (also could be proved by Theorem 1.5 or Theorem 1.6). The second part is the consequence of the first part. Now, we consider a generalization of Theorem 1.7. Let a_1, a_2, \ldots, a_n be any positive integers. Take

$$(a_1, a_2) = d_2, (d_2, a_3) = d_3, \ldots, (d_{n-1}, a_n) = d_n. \qquad (1.4)$$

Then we have the following theorem.

Theorem 1.8. $(a_1, a_2, \ldots, a_n) = d_n$.

Proof. We first show that d_n is a common factor of a_1, a_2, \ldots, a_n. Then we prove that any other common factor is less or equal to it. It follows from (1.4) that $d_n | a_n$ and $d_n | d_{n-1}$. Note that $d_{n-1} | a_{n-1}$ and $d_{n-1} | d_{n-2}$. Hence by Theorem 1.1, $d_n | a_{n-1}$ and $d_n | d_{n-2}$. In turn, we could deduce finally that $d_n | a_n, d_n | a_{n-1}, \ldots, d_n | a_1$. On the other hand, if d is any factor of a_1, a_2, \ldots, a_n, then $d | a_1$ and $d | a_2$. From Theorem 1.5, we deduce $d | d_2$. Go on in this way by using Theorem 1.4 repeatedly, we can deduce $d | d_n$, therefore $d \leq d_n$.

Next we introduce the concept of the least common multiple, which is symmetry to that of the greatest common factor, and also make the connection between the two. To this end, we first give a few lemmas. □

Lemma 1.1. *Let a, b, c be integers with $(a, c) = 1$. Then ab, c and b, c have the same common factors; if b, c is not both zero, then $(ab, c) = (b, c)$.*

Proof. We only need to prove the first part. By assumption and Theorem 1.6, there exist integers s and t such that

$$as + ct = 1.$$

Multiplying b to the both sides give that

$$(ab)s + c(bt) = b.$$

Let d be any common factor of ab and c. From the above equation and Theorem 1.2, $d|b$, hence d is a common factor of b and c. Conversely, any common factor of b and c is a common factor of ab and c. Thus we have proved Lemma 1.1. □

Lemma 1.2. *If $c|ab$ and $(a, c) = 1$, then $c|b$.*

Lemma 1.3. *Let a_1, a_2, \ldots, a_m and b_1, b_2, \ldots, b_n be two sets of integers, if $(a_i, b_j) = 1$ for any $1 \le i \le m, 1 \le j \le n$, then $(a_1 a_2 \ldots a_m, b_1 b_2 \ldots b_n) = 1$.*

Lemma 1.2 follows from Lemma 1.1, and by using Lemma 1.1 repeatedly we can deduce Lemma 1.3. Now we give the definition of the least common multiple.

Definition 1.3. Let a and b be any non-zero integers. If $a|m, b|m$, we say that m is a multiple of a and b. The least of all common multiples of a and b is called the least common multiple, denoted by $[a, b]$.

Similarly, the definition of common multiples and the least common multiple can be extended to any number of integers. By the

definition we can see, a, b and $|a|, |b|$ have the same common multiples, hence $[a, b] = [|a|, |b|]$. Therefore, we only discuss the least common multiple for positive integers.

Theorem 1.9. *Let a, b be any positive integers. Then all the common multiples of a and b are the multiples of $[a, b]$, and*

$$[a, b] = \frac{ab}{(a, b)}. \tag{1.5}$$

Proof. Suppose m is a multiple of a and b, so that $m = ak = bk'$ for some k and k'. Take $a = a_1(a, b)$, $b = b_1(a, b)$. Then $a_1 k = b_1 k'$. By Theorem 1.7, $(a_1, b_1) = 1$, hence it follows from Lemma 1.2 that $b_1 | k$. Let $k = b_1 t$. Then

$$m = ab_1 t = \frac{ab}{(a, b)} t. \tag{1.6}$$

Conversely, for any t, the number $\frac{ab}{(a,b)} t$ is a multiple of a and b, which means that all of multiples of a and b are in the form of (1.6). Taking $t = 1$, we get the least common multiple, therefore (1.5) holds. By (1.6), the first half of Theorem 1.9 is proved. \square

Corollary. *If c is a multiple of a and b, and $(a, b) = 1$, then $ab | c$.*

Finally, we consider the least multiple of more than two integers. Let a_1, a_2, \ldots, a_n be n positive integers. Take

$$[a_1, a_2] = m_2, [m_2, a_3] = m_3, \ldots, [m_{n-1}, a_n] = m_n. \tag{1.7}$$

Theorem 1.10. $[a_1, a_2, \ldots, a_n] = m_n$.

Proof. First, from (1.7) we can see that m_n is a multiple of a_1, a_2, \ldots, a_n. We only need to show that any common multiple is greater than or equal to it. Let m be any common multiple of a_1, a_2, \ldots, a_n. Then $a_1 | m, a_2 | m$. It follows from Theorem 1.9 that $m_2 | m$. Note that $a_3 | m$. Similarly, we could deduce that $m_3 | m$. In turn, we can prove finally that $m_n | m$. Therefore $m_n \leq m$.

In particular, if a_i $(1 \leq i \leq n)$ are pairwise coprime, then $[a_1, a_2, \ldots, a_n] = a_1 a_2 \ldots a_n$.

It is worth mentioning that in 1982, by using an analytic method, the English mathematician M. Nair (*Amer. Math. Monthly* **89**,

126–129) proved that if $a_i = i$ for any positive integer i, then $m_n \geq 2^n$ for $n \geq 7$. \square

Graham's conjecture

In 1970, the American mathematician R. L. Graham (1935–2020) raised a conjecture (*Amer. Math. Monthly* **77**, 775).

Let $A = \{a_1, a_2, \ldots, a_n\}$ be a set of distinct positive integers. Then $\max\limits_{1 \leq i,j \leq n} \frac{a_i}{(a_i, a_j)} \geq n$.

Graham is an expert in discrete mathematics and is also active in number theory. His wife Fan Chung (1949–) is a graph theory expert born in Taiwan. Many number theorists made effort to this conjecture, in particular they showed it is true if one of a_i is a prime number. In 1986, the Hungarian computer scientist Szegedy (*Combinatorica* **6** (1986) 67–71) showed that when n is sufficient large, the conjecture is true. Ten years later, two Indian mathematician Balasubramanian and Soundararajan proved Graham's conjecture (*Acta Arithmetica* **75** (1996) 1–38), but their method is quite complicated, so we still expect a simple proof. In 1999, American mathematician Granville and Roesler raised a new conjecture related with two sets of positive integers A and B, i.e.,

$$\max_{a \in A, b \in B} \left\{ \frac{a}{(a,b)}, \frac{b}{(a,b)} \right\} \geq \min\{|A|, |B|\}.$$

When $A = B$, this is Graham's conjecture. The author and Xiaodong Zhao (*J. Zhejiang Univ.* **33** (2005) 1–2) showed that if A or B contains a prime, then the conjecture is true.

In case of one set, suppose $1 \leq a_1 < a_2 < \cdots < a_n$, with $(a_1, a_2, \ldots, a_n) = 1$. For any positive integer k, we call the above set a good n-sequence or good sequence of order k, if $a_n < na_1$, and for any positive i and j, whenever $a_j \geq kn$, we have $(a_i, a_j) > k$.

Example. The sequences $\{10, 12, 15, 18, 30\}$ and $\{35, 45, 63, 75, 105\}$ are good 5-sequence of 1 order and 2 order, respectively; the sequence $\{35, 45, 63, 105\}$ is a good 4-sequence of 4 order, and $\{6, 12, 15, 20, 24, 30\}$ is a good 6-sequence of order 1. Obviously, a high order good n-sequence is also a low order good n-sequence.

Grahams and Erdős in Japan (1986)

Our problems are, for what n and k, there are good n-sequence of order k? If such sequences exist, how many are there?

Conjecture. *If* $\{a_1, a_2, \ldots, a_n\}$ *is a good sequence of order* k, *then*

$$\max_{1 \leq i,j \leq n} \frac{a_i + a_j}{(a_i, a_j)} \geq (k+1)n - k.$$

In particular, if $\{a_1, a_2, \ldots, a_n\}$ *is a good sequence of order 1, then*

$$\max_{1 \leq i,j \leq n} \frac{a_i + a_j}{(a_i, a_j)} \geq 2n - 1. \tag{G}$$

Obviously, it follows from (G) *that* $\max_{1 \leq i,j \leq n} \frac{a_i}{(a_i,a_j)} \geq n$.

Definition. Let $\{a_i\}$ $(1 \leq i \leq n)$ be a set of n positive integers with $a_1 < a_2 < \cdots < a_n$ and $a_n < na_1$, and let i, j and k be any integers. If $(a_i, a_j) > k$ whenever $a_j \geq kn$, then we call it an all-good sequence with n elements.

Theorem. *Graham's conjecture is true if and only if for any* $n > 1$, *there is no all-good sequence of* n *element.*

Proof. First we show the necessity. If Graham's conjecture is true and there is an all-good sequence of n elements, then there exist i

and j such that $\frac{a_i}{(a_i,a_j)} \geq n$. Obviously, $a_i \geq n$. Using the division algorithm, and letting $a_i = kn + r$ $(0 \leq r < n)$, we have $a_i \geq kn$. By assumption, this means $\frac{a_i}{(a_i,a_j)} = \frac{kn+r}{(a_i,a_j)} \leq \frac{kn+r}{k+1} < n$, which is a contradiction.

Next, we show the sufficiency. By assumption, for any distinct n positive integers $a_1 < a_2 < \cdots < a_n$, there exist i and j such that for some k, $a_i \geq kn$ and $(a_i, a_j) \leq k$. Therefore $\frac{a_i}{(a_i,a_j)} \geq \frac{kn}{k} = n$, so Graham's conjecture is true. $\qquad\square$

Remark. The above definition was raised by the author and Liuquan Wang, he was then a sophomore student. The two conditions are all necessary, otherwise, there are infinitely many counter examples. For example, if $a_n \geq na_1$, then we have two counterexamples $\{1, 2, 3, 4, 6\}$ or $\{2, 3, 4, 6, 12\}$ $(n = 5)$, the latter satisfying $(a_i, a_n) > 1$ for any $1 \leq i < n$; if $a_n < na_1$, and there exists $1 \leq i < n$ such that $(a_i, a_n) = 1$, then we have the counterexample $\{2, 3, 4, 5, \ldots, 35, 36, 38, 40, 42\}$ $(n = 38)$, where $(5, 42) = (25, 42) = 1$. Meanwhile, if $2n - 1$ is prime, then (G) holds unconditionally.

5. The Fundamental Theorem of Arithmetic

As previously stated, prime numbers play an extremely important role in the study of number theory, and it is the existence of prime numbers that makes this old subject attractive. One purpose of this section is to prove that, each positive integer greater than 1 can be uniquely expressed as the product of prime numbers, if the order of the factors is not considered. To this end, we need some lemmas,

Lemma 1.4. *Let a be any integer greater than 1 and q the minimum of all its factors other than 1, then q must be a prime number, and when a is composite, $q \leq \sqrt{a}$.*

Proof. By the method of contradiction and from Theorem 1.1, it is easy to show that q is a prime number. If a is a composite number, $a = bq$, $b > 1$. By assumption, $q \leq b$, hence $q^2 \leq qb = a$ and $q \leq \sqrt{a}$. $\qquad\square$

Lemma 1.5. *For any integer a and prime p, either $(p, a) = 1$ or $p | a$.*

The translators of the first Chinese version of *Elements*, Italian missionary Matteo Ricci and Chinese scholar Xu Guangqi in the Ming Dynasty

Lemma 1.6. *Let* a_1, a_2, \ldots, a_n *be integers and* p *is prime, if* $p | a_1 a_2 \ldots a_n$, *then* p *must divide some* a_i.

Note that $(p, a) = 1$ or p. Lemma 1.5 is obvious. Lemma 1.6 follows from Lemma 1.3, Lemma 1.5, and by the method of contradiction.

When $n = 2$, the proof of Lemma 1.6 was given already in the *Elements*, it is essentially equivalent to the fundamental theorem of arithmetic.

Portrait of Eratosthenes: the founder of the sieve method

Theorem 1.11. (Fundamental Theorem of Arithmetic). *Any integer a greater than 1 can be expressed as the product of primes,*

$$a = p_1 p_2 \ldots p_n. \tag{1.8}$$

And if not taking the order, the expression is unique.

Proof. We use induction on a. When $a = 2$, equation (1.8) is true obviously. Suppose (1.8) is true for all integers greater than 1 and less than a. If a is a prime, then (1.8) is obviously true; if a is a composite, then there exist two integers b and c such that

$$a = bc, \quad 1 < b < a, \quad 1 < c < a.$$

By assumption, $b = p_1 p_2 \ldots p_s$, $c = p_{s+1} p_{s+2} \ldots p_{s+t}$. Then $a = p_1 p_2 \ldots p_{s+t}$. So (1.8) is true.

Next we prove the uniqueness. Suppose

$$p_1 p_2 \ldots p_n = q_1 q_2 \ldots q_m.$$

And $p_1 \leq p_2 \leq \cdots \leq p_n$, $q_1 \leq q_2 \leq \cdots \leq q_m$. Since $q_1 | p_1 p_2 \ldots p_n$ and $p_1 | q_1 q_2 \ldots q_m$, by Lemma 1.6, there exist k and j such that $p_1 | q_j, q_1 | p_k$. Hence,

$$p_1 = q_j, \quad q_1 = p_k, \quad q_j = p_1 \leq p_k = q_1.$$

Therefore $p_1 = q_1$, $p_2 \ldots p_n = q_2 \ldots q_m$. Similarly, we can show that $p_2 = q_2$. Going on along the same line, we complete the proof of Theorem 1.11.

From Theorem 1.11 we deduce immediately that, any integer a greater than 1 could be expressed as

$$a = p_1^{\alpha_1} \cdots p_k^{\alpha_k}, \quad \alpha_i > 0, \ 1 \leq i \leq k,$$

where $p_1 < \cdots < p_n$ are distinct prime factors.

Furthermore, if a is expressed as above form, then all factors d of a could be expressed as

$$d = p_1^{\beta_1} \cdots p_k^{\beta_k}, \quad \alpha_i \geq \beta_i \geq 0, \ 1 \leq i \leq k.$$

The converse is true. Meanwhile, we also have the following theorem.

\square

Theorem 1.12. *Let a, b be any positive integers and*

$$a = p_1^{\alpha_1} \cdots p_k^{\alpha_k}, \quad \alpha_i \geq 0, \ 1 \leq i \leq k,$$

$$b = p_1^{\beta_1} \cdots p_k^{\beta_k}, \quad \beta_i \geq 0, \ 1 \leq i \leq k.$$

then

$$(a, b) = p_1^{\mu_1} \cdots p_k^{\mu_k}, \quad [a, b] = p_1^{\nu_1} \cdots p_k^{\nu_k},$$

where $\mu_i = \min(\alpha_i, \beta_i), \nu_i = \max(\alpha_i, \beta_i), \ 1 \leq i \leq k.$

We have proved that any positive integer greater than 1 can be uniquely decomposed into the product of prime numbers, but in the actual calculation, to determine whether a given integer is prime or not is never easy. Here, we introduce a simple method, which was established by of the ancient Greek mathematician Eratosthenes (c. 276–194 BC) who lived in Cyrene (now Libya). It is stated as follows:

Given any positive integer n, we use the following procedure to find all the prime numbers less than n: let the n natural numbers be placed in a row as

$1, 2, 3, \ldots, n.$

First rule out 1, and leave a but rule out the rest multiples of 2, so we have

$2, 3, 5, \ldots, n'$, where n' is n if n is odd and $n-1$ if n is even.

Then leave the first number of the remaining, i.e., 3 (must be prime), and rule out the rest multiples of 3; and then leave the first

number of the remaining, i.e., 5 (must be prime), and rule out the rest multiples of 5; continuing in this way, we have left all primes less than n and ruled out all the non-prime numbers.

By Lemma 1.4, we need only rule out the true multiples of primes less than or equal to \sqrt{n}. This method is like a sieve to rule out the composite numbers, so it is called Eratosthenes sieve method. Take $n = 15$ as an example. Since $\sqrt{15} < 4$, we need only to use the above method twice, then we get all the prime numbers no larger than 15, which are $2, 3, 5, 7, 11$, and 13. It is worth mentioning that Eratosthenes is "the father of geography", "the namer of geography", etc. He was the first to accurately measure the circumference of the earth. Through the observation of the ebb and flow of the Nile and the Red Sea, he concluded that the Atlantic and India oceans are connected. This conclusion inspired the navigation of the Italian explorer Christopher Columbus (c. 1451–1506).

However, we cannot get all the primes by the sieve method or any other method, because of the following theorem.

Theorem 1.13. (Euclid). *There are infinitely many prime numbers.*

Proof. The proof is by contradiction. Suppose that there are only a finite number of prime numbers, i.e., p_1, p_2, \ldots, p_k. Take $p_1 p_2 \ldots p_k + 1 = N$. Then $N > 1$. The number N must have a prime factor p by Lemma 1.4. It is easy to see that $p \neq p_i$ for $1 \leq i \leq k$ since otherwise $p | p_1 p_2 \ldots p_k + 1$ and so $p | 1$, and contradicts the fact that p is prime. So p is different from the k prime numbers. Theorem 1.3 is thus proved. $\qquad\square$

To a large extent we can say that the secret of the natural numbers or number theory lies in the irregular distribution of prime numbers and some of the rules. This kind of rule is hidden, even if there is. For example, in 1845, the French mathematician Joseph Louis Bertrand (1822–1900) raised the so-called Bertrand's hypothesis or Bertrand's postulate.

Bertrand's Postulate. For any positive integer n, there exists a prime p such that $n < p \leq 2n$.

In 1850, the Russian mathematician Pafnuty Chebyshev (1821–1894) gave a non-constructive proof. He is considered to be a founding father of Russian mathematics. According to the Mathematics Genealogy Project, Chebyshev has 7483 mathematical "descendants" as of 2010. In 1892, the British mathematician J. J. Sylvester (1814–1897) strengthened the Bertrand postulate, he raised the following postulate.

Sylvester's Postulate. For any positive integer n, k with $2k \leq n$, there exists a prime $p > k$ among the $n, n-1, \ldots, n-k+1$.

When $n = 2k$, this is the Bertrand postulate. In 1934, by using the elementary method, the 21-year-old Erdős gave a proof of the Sylvester postulate (*J. London Math. Soc.* **9**, 282–288). The Bertrand postulate has been improved. For example, in 1952, the Japanese mathematician Nagura proved that, when $n > 25$, there is a prime number p such that $n < p < \frac{6}{5}n$. This is to say that for $n > 30$, there exists a prime p such that $\frac{5}{6}n < p < n$. Here make a note, it is also true that for $n > 1$ there exists a prime p such that $\frac{2}{3}n < p < n$.

By Bertrand postulate, we can give the third example of Section 3 a simple proof.

If p is the largest prime less than or equal to n, then $2p > n$. Otherwise, there will be a contradiction with the Bertrand postulate. Therefore, in $1, 2, \ldots n$, only p is multiple of p. Multiplying the harmonic sum $\sum_{i=1}^{n} \frac{1}{i}$ by $\frac{n!}{p}$ and making use of the corollary of Theorem 1.9, we can deduce that only $\frac{n!}{p^2}$ is not an integer, then we get the conclusion.

Using the symbol of binomial coefficients, the Sylvester postulate can be equivalently described as below; this statement will be used in Section 13:

Let n and k be positive integers. If $2k \leq n$, then there exists a prime $p > k$ in the binomial coefficients

$$\binom{n}{k} = \frac{n(n-1)\cdots(n-k+1)}{k!}.$$

Now, there is still one question that no one can answer: does there exist one prime number in between n^2 and $(n+1)^2$?

The above question was raised by the French mathematician Adrien-Marie Legendre (1752–1833). He guessed that there must be a prime. In 1798, Legendre published a book named *Essai sur la Théorie des Nombres* the title of which was significantly affected by the senior French writer Michel de Montaigne (1533–1592), who is famous for "*Essais*". But when Legendre's book was reprinted, "*Essai*" was deleted from the title. Legendre is recognized as the most outstanding successor of the 18th century French mathematician Lagrange, better than each disciple of Euler. If $\pi(x) = \sum_{p \leq x} 1$ represents the number of primes less than x (known as the prime function), Legendre and Gauss made following conjecture at the end of 18th century respectively:

$$\pi(x) \sim \frac{x}{\log x}.$$

This statement is known as the prime number theorem. Here the symbol similarity symbol means that the ratio of the two sides tends to 1 when x tends to infinity.

In 1896, using the complex variable function method, the French mathematician Jacques Hadamard (1865–1963) and the Belgian mathematician Nicolas Poussin (1866–1962) proved the above result independently. The latter became a baron therefore. The meaning of this result for the 19th century is like the proof of Fermat's Last Theorem for the 20th century. In 1949, the Norway born American mathematician Atle Selberg (1917–2007) and Erdős proved the prime number theorem independently by the elementary method. The following year, the Selberg won the Fields award, while Erdős won the Wolf prize after 35 years.

Let p_n be the nth prime. By the prime number theorem, we have

$$p_n \sim n \log n.$$

Gauss gave a more accurate estimate

$$\pi(x) \sim \lim(x) = \int_2^x \frac{dt}{\log t}.$$

Using this result and integration by parts, it is easy to deduce the prime number theorem.

A chart depicting $\pi(x)$, the upper $\text{Li}(x)$ is more accurate compared to the lower $\frac{x}{\ln x}$

On the other hand, as early as in 1735, Euler proved that the infinite series

$$\frac{1}{2} + \frac{1}{3} + \frac{1}{5} + \cdots + \frac{1}{p} + \cdots$$

is divergent, which gave an analytic proof that there are infinitely many primes. In 1938, 24-year-old Erdős gave a beautiful elementary proof, which we will show it in the 7th section.

Hilbert's 8th problem

The fundamental theorem of arithmetic shows the multiplication structure of integers, every integer greater than 1 can be expressed as a product of prime numbers; for addition, there should also be a corresponding proposition on prime numbers, which is still a conjecture that we are not able to prove. In 1742 Euler, who just moved from St Petersburg Academy of Sciences to the Berlin Academy of Sciences, wrote in a letter answer to Goldbach:

> *Any even number greater than 4 could be expressed as the sum of two odd primes.*

Earlier, Goldbach wrote to Euler to announce his discovery that any odd number greater than or equal to 9 can be expressed as the

Statue of Hilbert in Göttingen. Photo taken by the author

Portrait of Goldbach, an amateur mathematician

sum of three odd primes. It is not difficult to see, Goldbach's conjecture can be derived directly from Euler's conjecture. Later both of them are called the Goldbach conjecture. The reason might be that in the history of mathematics, there are already so many theorems and formulas named after Euler. But later, people found in the lost

work of Descartes that as early as in the 17th century, the French observed this secret of natural numbers.

Although each pupil can verify by $6 = 3 + 3$, $8 = 3 + 5$, $10 = 3 + 7 = 5 + 5, \ldots$, so far still no one can prove or disprove Goldbach's conjecture. The result closest to it is by the Chinese mathematician Chen Jingrun (1933–1996), who proved in 1966 that every sufficiently large even integer can be expressed as the sum of an odd prime and a positive integer which is the product of at most 2 prime factors. For example, $100 = 23 + 7 \cdot 11$. Chen used a new weighted sieve method, this is a variant of the old sieve method by Eratosthenes. On the other hand, as early as in 1937, the former Soviet Union mathematician I. M. Vinogradov (1891–1983) proved that each sufficiently large odd integer is a sum of three old primes, he used a circled method which is different from the sieve method.

We need to point out that many result on even Goldbach conjecture can be obtained to the twin prime conjecture. The so-called twin prime means two prime numbers with the difference 2, such as $(3, 5), (5, 7), (11, 13), (17, 19)$. The twin prime conjecture says that there are infinitely many twin primes. Although historians of mathematics have been unable to find out when, where and who first

The number of ways an even number can be represented as the sum of two primes

conjectured that there are infinitely many twin primes (generally believed that the ancient Greeks knew it). One thing is sure in 1849, the French mathematician Alphonse de Polignac (1826–1863) proposed the following conjecture.

Conjecture (de Polignac). *For any positive integer k, there exist infinitely many primes p and q such that $p - q = 2k$.*

When $k = 1$, this is the twin prime conjecture. De Polignac's father was the prime ministry of King Charlie 10, he proposed that conjecture when he was 23-year-old student of Paris Polytechnic School. Here, we point out a k-power twin prime conjecture.

Conjecture A. *For any fixed positive integer k, there exist infinitely many prime p such that $p^k - 2$ is also prime; if k is odd, then there exist infinitely many prime p such that $p^k + 2$ is also prime.*

When $k = 1$, this is the twin prime conjecture. When $k = 2$, the primes here are even more than the twin primes. After calculation, we found that, for the odd k and even k, the number of prime numbers here decrease as k increases. The reason to use the word k-power is due to the finite number of solutions for the equation $p^m - q^n = 2^h$ with $m, n > 1$, and for given m, nm and h, there is at most only one solution. At present, the known solutions are $3^2 - 2^3 = 2^0$, $3^3 - 5^2 = 2^1$, $5^3 - 11^2 = 2^2$, $5^2 - 3^2 = 2^4$, $3^4 - 7^2 = 2^5$.

In Section 15, we will also raise another form of the twin prime conjecture — the k-prime number conjecture.

In 2004, by the analysis of ergodic theory and the Ramsay theory in combinatorics, the Chinese Australian mathematician Terence Tao (1975–) and the British mathematician Ben Green (1977–) proved the following result.

Theorem (Green–Tao). *There are arbitrarily long arithmetic progressions of prime numbers.*

Here the so-called length means the number of elements in the arithmetic progressions, for example, 3, 5, 7 is an arithmetic progression of prime numbers with length 3 and common difference 2; and 109, 219, 329, 439, 549 is an arithmetic progression of prime numbers with length 5 and common difference 110. In 2007, the Poland mathematician Jarosław Wróblewski found an arithmetic progression

of prime numbers with length 24 (in 2008 and 2010, arithmetic progressions of prime numbers with length 25 and 26 are also found).

$$468395662504823 + 45872132836530n \quad (0 \leq n \leq 23).$$

The Green–Tao theorem is very strong, since previously we even did not know if there are infinitely many arithmetic progressions of prime numbers with length 3. Because of this work and other work, Tao won the Fields Medal in 2006. His parents came from Hong Kong and graduated from the University of Hong Kong. His father was a pediatrician and his mother was a high school math teacher. In 1972, the couple moved to Australia. Unfortunately, the proof of Tao and Green is non-constructive and the common difference of primes is not fixed. Therefore, one cannot deduce the ancient twin prime conjecture.

On the other hand, in 1940, Erdős proved that there exist a constant $c < 1$ and infinitely many pairs of consecutive primes p and p' such that $p' - p < c \ln p$. In 2005, the American Daniel Goldston, Hungarian János Pintz and Turkish Cem Yildirim proved that such $c > 0$ could be arbitrarily small. In 2013, the American Chinese mathematician Zhang Yitang (1955–) proved the following theorem.

Theorem (Zhang Yitang). *There are infinitely many pairs of primes p, q such that their difference is less than 70 million.*

This means an important step towards the proof of the twin prime conjecture. Zhang was born in Shanghai and graduated from Peking University, and later got his Ph.D. degree from Purdue University. He worked as a waiter, accounting, and delivery person. When he was 58, he was still a temporary lecturer at the University of New Hampshire. It is worth mentioning that, by the program of Polymath 8 later proposed by Tao. The bound in Zhang's theorem is declining, it became 246 in March 2014.

It is interesting that Goldbach's occupation was not a mathematician, but a rich person who loved mathematics. Born in the Prussian city of Konigsberg (now in Russia), he liked to travel, make friends with top mathematicians, and communicate with them. Goldbach worked as a high school math teacher, and served as a secretary of the Petersburg Academy of Sciences and the tutor for Tsar Peter II.

In 1742, he entered the Russian Ministry of Foreign Affairs (until his retirement). It was in that year when he proposed the Goldbach conjecture.

Another interesting thing is when Zhang announced his amazing result in Boston, at the same time on the other side of the Atlantic, Lima born Peru mathematician Harald Helfgott (1977–) who works at the École Normale Supérieure in Paris, announced on the Internet that with the help of the computer, he has fully proved Goldbach's conjecture that every odd number greater than 7 can be expressed as the sum of three odd primes.

Now, we must point out that, prime numbers are used for the decomposition of natural numbers according to multiplication, and they are not advantage for the construction of natural numbers with respect to addition. Moreover, that even numbers are the sum of two prime numbers while odd numbers are the sum of three primes does not appear consistent and beautiful. This might be the pity of Goldbach's conjecture. Therefore, we give binomial coefficients a new concept and define figurate prime number as

$$\binom{p^i}{j},$$

where p is a prime, i and j are positive integers. These are special binomial coefficients, having the property of both prime numbers and figurate numbers. These numbers include 1, all prime numbers and their powers, also fewer but infinitely many even numbers. However, it is easy to see that, the number of figurate primes not exceeding x is the same order as the number of primes not exceeding x. We have verified to 10^7.

Conjecture B. *Any integer greater than 1 can be expressed as the sum of two figurate prime numbers.*

Further, if we define figurate numbers which are not prime as true figurate prime, almost half of them are even. Then we have the following stronger statement.

Conjecture C. *Any integer greater than 5 can be expressed as the sum of a prime number and a true figurate prime number.*

Here is the statistical table we make with $C = 2 + 2\sqrt{2}$:

Range	Primes	Figurate primes	True figurate primes
≤ 100	25	47	22
≤ 1000	168	226	58
≤ 10000	1229	1355	126
≤ 100000	9592	9866	274
≤ 1000000	79498	79096	598
order	$x/\log x$	$x/\log x$	$c\sqrt{x}/\log x$

Meanwhile, it is natural for us to raise a conjecture more subtle and maybe more difficult than twin primes conjecture.

Conjecture D. *There are infinitely many pairs of adjacent figurate prime numbers.*

Obviously, if p and $\frac{p^2-p+2}{2}$ are both primes, then $\binom{p}{2}$, $\binom{p}{2}+1$ is a pair of adjacent figurate prime numbers. This sequence of primes could be found in OEIS A164620, the first few are 2, 5, 13, 17, 41, 61, 89, 97, and the 1000th prime $p = 116797$. If the sequence has infinite number of elements, then Conjecture D is true. However, even if the following conclusion is not easy to prove or to give a counter example:

Conjecture E. *Except for 1 and trivial ones, figurate prime numbers are distinct.*

Conjecture E seems much easier than any other one. But we still cannot prove it, here is the result we obtain:

Proposition. Suppose $\binom{p^\alpha}{i} = \binom{q^\beta}{j}$ not trivial, α and β are positive integers, $0 < i < p^\alpha$, $0 < j < q^\beta$, then $p|i$, or $q|j$.

This was raised by Shane Chern in our number theory seminar in 2015. In order to prove it, we need a lemma established by E. E. Kummer in 1952, which was rediscovered by Lucas in 1878.

Lemma 1. *Let a and b be positive integers, m is the exact power of p dividing $\binom{a+b}{\alpha}$, i.e., $p^m \| \binom{a+b}{\alpha}$. Then m is equal to the number*

if "carry-overs" when performing the addition of a and b, written in the base p.

For the proof see Paulo Ribenboim, The New Book of Prime Number Records (Springer, New York, 1995). It also follows from the proof of Lemma 2.

Lemma 2. *Let $p^m \| \binom{n}{k}$, $1 \leq k \leq n$. Then $p^m \leq n$.*

Proof of Proposition. If $p \nmid i$, $q \nmid i$, for any $1 \leq i_0 < i$, let $p^0 \| i_\alpha$, $0 \leq v < \alpha$, then $p^v \| p^a - i_0$. Hence in the fraction

$$\binom{p^\alpha}{i} = \frac{p^\alpha(p^\alpha - 1) \cdots (p^\alpha - (i-1))}{1 \cdots (i-1)l}$$

from the second term to the ith one in the numerator has the same power of p as terms from the first to the i-1th in the denominator, respectively, therefore $p^\alpha \| \binom{p^\alpha}{i}$. Similarly, we can show that $q^\beta \| \binom{q^\beta}{i}$. By Lemma 2, $p^\alpha \leq q^\beta$, $q^\beta \leq p^\alpha$, we deduce $p^\alpha = q^\beta$, hence $i = j$ or $p^\alpha - j$. A contradiction! This completes the proof of the proposition.

Look back Hilbert's 8th problem. What he raised is, "After the thorough discussion on the Riemann formula, we might be able to solve the Goldbach problem strictly, i.e., whether every even number can be expressed as the sum of two prime numbers, and further more to solve the problem whether there exist infinitely many pairs of primes with difference 2, and even the more general problem, namely linear Diophantine equation $ax + by + c = 0$, $(a, b) = 1$ if there always exists prime solutions x and y."

After introducing the concept of figurate prime numbers, we now try to make the Diophantine equation that Hilbert worried about more meaningful, at the same time make the Goldbach conjecture linked with the twin prime conjecture. We have the following conjecture (the second part can be deduced by the properties of the Diophantine equation under Schinzel's Hypothesis).

Conjecture F. *Let a and b be positive coprime integers. Then for each $n \geq (a-1)(b-1)$, the equation*

$$ax + by = n$$

always has figurate prime number solutions (x, y); and as long as $n \equiv a + b \pmod 2$, the equation

$$ax - by = n$$

has infinitely many prime solutions (x, y).

Exercises 1

1. Prove that each odd number can be the difference of two squares.
2. Prove that for any integer n, we have $3|n^3 - n$, $5|n^5 - n$, and $7|n^7 - n$. How about $9|n^9 - n$?
3. Prove that if both p and $p + 2$ are primes greater than 3, then $6|p + 1$.
4. Show that $\sqrt{2}$ and $\sqrt{6}$ are not rational.
5. Show that for any positive integer, there are four consecutive integers which are composite.
6. Determine all the positive integer n such that $7|2^n - 1$.
7. Let n be odd. Find the number of ways that n can be the difference of two square numbers.
8. Let a, b, c be positive integers, $(a, c) = 1$, and $\frac{1}{a} + \frac{1}{b} = \frac{1}{c}$. Show that $a + b$ is square.
9. Show that there are infinitely many primes in the form of $4n + 3$.
10. Prove that there are no consecutive odds which are both the sum of two non-zero squares.
11. Prove that for any positive integer n, the number $\frac{1}{3} + \frac{1}{5} + \cdots + \frac{1}{2n+1}$ can be an integer.

Chapter 2

The Concept of Congruence

6. The Concept of Congruence

The Euclidean division tells us that for any given integer $b > 1$, when b divides an integer a, it produces a remainder r such that $0 \le r < b$. This concept is also quite common in daily life, for example, when we ask what time it is now, in fact it is a remainder of a total number of the time divided by 24. Another example is that when we ask what day today is, it is equivalent to the remainder of a total number of the days divided by 7. Further, if we know the 13th day of some month and some year is Wednesday, it is easy to know that the 6th, the 20th and the 27th days of that month are Wednesdays, too.

The face of a clock is a complete residue system modulo 12

It is worth mentioning that, either the concept "24 hours per day" or "7 days per week" was derived from ancient Babylon. The latter one might be related to the astronomical knowledge of those days, namely the moon circles around the earth in 28 days, which means

57

there are exactly 4 weeks per month. This argument had been extended to the ancient Greeks. According to Section1, the numbers 6 and 28 are perfect numbers. In this way, we have the concept of congruence in mathematics, which is defined as follows.

Definition 2.1. Given positive integer $m > 1$, for any two integers a and b, if they have the same remainders when divided by m, then we say a is congruent to b modulo m, denoted as $a \equiv b \pmod{m}$, and m is called the modulus of the congruence. If the remainders are different, then a and b are incongruent modulo m, denoted as $a \not\equiv b \pmod{m}$.

Comparing the above definition with Definition 3.1 in Section 3, we find that a and b are congruent modulo m if and only if $m|(a-b)$. On the other hand, a is divisible by b if and only if $a \equiv 0 \pmod{b}$. In other words, the division can be defined by congruence.

Immediately we get the following properties from Definition 2.1,

(1) reflexivity: $a \equiv a \pmod{m}$;
(2) symmetry: if $a \equiv b \pmod{m}$, then $b \equiv a \pmod{m}$;
(3) transitivity: if $a \equiv b \pmod{m}$, $b \equiv c \pmod{m}$, then $a \equiv c \pmod{m}$.

The following properties about congruence are obvious.

Theorem 2.1. *If $a \equiv b \pmod{m}$, $a' \equiv b' \pmod{m}$, then $a \pm a' \equiv b \pm b' \pmod{m}$ and $aa' \equiv bb' \pmod{m}$. Further, let $f(x)$ be any polynomial with integer coefficients. If $a \equiv b \pmod{m}$, then $f(a) \equiv f(b) \pmod{m}$.*

Theorem 2.2. *If $a \equiv b \pmod{m}$, $a = a_1 d, b = b_1 d$, and $(d, m) = 1$, then $a_1 \equiv b_1 \pmod{m}$. If $a \equiv b \pmod{m}$, $d|m$, and $d > 1$, then $a \equiv b \pmod{d}$.*

Theorem 2.3. *If $a \equiv b \pmod{m}$, then $(a, m) = (b, m)$.*

Using Theorem 2.1, we obtain the following two methods that easily determine divisibility. It is necessary to point out that such tests are likely to have come from India originally, and later they were first written into their books by Arabian mathematicians Al-Khwarezmi

(c. 783–850) and Al-Karaji (c. 953–c. 1029) and then transferred to the West. Al-Khwarezmi coined the term of algebra, and a transliteration of his name later gradually evolved to the mathematical term algorithm.

Example 2.1. A natural number is divisible by 3 or 9 if and only if the sum of its digits in decimal system is divisible by 3 or 9.

Example 2.2. Let a natural number a be expressed as

$$a = a_n 1000^n + a_{n-1} 1000^{n-1} + \cdots + a_0, \quad 0 \le a_i < 1000.$$

Then it is divisible by 7, 11 or 13 if and only if

$$(a_0 + a_2 + \cdots) - (a_1 + a_3 + \cdots) = \sum_{i=0}^{n} (-1)^i a_i$$

is divisible by 7, 11 or 13.

Example 2.3. By using induction, we can prove that for any odd number a,

$$a^{2^n} \equiv 1 \pmod{2^{n+2}}, \quad n \ge 1.$$

Example 2.4. By using the properties of congruence and some calculation, we can get the conclusion about Fermat's numbers in Section 3, namely

$$F_5 = 2^{2^5} + 1 \equiv 0 \pmod{641}.$$

Example 2.5. Let m be a positive integer and let an integer a be such that $(a, m) = 1$. Then there exits an integer x such that

$$ax \equiv 1 \pmod{m}.$$

Obviously, $(x, m) = 1$, we call x the inverse of a, noted as $a^{-1} \pmod{m}$ or a^{-1}. Theorem 1.6 in Section 4 shows that there exist integers x_0, y_0 such that $ax_0 + my_0 = 1$. Letting $x = x_0$ gives the conclusion. As a matter of fact, this is Qin Jiushao's *Da Yan Shu*, which is the key step towards the Qin Jiushao theorem (the Chinese Remainder Theorem) in Section 11. And it is also the starting point of the RSA system in Section 10.

Numbers and Roses, the covers of the first and revised Chinese editions

Example 2.6. The following congruences

$$a \equiv b \ (\text{mod } m_j), \quad 1 \le j \le k$$

are true if and only if

$$a \equiv b \ (\text{mod}[m_1, m_2, \dots, m_k]).$$

In particular, if $m_i (1 \le i \le k)$ are coprime to each other, then the above formula is equivalent to

$$a \equiv b \ (\text{mod } m_1 m_2 \dots m_k).$$

Here, we introduce a small secret about the ISBN, which is closely related to congruence. As we know, every book has the international standard book number (ISBN). Before 2004, ISBN generally consisted of 10 digits, which are divided into four parts, separated by a small dash. The first part is the country code, the second is the press code, the third is the book code, and the last is the checking code. The country code usually has only 1 digit, for example, for English

speaking countries it is 0 and 1. The code 2 is for the French speaking countries, 3 for the German speaking countries, 4 is for Japan, 5 for Russian speaking countries, 7 for the Chinese publications in the mainland. The country codes for less populous countries or regions generally have more than 1 digit, for example, Sweden is 91, Croatia is 953, Taiwan is 957 or 986. There is no unified code in the Spanish speaking countries, Mexico is 970 and Columbia is 958.

Suppose the digits from the left to the right are $\{a_1, a_2, \ldots, a_{10}\}$ in ISBN. Then the last code, called the checking code, satisfies the following relation:

$$a_{10} \equiv \sum_{i=1}^{9} i a_i \pmod{11}.$$

For example, the ISBN of the author's book *Numbers and Roses* (SDX Joint Publishing Company, Beijing) published in 2003 is 7-108-01706-7, where the first 7 is the country code for China, 108 is the code for SDX Joint Publishing Company, 01706 is the code for the book and

$$7 \equiv 1 \times 7 + 2 \times 1 + 3 \times 0 + 4 \times 8 + 5 \times 0 + 6 \times 1 + 7 \times 7$$
$$+ 8 \times 0 + 9 \times 6 \pmod{11}.$$

The number of digits for the ISBN of the books published after 2007 is increased from 10 to13. The first 3 are always 978, and the 4-12th are the same as the 1-9th before, only the calculation of the last code a_{13} is changed to

$$10 - a_{13} \equiv (a_1 + a_3 + \cdots + a_{11}) + 3(a_2 + a_4 + \cdots a_{12}) \pmod{10}.$$

For example, the ISBN of the revised edition of *Numbers and Roses* published by Commercial Press in 2012 is 978-7-100-09319-4, and 100 is the code for the Commercial Press, and

$$10 - 4 \equiv 9 + 8 + 1 + 0 + 9 + 1 + 3(7 + 7 + 0 + 0 + 3 + 9) \pmod{10}.$$

Gauss's Disquisitiones Arithmeticae

In 1801, at the age of 24, Gauss published his first book *Disquisitiones Arithmeticae*, written in Latin and published in Leipzig,

which is more than two hundred kilometers away from the University City Göttingen where Gauss taught. As the doctoral dissertations Gauss had previously completed, *Disquisitiones Arithmeticae's* printing fee were also covered by Ferdinand, the Duke of his hometown Brunswick. Earlier he had sent the book to the French Academy of Sciences but it was rejected. The Duke had been the patron of Gauss who was born poor and precocious. His statue still stands in the central square in downtown Brunswick today. In the dedication of the cover page, Gauss wrote.

> *Were it not for your (Ferdinand) unceasing benefits in support of my studies, I would not have been able to devote myself totally to my passionate love, the study of mathematics.*
>
> *Disquisitiones Arithmeticae* is divided into seven sections, they are, respectively, *Congruence in General, Congruence*

DISQVISITIONES

ARITHMETICAE

AVCTORE

D. CAROLO FRIDERICO GAVSS

LIPSIAE

IN COMMISSIS APVD GERH. FLEISCHER, JVN.

1801.

Title page of the first edition of Gauss' *Disquisitiones Arithmeticae* (1801)

of the First Degree, Residues of Powers, Congruence of the Second Degree, Forms and Indeterminate Equations of the Second Degree, Various Applications of the Preceding Discussions, and *Equations Defining Sections of a Circle.* The book includes the first proof of the quadratic reciprocity law known as the stone of arithmetic by Gauss, the discussion of primitive roots and indices, and the Euclidean mapping theory for regular polygons, which connects geometry, algebra and arithmetic together. It's worth mentioning that Gauss introduced the congruent notation "≡".

This book was so hard that few people understood it at the time. However, after reading it, elder Lagrange in Paris sent a letter to congratulate Gauss, "your *Disquisitiones Arithmeticae* has immediately made you a first-class mathematician". The younger German mathematician, intuitionistic pioneer Kronecker (1823–1891), regarded it as "the king of all the books". At the end of the century, the German historian of mathematics Moritz Cantor (1829–1920) commented,

Disquisitiones Arithmeticae *is the charter of arithmetic. Gauss was always reluctant to publish his book, and the benefit it brings to science is that today his printed work is still as important as its first publication. And his publication is the code better than any other codes of humans, because we can find no mistakes in it no matter when and where, which helps us understand what Gauss talked about his first masterpiece in the early years:* "Disquisitiones Arithmeticae *is a historical wealth". It is quite reasonable that he was so proud at that time.*

Before Gauss, Fermat, Euler and Lagrange (1738–1813) and others all had the chance to complete the foundation of number theory. However, Fermat did not like to show his own discovery. Instead, he only wrote down his own work piecemeal in the notebook. The interests of Euler and Lagrange in mathematics were too broad, and they had long-term work in the Academy of Sciences. Of course, Gauss also had many talents and great achievements. Fortunately, in his youth he devoted himself to number theory. When he completed the major work on number theory, he was working in the university,

so he completed this work naturally, and then he turned to other areas of research and discovery.

It is worth mentioning that, Gauss's doctoral degree was not awarded by the University of Göttingen, but by the University of Helmstedt in the Lower Saxony town that is in the east of Brunswick, where there was also Pfaff (1765–1825), the most famous mathematician in Germany at that time. And Gauss benefitted from his careful guidance. Therefore, as a student Gauss often walked through the three cities of Göttingen, Brunswick and Helmstedt. In 1799, when he was only 22 years old, Gauss got his doctoral degree from the University of Helmstedt, his thesis topic was *A New Proof of That Every Single Variable Polynomial Can be Decomposed into A Primary Polynomial or A Quadratic Polynomial*. In 1810, the University of Helmstedt merged into the University of Göttingen, but Pfaff went to the University of Halle.

By the end of 2011, the Chinese version of *Disquisitiones Arithmeticae* was published, and the translators are Chengbiao Pan and Mingyao Zhang.

7. Residue Classes and Residue Systems

We have introduced the concept of congruence in the previous section. For any given module, if we put the numbers with the same remainder together, we can get the concept of residue class. In this section, we mainly discuss the properties of residue classes and residue systems that are closely related to residue classes. By the Euclidean division and the definition of congruence, we easily get the following theorem.

Theorem 2.4. *Let $m > 1$ be an arbitrary positive integer. Then all integers can be divided into m sets, denoted as $\{[0], [1], \ldots, [m-1]\}$, where $[i](0 \leq i \leq m-1)$ is composed of all integers of the form $qm + i(q = 0, \pm 1, \pm 2 \ldots)$. These sets satisfy the following properties:*

(1) *Every integer is included in only one such set.*
(2) *Two integers are in the same set if and only if they are congruent modulo m.*

Now, we can give the definition of residue classes and complete residue systems.

Definition 2.2. Each $[i]$ in Theorem 2.4 is called a residue class modulo m. The set of all residue classes is written by Z_m, i.e., $Z_m = \{[0], [1], \ldots, [m-1]\}$; if $a_0, a_1, \ldots, a_{m-1}$ are integers, and any two of them are not in the same residue class, then $\{a_0, a_1, \ldots, a_{m-1}\}$ is called a complete residue system modulo m.

This definition and Theorem 2.3 show that if there exists a certain number in a residue class modulo m, which is coprime with m, then all numbers in that residue class are coprime with m. From Theorem 2.4, it's easy to get the following corollary.

Corollary 2.1. *A set of m integers forms a complete residue system modulo m if and only if each two of them are not congruent modulo m.*

Example 2.7. The following sequences are all special complete residue systems modulo m:

$$0, 1 \ldots m - 1 \text{ (The least nonnegative complete residue system)};$$

$$1, 2 \ldots m \text{ (The least positive complete residue system)};$$

$$-\frac{m-1}{2}, \ldots, -1, 0, 1, \ldots, \frac{m-1}{2} \text{ (if } m \text{ is odd)}$$

$$-\frac{m}{2}, \ldots, -1, 0, 1, \ldots, \frac{m}{2} - 1 \text{ (if } m \text{ is even)}$$

$$-\frac{m}{2} + 1, \ldots, -1, 0, 1, \ldots, \frac{m}{2} \text{ (if } m \text{ is even)}.$$

The last three complete residue systems are called the absolute minimum complete residue systems modulo m.

Theorem 2.5. *Let m be a positive integer and b an integer, $(a, m) = 1$. If x passes through a complete residue system modulo m, then $ax + b$ also passes through a complete residue system modulo m.*

Proof. By Corollary 2.1, we only need to show that when each two of x_1, x_2, \ldots, x_m are incongruent modulo m, so are each two of

$ax_1 + b, ax_2 + b, \ldots, ax_m + b$. Suppose $ax_i + b \equiv ax_j + b \pmod{m}$ $(i \neq j)$. From Theorem 2.1 we know that $ax_i \equiv ax_j \pmod{m}$. Because $(a, m) = 1$, we get $x_i \equiv x_j \pmod{m}$ from Theorem 2.2. This completes the proof of Theorem 2.5. □

Theorem 2.6. *Let m_1, m_2 be two coprime positive integers. If x_1, x_2 pass through the complete residue systems modulo m_1, m_2, respectively, then $m_2 x_1 + m_1 x_2$ passes through the complete residue system modulo $m_1 m_2$.*

Proof. According to the assumption, we know that when x_1, x_2 passes through m_1, m_2, respectively, $m_2 x_1 + m_1 x_2$ passes through $m_1 m_2$ numbers. By Theorem 2.4, we only need to prove that these integers $m_1 m_2$ are incongruent modulo $m_1 m_2$ each other.

Conversely, suppose $m_2 x_1' + m_1 x_2' \equiv m_2 x_1'' + m_1 x_2'' \pmod{m_1 m_2}$, where x_1', x_1'' and x_2', x_2'' are taken from the complete residue system modulo m_1, m_2, respectively. By Theorem 2.1 and Theorem 2.2, we have,

$$x_1' \equiv x_1'' \pmod{m_1}, \quad x_2' \equiv x_2'' \pmod{m_2}.$$

This completes the proof of Theorem 2.6. □

Example 2.8. If s, t are two arbitrary positive integers. If u, v pass through the complete residue systems modulo p^{s-t} and p^t, respectively, then $x = u + p^{s-t} v$ passes through the complete residue system modulo p^s.

Example 2.9. Suppose the positive integers m_1, m_2, \ldots, m_k are coprime each other, $m_1 m_2 \ldots m_k = m$ and $m = m_i M_i$ with $1 \leq i \leq k$. If x_1, x_2, \ldots, x_k pass through the complete residue system modulo m_1, m_2, \ldots, m_k, respectively, then $M_1 x_1 + M_2 x_2 + \cdots + M_k x_k$ passes through the complete residue system modulo m.

Example 2.10. Suppose m_1, m_2, \ldots, m_k are positive integers. If x_1, x_2, \ldots, x_k pass through the complete residue system modulo m_1, m_2, \ldots, m_k, respectively, then

$$x_1 + m_1 x_2 + m_1 m_2 x_3 + \cdots + m_1 m_2 \cdots m_{k-1} x_k$$

passes through the complete residue system modulo $m_1 m_2 \ldots m_k$.

Example 2.11. Let $m > 2$ if x and y pass through the complete residue system modulo m, respectively, could it be possible for $x + y$ or $x - y$ to pass through the complete residue system modulo m?

When m is odd, the answer is positive. In fact, taking $x = \{0, 1, 2, \ldots, m - 1\}$, $y = \{1, 2, \ldots, m - 1, 0\}$ then $x + y = \{1, 3, \ldots, m - 2, 0, 2, \ldots, m - 1\}$ forms a complete residue system modulo m. And if we take $y = \{0, m - 1, m - 2, \ldots, 1\}$, then $x - y = \{0, 2, \ldots, m - 1, 1, 3, \ldots, m - 2\}$ forms a complete residue system modulo m. When m is even, the answer is negative. In fact, let x and y be two sets of complete residue systems modulo m, then the sum of either sums or differences between corresponding elements are multiples of m, while if $x + y$ or $x - y$ forms a complete residue system modulo m, then those sums are both congruent to $m/2$ modulo m. A contradiction!

The above example is about the sum and the difference, in Section 14, concerning covering systems, we will discuss the multiplication problem. Among the following two problems, the latter is a little more difficult:

Let $n > 1$, if and only if n is odd, we can choose and make pairs for the set $\{0, 1, 2, \ldots, 2n - 1\}$, their sum passes through a complete residue system modulo m. The necessity: first of all, the sum is $n(2n - 1)$, if we make the pairs successfully, then the sum of elements of the least residue system modulo is $n(n + 1)/2$. Hence,

$$n(2n - 1) \equiv \frac{n(n + 1)}{2} \pmod{n}.$$

When n is even, the above congruence is impossible. When n is odd, $(4, n) = 1$, then if x passes through the complete residue system modulo m, so does $4x + 1$. Adding the elements of $\{0, 1, 2, \ldots, 2n - 1\}$ in pairs in order, then we get a set of n elements $\{1, 1 + 4, 1 + 8, \ldots, 1 + 4(n - 1)\}$. This completes the proof.

Let $n > 1$, if and only if $n \equiv 0$ 或 $1 \pmod{4}$, we can choose and make pairs for the set $\{0, 1, 2, \ldots, 2n - 1\}$, their difference is exactly $\{1, 2, \ldots, n\}$. The necessity: Let S and T be the sums of minuends and subtractors, respectively, after making pairs, then $S + T = n(2n - 1)$, $S - T = n(n + 1)$, hence $4S = n(5n - 1)$, taking into congruence modulo 4, we deduce the result.

For the cases without solution, we could remove an element from the set $\{0, 1, 2, \ldots, 2n\}$ then make the pairs successfully for both addition and subtraction. Moreover, if there are solutions, how about the number of solutions?

Now, we need to define a special kind of residue class — reduced residue classes and the reduced residue system that consist of reduced residue classes, they will play an important role in the future.

Definition 2.3. If the number in a certain residue class modulo m is coprime with m, then we call it a residue class coprime with m. If we pick out one number from each residue class coprime with m, then we get a set that is called the reduced system modulo m.

Example 2.12. If p is a prime number, then $1, 2 \ldots p - 1$ form a reduced residue system modulo p.

In order to further discuss the properties of reduced residue systems, we need to introduce a significant arithmetical function. The so-called arithmetical function refers to any function with its independent variables taking values among positive integers or integers. Such functions are indispensable in the study of number theory.

Definition 2.4. Euler's totient function $\phi(n)$ is a function defined on positive integer n, whose value is equal to the number of the integers in the sequence $1, 2 \ldots n$, coprime with n, i.e.,

$$\phi(\text{n}) = \sum_{\substack{1 \leq i \leq n \\ (i,n)=1}} 1.$$

For example, $\phi(1) = \phi(2) = 1, \phi(3) = \phi(4) = \phi(6) = 2, \phi(5) = \phi(8) = \phi(10) = 4$, and $\phi(7) = \phi(9) = 6$. This function was defined by Euler in 1760, and he denoted it by $\pi(\text{n})$. In 1801, Gauss used $\phi(\text{n})$ for the first time in *Disquisitiones Arithmeticae*. It is also known as the Euler phi function. In 1879, Sylvester invented an English word *totient* to represent this function, so it is also called Euler's totient function. The author guesses that this is because *totter* has a meaning toddle, while *tient* is a suffix of the word quotient.

It is easy to see that $\phi(n)$ is even when $n > 2$. The least even nontotient is 14, which means that there is no solution for $\phi(x) = 14$. In 1993, Mingzhi Zhang (*J. Number Theory*, 43, 168–173) showed that, for any positive integer m, there exists a prime p such that mp is nontotient. This result shows that there exist infinitely many nontotients. However, Erdős has the following conjecture.

Conjecture (Erdős). *For any positive integer n, the equation $2n = \phi(x) + \phi(y)$ has a solution.*

Obviously, if the even Goldbach conjecture is true, then the above equation always has prime solutions, so it is a weak Goldbach conjecture.

The first 1000 numbers of Euler's function

Function $[x]$ and the $3x + 1$ problem

In the study of number theory, we also need to consider real numbers in addition to integers and rational numbers as well as functions defined on them. Functions $[x]$ and $\{x\}$ are two examples; they play an important role in some topics of number theory. Here x is an arbitrary real number, $[x]$ represents the maximum integer not

exceeding x, or the integer part or x, $\{x\} = x - [x]$ is the decimal part of x. For example,

$$[e] = 2, [-\pi] = -4, \left\{-\frac{1}{3}\right\} = \frac{2}{3}, \{\sqrt{2}\} = 0.414\cdots.$$

We deduce the following properties from the definition:

(1) $x = [x] + \{x\}, x - 1 < [x] \le x$;
(2) $[x] + [y] \le [x + y], \{x\} + \{y\} \ge \{x + y\}$;
(3) In the Euclidean division of two integers a, $b(>0)$, i.e., $a = bq + r(0 \le r < b)$,

$$q = \left[\frac{a}{b}\right], \quad r = \left\{\frac{a}{b}\right\} b.$$

(4) If a and b are arbitrary positive integers, then the number of multiples of b not greater than a is $[\frac{a}{b}]$. Further, we have

$$\left[\frac{a}{b}\right] \le \frac{a}{b} \le \left[\frac{a}{b}\right] + 1 - \frac{1}{b}.$$

First of all, as an application of $[x]$, we obtain a standard factor decomposition for $n!$. Here $n!$ represents the product of all the natural numbers from 1 to n, i.e., $n! = 1 \times 2 \times \cdots \times n$.

Lagrange Theorem. For any integer $n > 1$, we have $n! = \prod_{p \le n} p^{\alpha_p}$, where $\alpha_p = \sum_{r=1}^{\infty} [\frac{n}{p^r}]$.

Proof. Obviously, α_p is the sum of indices of p in the standard factor decomposition of $2, 3, \ldots, n$. Suppose the number of integers with the index of p numbering r is n_r $(r \ge 1)$. Then

$$\alpha_p = n_1 + 2n_2 + 3n_3 + \cdots$$
$$= n_1 + n_2 + n_3 + \cdots$$
$$+ n_2 + n_3 + \cdots$$
$$+ n_3 + \cdots$$
$$+ \cdots$$
$$= N_1 + N_2 + N_3 + \cdots$$

where $N_r = n_r + n_{r+1} + \cdots$, is exactly the number of integers divisible by p^r in $2, 3, \ldots, n$. According to above Property 4, we have $N_r = [\frac{n}{p^r}]$. This completes the proof of theorem. \square

Next, we show the genius proof on the divergence of the series $\sum_p \frac{1}{p}$ by Erdős in 1938 when he was 25.

Proof. Suppose that the series is convergent. Let p_i be the ith prime, then there exists j, such that $\sum_{i \geq j+1} \frac{1}{p_i} < \frac{1}{2}$. We say that p_1, \ldots, p_j are small primes, and p_{j+1}, p_{j+2}, \ldots are big primes. For any positive integer N, denote by N_b the number of the integers $n \leq N$ which are divisible by at least one big prime and N_s the number of the integers $n \leq N$ whose factors are all small primes. Then it is easy to see that $N = N_b + N_s$.

Since $[\frac{N}{p_i}]$ represents the number of all the multiples of p_i which are less or equal to N, hence

$$N_b \leq \sum_{i \geq j+1} \left[\frac{N}{p_i}\right] \leq \sum_{i \geq j+1} \frac{N}{p_i} < \frac{N}{2}.$$

Now we try to get an upper bound of N_s. Each $n \leq N$ with only small prime factors can be expressed as $n = a_n b_n^2$, where a_n is square-free and each a_n is the product of distinct small primes. All together there are exactly 2^j choices for the square-free part. Meanwhile, since $b_n \leq \sqrt{n} \leq \sqrt{N}$, there are no more than \sqrt{N} choices for b_n, therefore

$$N_s \leq 2^j \sqrt{N}.$$

In order to make $2^j \sqrt{N} \leq \frac{N}{2}$, we only need to take $N = 2^{2j+2}$. Then, we have $N_b + N_s < N$. A contradiction!

Finally, we want to talk about the famous $3x+1$ problem. In 1937, the German mathematician Lothar Collatz (1910–1990) considered the following arithmetical function:

$$f(x) = \begin{cases} \dfrac{x}{2}, & (x \equiv 0 \pmod 2) \\ \dfrac{3x+1}{2}, & (x \equiv 1 \pmod 2) \end{cases}.$$

He guessed that, no matter what number you start with, you will always eventually reach 1, and the recursive times for n to reach 1 is called the stopping time of n. It is called the Collatz conjecture, but there are some other names, such as Kakutani conjecture. Shizuo Kakutani (1911–2004) was a Japanese-born American mathematician; his daughter Michi Ko is a literary critic, who won the Pulitzer prize, and she works for the *New York Times*. Although some people have checked that while x is less than 3×2^{50}, the conjecture is true, so far no one can prove or disprove it. Erdős said about the Collatz conjecture: "Mathematics may not be ready for such problems." And some people excluded the possibility of the $qx + 1$ problem (q is an odd number greater than 3), that is to say a natural generalization of the conjecture does not exist. Recently, thanks to the communication with middle school teacher Shengli Xu who lives in the mountain area in southwest Zhejiang province and amateur Lixing Shen who works in an international cooperation organization in Cambodia, we have made some new attempts.

Firstly, we change $f(x)$ into the following equivalent arithmetical function:

$$g(x) = \begin{cases} \dfrac{x}{2}, & x \equiv 0 \ (\mathrm{mod}\ 2) \\ \left[\dfrac{3}{2}x\right] + 1, & x \equiv 1 \ (\mathrm{mod}\ 2) \end{cases}.$$

Here $[x]$ is the integer part of x, this is because if we write $n = 2s+1$, then, $[\frac{3n}{2}] + 1 = [\frac{6s+3}{2}] + 1 = 3s + 2 = \frac{3n+1}{2}$. For any positive integer c, we set

$$g_c(x) = \begin{cases} \dfrac{x}{2}, & x \equiv 0 \ (\mathrm{mod}\ 2) \\ \left[\dfrac{c+2}{c+1}x\right] + 1, & x \equiv 1 \ (\mathrm{mod}\ 2) \end{cases}.$$

When $c = 1$ we get $g_1(x) = g(x)$, which is the $3x+1$ problem. When c is even then $c+2$ does not reach 1. When $c = 3$, there exists a loop $\{37, 47, 59, 74\}$. While c is an odd greater than 4, the number $g_c(x)$ can still eventually reach 1. We have verified this for even $c \leq 50000$ and $x \leq 10^7$.

Directed graph showing the orbits of the first 1000 numbers

As for the corresponding $3x-1$ problem, it's known that it cannot be always induced into 1. In fact, this is equivalent to

$$h(x) = \begin{cases} \dfrac{x}{2}, & x \equiv 0 \pmod 2 \\[2mm] \left[\dfrac{3}{2}x\right], & x \equiv 1 \pmod 2 \end{cases},$$

which is expected to be turned into 1 or one of the following two loops, namely, $\{5, 7, 10\}$ and $\{17, 25, 37, 55, 82, 41, 61, 91136, 68, 34\}$. However, if we define

$$i(x) = \begin{cases} \dfrac{x}{2}, & x \equiv 0 \pmod 2 \\[2mm] \left[\dfrac{4}{3}x\right], & x \equiv 1 \pmod 2 \end{cases},$$

it is still possible to change any positive integer into 1 (it has been verified by us for $x \le 10^{12}$).

As for the original $3x + 1$ problem, we also have the following extension. Let k be any non-negative integer, considering the function

$$f_k(x) = \begin{cases} \dfrac{x}{2}, & \text{if } x \text{ is even,} \\[2mm] \dfrac{3x + 3^k}{2}, & \text{if } x \text{ is odd.} \end{cases} \qquad \square$$

We conjecture that for any positive integer x, $f_k(x)$ can eventually reach 3^k. In particular, if $k = 0$, this is the $3x + 1$ problem.

Remark. In 1988, B. C. Wiggin (Conjecture about the $3x+1$ family, *J. Recreational Math.*, 20.2, 52–56) proposed a new kind of function, he conjectured that for any positive integer x, each of them can eventually reach 1. For example,

$$f_3(x) = \begin{cases} \dfrac{x}{3}, & \text{if } x \equiv 0 \pmod 3, \\[2mm] \dfrac{4x \mp 1}{3}, & \text{if } x \equiv \pm 1 \pmod 3, \end{cases}$$

$$f_4(x) = \begin{cases} \dfrac{x}{2}, & \text{if } x \equiv 0 \ (\text{mod}\,2), \\[2mm] \dfrac{5x \mp 1}{4}, & \text{if } x \equiv \pm 1 \ (\text{mod}\,4), \end{cases}$$

$$f_5(x) = \begin{cases} \dfrac{x}{5}, & \text{if } x \equiv 0 \ (\text{mod}\,5), \\[2mm] \dfrac{6x + 1}{5}, & \text{if } x \equiv \pm 1 \ (\text{mod}\,5), \\[2mm] \dfrac{6x \mp 2}{5}, & \text{if } x \equiv \pm 2 \ (\text{mod}\,5). \end{cases}$$

8. The Fermat–Euler Theorem

Let $m > 1$ be an integer, there are a total $\phi(m)$ numbers in each reduced residue systems modulo m. Denote all the elements in a reduced residue system modulo m as Z_m^*. For any given $\phi(m)$ integers $a_1, a_2, \ldots, a_{\phi(m)}$ that are incongruent modulo m each other, if for any $1 \le i \le \phi(m)$, there holds $(a_i, m) = 1$, then they form a reduced residue system modulo m.

Numbers below 10000 and their corresponding total stopping time

Similar to Theorems 2.5 and 2.6, we have

Theorem 2.7. *Let* $(a, m) = 1$*. If* x *passes through a reduced residue system modulo* m*, then* ax *also passes through the reduced residue system modulo* m*.*

Proof. First, ax also passes through $\phi(m)$ numbers, due to $(a, m) = 1$, $(x, m) = 1$, and so $(ax, m) = 1$. If $ax_1 \equiv ax_2 \pmod{m}$, then by Theorem 2.2, it can be derived that $x_1 \equiv x_2 \pmod{m}$, which is a contradiction. □

Example 2.13. Let $m > 1$ and $(a, m) = 1$. If x passes through the reduced residue system modulo m, then

$$\sum_x \left\{\frac{ax}{m}\right\} = \frac{1}{2}\phi(m).$$

Proof. For $(a, m) = 1$, it's easy to see that $(m - a, m) = 1$. By Theorem 2.7, if x passes through the reduced residue system modulo m, then so do ax and $(m - a)x$. Therefore,

$$\sum_x \left\{\frac{ax}{m}\right\} = \sum_x \left\{\frac{(m - a)x}{m}\right\}.$$

On the other hand, note that $\left\{\frac{ax}{m}\right\} + \left\{\frac{(m-a)x}{m}\right\} = 1$ for each integer x. Then

$$\sum_x \left\{\frac{ax}{m}\right\} + \sum_x \left\{\frac{(m - a)x}{m}\right\} = \phi(m).$$

Add the two sides of the above two equations, we get the required result. □

Theorem 2.8. *Let* m_1, m_2 *be two positive integers coprime with each other. If* x_1, x_2 *pass through the reduced residue system modulo* m_1, m_2*, respectively, then* $m_2 x_1 + m_1 x_2$ *passes through the reduced residue system modulo* $m_1 m_2$*.*

Proof. Since a reduced residue system consists of all the integers in the complete residue system that are coprime with the module, we only need to prove: if x_1, x_2 passes through the reduced

residue system modulo m_1, m_2, respectively, then $m_2x_1+m_1x_2$ passes through the integers in the complete residue system modulo m_1m_2 that are coprime with m_1m_2.

By Theorem 2.6, if x_1, x_2 pass through the complete residue system modulo m_1, m_2, respectively, $m_2x_1 + m_1x_2$ passes through the complete residue system modulo m_1m_2. And if $(x_1, m_1) = (x_2, m_2) = 1$, noting that $(m_1, m_2) = 1$, we have $(m_2x_1, m_1) = (m_1x_2, m_2) = 1$. Then, $(m_2x_1 + m_1x_2, m_1) = (m_2x_1 + m_1x_2, m_2) = 1$. According to Lemma 4.3 in Section 4, $(m_2x_1 + m_1x_2, m_1m_2) = 1$. On the other hand, if $(m_2x_1 + m_1x_2, m_1m_2) = 1$, it can also be gradually reasoned backward, so that $(x_1, m_1) = (x_2, m_2) = 1$. This completes the proof of Theorem 2.8.

The following corollary comes from Theorem 2.8 and its proof directly. □

Corollary 2.2. *If* $(m_1, m_2) = 1$, *then* $\phi(m_1m_2) = \phi(m_1)\phi(m_2)$.

Defined on the positive integers, an arithmetical function not vanishing identically and meeting the above property is called multiplicative function. If it is without the coprime condition, then it is called a complete multiplicative function. The Euler totient function is a multiplicative function, but not a complete multiplicative function.

Example 2.14. Suppose positive integers m_1, m_2, \ldots, m_k are coprime with each other, $m_1m_2\ldots m_k = m$, and $m = m_iM_i$ with $1 \le i \le k$. If x_1, x_2, \ldots, x_k pass through the reduced residue system modulo m_1, m_2, \ldots, m_k, respectively, then $M_1x_1 + M_2x_2 + \cdots + M_kx_k$ passes through the reduced system modulo m.

Theorem 2.9. *Let* $n = p_1^{\alpha_1}p_2^{\alpha_2 0} \cdots p_k^{\alpha_k} > 1$. *Then*

$$\phi(n) = n \left(1 - \frac{1}{p_1}\right)\left(1 - \frac{1}{p_2}\right)\cdots\left(1 - \frac{1}{p_k}\right).$$

Proof. Corollary 2.2 shows that $\phi(n) = \phi(p_1^{\alpha_1})\phi(p_2^{\alpha_2})\cdots\phi(p_k^{\alpha_k})$. We only need to prove that $\phi(p^\alpha) = p^\alpha - p^{\alpha-1}$. By the definition, $\phi(p^\alpha)$ equals p^α, subtracting the numbers of those numbers in $1, 2\ldots p^\alpha$ not coprime with p^α, namely $p^{\alpha-1}$. By the standard factor decomposition of n, we obtain Theorem 2.9.

It is easy to see from Theorem 2.9 that, for any positive integers m, n, if $m|n$, then $\phi(m)|\phi(n)$. □

Example 2.15 (Gauss). For any positive integer n, it is true that $\sum_{d|n} \phi(d) = n$.

Proof. Classifying the set of positive integers $1, 2 \ldots n$ according to their greatest common factor with n, we find that different positive factor of n corresponds to the disjoint subsets in this set. We shall determine this subset,

$$(j, n) = d, 1 \leq j \leq n.$$

Let $j = dh$. Then the above formula is equivalent to

$$\left(h, \frac{n}{d}\right) = 1, 1 \leq h \leq \frac{n}{d}.$$

The number of h that meet this condition is $\phi(\frac{n}{d})$. So $\sum_{d|n} \phi(\frac{n}{d}) = n$. Moreover, when d passes through all positive divisors of n, so does $\frac{n}{d}$, this helps us to get the conclusion.

Now we introduce two famous congruences in number theory, namely Fermat's Little Theorem and Euler's Theorem. In 1640, Fermat pointed out the following result in a letter to a friend Bernard Frénicle de Bessy (c. 1605–1675). □

Theorem 2.10 (Fermat). *For any integer a and prime number p,*

$$a^p \equiv a \pmod{p};$$

In particular, if $(a, p) = 1$,

$$a^{p-1} \equiv 1 \pmod{p}.$$

It is obvious that these two congruences in Theorem 2.10 can be derived by each other, which means they are equivalent, known as Fermat'slittle theorem. This name was given by Hensel in 1913, who found p-adic number. An interesting generalization of Fermat's little theorem is: Let p be prime, and m, n positive integers, if $m \equiv n \pmod{p-1}$, then for any integer a coprime with p, $a^m \equiv a^n$ \pmod{p}.

Now we give a proof by using Lagrange theorem in abstract algebra (the order of a finite group is a multiple of the order of its

arbitrary subgroup): let U_P be the multiplicative group consisting of invertible elements in quotient group Z/Z_P, whose unit is 1. Obviously, $|U_p| = p - 1$. If $a \in U_p$, then the order of the subgroup generated by a is l. By the Lagrange theorem, $l|p - 1$, so $a^{p-1} = (a^l)^{(p-1)/l} \equiv 1 \pmod{p}$.

It is worth mentioning that Fermat himself did not give a proof. Fortunately, we have Euler, who never let any question Fermat studied go easily. In 1736, the first proof of Euler's was published. After a long time, people found in Leibniz's posthumous manuscript that not later than 1683 he had given the same proof. But in 1760, Euler proved a stronger result, now called Euler's theorem.

The King of amateur mathematicians, Pierre de Fermat

Theorem 2.11 (Euler). *If $m > 1$ is an integer and $(a, m) = 1$, then*

$$a^{\phi(m)} \equiv 1 \pmod{m}.$$

Proof. Let $x_1, x_2, \ldots, x_{\phi(m)}$ be a reduced residue system modulo m, by Theorem 2.7, $ax_1, ax_2, \ldots, ax_{\phi(m)}$ is also a reduced residue system

modulo m, so

$$(ax_1)(ax_2)\cdots(ax_{\phi(m)}) \equiv x_1x_2\ldots x_{\phi(m)} \pmod{m}.$$

That is,

$$a^{\phi(m)}(x_1x_2\ldots x_{\phi(m)}) \equiv x_1x_2\ldots x_{\phi(m)} \pmod{m}.$$

By Lemma 1.3 in Section 4 and $(x_1, m) = (x_2, m) = \cdots = (x_{\phi(m)}, m) = 1$, we have $(x_1x_2\ldots x_{\phi(m)}, m) = 1$. Theorem 2.11 follows from Theorem 2.2.

Earlier in the year that Fermat discovered his little theorem, he wrote to Mersenne saying that: if n is a composite, then $2^n - 1$ must be a positive, too, and if n is a prime, then $2n|2^n - 2$. From this we can infer that Fermat may have got the idea for his little theorem when he studied the problem of perfect numbers (Mersenne prime).

Fermat's little theorem follows immediately from the Euler theorem. We show that Euler's theorem can be derived from Fermat's little theorem, too.

By Fermat's little theorem, there is an integer t_1 such that $a^{p-1} = 1 + pt_1$. Therefore, there is an integer t_2, such that $a^{(p-1)p} = (1 + pt_1)^p = 1 + p^2t_2$. We continue it until there is t_α such that $a^{\phi(p^\alpha)} = 1 + p^\alpha t_\alpha$. Therefore, $a^{\phi(n)} \equiv 1 \pmod{p^\alpha}$. By the conclusion of Example 2.6, $a^{\phi(n)} \equiv 1 \pmod{n}$. So Euler's theorem is proved.

In 1909, the German mathematician Arthur Wieferich (1884–1954) asked whether there is a prime p such that

$$2^{p-1} \equiv 1 \pmod{p^2}.$$

Such a prime is known as Wieferich prime. So far, in the range less than 1.7×10^{16}, people only found two Wieferich primes, namely 1093 and 3511, respectively, in 1913 and 1922. Even so, we cannot prove that there exist infinitely many non Wieferich primes.

There are also so-called Wieferich pairs, namely two distinct primes p and q, such that

$$p^{q-1} \equiv 1 \pmod{q^2}, q^{p-1} \equiv 1 \pmod{p^2}.$$

So far, only seven Wieferich pairs are known, i.e., (2, 1093), (3, 1006003), (5, 1645333507), (5, 188748146801), (83, 4871), (911, 318917), (2903, 18787). □

Finally, we introduce an application of Fermat's little theorem. In 1961, by using the method of algebra and combination, Erdős *et al.* showed the following result:

Theorem 2.12 (Erdős-Ginzburg-Ziv). *One can always take n from arbitrary $2n - 1$ integers such that their sum is a multiple of n.*

There are several proofs of Theorem 2.12 by the elementary method. Here we present one by Gao Weidong (*Journal of Northeastern Normal University*, 4.1985, 15–17). His proof consists of two Lemmas.

Lemma 2.1. *Let p be prime. Then one can always take p numbers from arbitrary $2p - 1$ integers such that their sum is a multiple of p.*

Proof. For any $2p-1$ integers $a_1, a_2, \ldots, a_{2p-1}$, we can make totally $C_{2p-1}^p = \binom{2p-1}{p}$ sums:

$$S_1 = a_1 + a_2 + \cdots + a_p, \ldots, S_{C_{2p-1}^p} = a_p + a_{p+1} + \cdots + a_{2p-1}.$$

Suppose $(S_i, p) = 1 (1 \leq i \leq C_{2p-1}^p)$. Then by Fermat's little theorem,

$$S_i^{p-1} \equiv 1 \pmod{p}, 1 \leq i \leq C_{2p-1}^p,$$

hence

$$\sum_{i=1}^{C_{2p-1}^p} S_i^{p-1} \equiv \sum_{i=1}^{C_{2p-1}^p} 1 = C_{2p-1}^p = \frac{(2p-1)!}{(p-1)!p!} \pmod{p},$$

which is not a multiple of p.

On the other hand, consider the expressions of the left sum in the above congruence. The term containing

$$\frac{(p-1)!}{r_1! r_2! \cdots r_k!} a_{i_1}^{r_1} a_{i_2}^{r_2} \cdots a_{i_k}^{r_k} (r_1 + r_2 + \cdots + r_k$$
$$= p - 1, 1 \leq k \leq p - 1, r_i > 0, 1 \leq i \leq k)$$

appears

$$C_{2p-1-k}^{p-k} = \frac{(2p-1-k)!}{(p-k)!(p-1)!}$$

times. Obviously, each such binomial coefficient is a multiple of p. A contradiction! □

Lemma 2.2. *If Theorem 2.12 is true for both $n = k$ and $n = l$, then it is also true for $n = kl$.*

Proof. By assumption, one can select $2l - 1$ groups from $2kl - 1$ numbers, each group contains k numbers such that their sum is a multiple of k, i.e.,

$$kr_1, kr_2, \ldots, kr_{2l-1}.$$

Again by assumption, from $r_1, r_2, \ldots, r_{2l-1}$ we can select l numbers such that their sum is a multiple of kl. This completes the proof of Lemma 2.2.

In the spring of 2014, the author raised this problems in the class of Elementary Number Theory, and together we improved the result to the following form:

> *If d is a positive factor of n, then we can take n numbers from arbitrary $n + d - 1$ integers such that their sum is a multiple of d.*
>
> *If s is a positive integer, then we can take sn from arbitrary $(s + 1)n - 1$ integers such that their sum is a multiple of sn.*

When $d = n$ or $s = 1$, this is Theorem 2.12. Next we use a congruent linear transformation to give a proof of the latter.

As a matter of fact, let $a_i \equiv j \pmod{s}$, $1 \leq i \leq (s + 1)n - 1$. Taking $b_i = \frac{a_i - j}{s}$, we get $(s + 1)n - 1$ integers. By using Theorem 2.12 repeatedly, we can take s groups, each containing n integers such that their sums are nr_1, nr_2, \ldots, nr_s, respectively. Adding all the sn numbers, we have

$$\sum_{i(sn\uparrow)} a_i = \sum_{i(sn\uparrow)} (sb_i + j) = s \sum_{i(s\uparrow)} nr_i + snj = sn \left(\sum_{i(s\uparrow)} r_i + j \right).$$

Finally we have to say that $(s+1)n - 1$ cannot be smaller, since we can take $(s+1)n - 2$ numbers as

$$a_1 = a_2 = \cdots = a_{sn-1} = j, b_1 = b_2 = \cdots = b_{n-1} = -s + j.$$

Take k numbers as j and $sn - k$ numbers as $-s + j$. Their sum is $kj + (sn - k)(-s + j) = sk + sn(-s + j)$.

Note that $(s - 1)n + 1 \leq k \leq sn - 1$, therefore the above sum could not be a multiple of sn. \square

The Euler number and the Euler prime

By the definition of Euler's function, $\phi(a) < a$ for any integer $a > 1$. Suppose n is an arbitrary natural number, and let $\phi^n(a) = \phi(\phi^{n-1}(a))$. Then for any a, there must exist a positive integer $h(a)$ that meets $\phi^{h(a)}(a) = 1$. We call it the height of a. For example, $h(2) = 1, h(3) = h(4) = h(6) = 2, h(5) = 3, h(11) = 4, \ldots$. In fact, with the help of the Euler function, the set of all positive integers constitute a tree in graph theory. For a non-negative integer n, let $E_0 = 2$, and when $n \geq 1$, we let $E_n = \min\{a > 1 \,|\, \varphi^n(a) = 2\}$, which is called the nth Euler number. If an Euler number is prime, then we call this number Euler's prime. These two terms are named by the author. Before that T. D. Noe used the computer to calculate the first 1000 Euler numbers. In particular, we have

n	E_n	n	E_n
0	2	11	$2329 = 17 \times 137 = E_4 \times E_7$
1	$3 = F_0$	12	$4369 = 17 \times 257 = E_4 \times E_8$
2	$5 = F_1$	13	$10537 = 41 \times 257 = E_5 \times E_8$
3	11	14	17477
4	$17 = F_2$	15	$35209 = 137 \times 257 = E_7 \times E_8$
5	41	16	$65537 = F_4$
6	83	17	140417
7	137	18	$281929 = 257 \times 1097 = E_8 \times E_{10}$
8	$257 = F_3$	19	557057
9	641	20	$1114129 = 17 \times 65537 = E_4 \times E_{16}$
10	1097		

The Euler numbers without the factor decomposition above are Euler's primes. It's known that Euler number must be odd and Fermat's primes must be Euler's primes. Therefore, if there is the sixth Fermat prime, it must be in the Euler numbers (Shapiro, *Amer., Math. Monthly* 50.1943, 18–30). For an odd Euler number E_n, if E_n is an Euler prime, the prime factors for $E_n - 1$ are Euler's primes, where the sum of the subscripts (containing the multiplicities) of odd Euler primes and power of 2 is equivalent to n; if E_n is a composite number, then the prime factors for it are Euler's primes, and the sum of its subscripts (containing the multiplicities) equals to n (Noe, *J. Integer Sequences*, 11.2008, Article 08.1.2).

We have the following conjectures: the set of Euler's numbers is infinite; other than 3, an Euler number no longer equals to a Mersenne prime. In addition, since Euler's numbers are much more than Fermat's primes and Mersenne primes (among the first 100 Euler's numbers, 288 are Euler's primes), can we find the new largest prime in this way?

In 2001, the Romanian mathematician Florian Luca (*Divulgaciones Matemáticas*, 9, 191–195), who lives in Mexico, proved that:

$$F_m \neq \binom{m}{k}, \quad 2 \leq k \leq m - 2.$$

In 2014, Shane Chern (*J. Integer Sequences*, 17.2014. Article 14.3.2) generalized the result to multinomial coefficients (see the definition in Section 30). Our problem is, how about the following?

$$E_m \neq \binom{m}{k}, \quad 2 \leq k \leq m - 2.$$

9. The Fraction Expressed as Repeating Decimal

In this section, we shall discuss the problem of the rational number expressed as a repeating decimal; this is an ingenious application of Euler's theorem.

Let $a, b(> 0)$ be two arbitrary integers. By the Euclidean division, there exist q and r such that $a = bq + r, 0 \leq r < b$. Therefore

$$\frac{a}{b} = q + \frac{r}{b}, 0 \leq \frac{r}{b} < 1.$$

We only need to consider the fraction between 0 and 1. In *Disquisitiones Arithmeticae*, Gauss had given the repeating decimal representation and the length of repetend of $\frac{1}{n}$ for $n \leq 1000$.

First of all, we define the repeating decimal.

Definition 2.5. The so-called repeating decimal refers to an infinite decimal $0.a_1 a_2 \ldots a_n \ldots$, where there exist $s \geq 0, t > 0$ such that $a_{s+i} = a_{s+kt+i}, 1 \leq i \leq t, k \geq 0$, noted as $0.a_1 a_2 \ldots a_s \dot{a}_{s+1} \ldots \dot{a}_{s+t}$. For repeating decimals, such s and t may be more than one pair, and if we can find the smaller such t, we call the corresponding $a_{s+1} a_{s+2} \ldots a_{s+t}$ the repetend, and t the length of the repetend. If the minimum s is zero, it is called a pure repeating decimal, otherwise a mixed repeating decimal.

By using the basic properties of divisibility and congruence, we can prove the following theorem.

$$\frac{\mathsf{I}}{\mathsf{II}} = \frac{1}{7}$$

The ancient Chinese fraction, without the small line in the middle

Theorem 2.13. *A rational number $\frac{a}{b}, 0 < a < b, (a, b) = 1$ can be expressed as a pure repeating decimal if and only if $(b, 10) = 1$. And in such a case, the length of the repetend t is the minimum positive integer satisfying the congruence $10^t \equiv 1 \pmod{b}$.*

Proof. Sufficiency: If $\frac{a}{b}$ can be expressed as a pure repeating decimal, by the assumption and the definition,

$$\frac{a}{b} = 0.a_1 a_2 \ldots a_t a_1 a_2 \ldots a_t \ldots$$

Thus,

$$10^t \frac{a}{b} = 10^{t-1}a_1 + 10^{t-2}a_2 + \cdots + 10a_{t-1} + a_t$$

$$+ 0.a_1 a_2 \ldots a_t a_1 \ldots a_t \ldots$$

$$= q + \frac{a}{b}$$

where q is a positive integer. Therefore $\frac{a}{b} = \frac{q}{10^t-1}$, i.e., $a(10^t - 1) = bq$. For $(a, b) = 1$, we have $b|(10^t - 1)$, thereby $(b, 10) = 1$.

Necessity: If $(b, 10) = 1$, by Euler's theorem, we know that there is a positive integer t such that

$$10^t \equiv 1 \pmod{b}, \quad 0 < t \le \phi(b)$$

holds. Therefore $10^t a = qb + a$, i.e.,

$$10^t \frac{a}{b} = q + \frac{a}{b},$$

where $0 < q < 10^t \frac{a}{b} \le 10^t(1 - \frac{1}{b}) < 10^t - 1$, so we may set

$$q = 10^{t-1}a_1 + \cdots + 10a_{t-1} + a_t, \quad 0 \le a_i \le 9, \ 1 \le i \le t.$$

Here a_1, \ldots, a_t are not all zero nor all 9. Therefore $\frac{q}{10^t} = 0.a_1 a_2 \ldots a_t$, so

$$\frac{a}{b} = 0.a_1 a_2 \ldots a_t + \frac{1}{10^t} \frac{a}{b}.$$

Repeating the above process, we have

$$\frac{a}{b} = 0.a_1 a_2 \ldots a_t a_1 a_2 \ldots a_t \ldots = 0.\dot{a}_1 \dot{a}_2 \ldots \dot{a}_t.$$

The first half part of Theorem 2.13 has been proved, and then we focus on the second part. If t is the length of the repetend for $\frac{a}{b}$, then by the proof of sufficiency, $10^t \equiv 1 \pmod{b}$. And if there exists $0 < t_0 < t$ such that $10^{t_0} \equiv 1 \pmod{b}$, then by the proof of necessity, $t \le t_0$. A contradiction! So t is the smallest positive integer satisfying the congruence $10^t \equiv 1 \pmod{b}$. Conversely, if t is the smallest positive integer satisfying $10^t \equiv 1 \pmod{b}$, then the proof of necessity shows that $\frac{a}{b} = 0.\dot{c}_1 \cdots \dot{c}_t$; if there exists $0 < t_0 < t$ such that $\frac{a}{b} = 0.\dot{d}_1 \cdots \dot{d}_{t_0}$, then by the proof of sufficiency, $10^{t_0} \equiv 1 \pmod{b}$. Another contradiction! So t is the length of the repetend for $\frac{a}{b}$. $\qquad\square$

Theorem 2.14. If $\frac{a}{b}$ is a rational number such that $0 < a < b, (a, b) = 1, b = 2^\alpha 5^\beta b_1, (b_1, 10) = 1$, and α, β are not both zero, then $\frac{a}{b}$ can be expressed as a mixed repeating decimal, where the digits that are not cyclic are $\max(\alpha, \beta)$. In such case, the length of the repetend t is the smallest positive integer satisfying the congruence $10^t \equiv 1 \pmod{b_1}$.

Proof. Let $\mu = \beta \geq \alpha$, Then

$$10^\mu \frac{a}{b} = \frac{2^{\beta-\alpha}a}{b_1} = M + \frac{a_1}{b_1}$$

where $0 < a_1 < b_1, 0 \leq M < 10^\mu$ and $(a_1, b_1) = (2^{\mu-\alpha}a - Mb_1, b_1) = (2^{\mu-\alpha}a, b_1) = 1$. By Theorem 2.13, $\frac{a_1}{b_1}$ can be expressed as a pure repeating decimal:

$$\frac{a_1}{b_1} = 0.\dot{c}_1 \ldots \dot{c}_t.$$

Set $M = m_1 10^{\mu-1} + \cdots + m_\mu (0 \leq \mu \leq 9)$. Then

$$\frac{a}{b} = 0.m_1 \ldots m_\mu \dot{c}_1 \ldots \dot{c}_t.$$

Now, we shall prove that non repeating digits are not less than μ. If $\frac{a}{b}$ can also be expressed as

$$\frac{a}{b} = 0.n_1 \ldots n_\nu \dot{d}_1 \ldots \dot{d}_s (\nu < \mu),$$

then by Theorem 2.13,

$$10^\nu \frac{a}{b} - \left[10^\nu \frac{a}{b}\right] = 0.\dot{d}_1 \ldots \dot{d}_s = \frac{a_2}{b_2}$$

where $(b_2, 10) = 1$, therefore $10^\nu \frac{a}{b} = \frac{a'}{b_2}$, i.e., $10^\nu ab_2 = a'b$. The right side can be divided by 5^μ, while ab_2 in the left side is coprime with 5. Hence, $5^\mu | 10^\nu$, which is a contradiction with $\mu > \nu$. The first half part of Theorem 2.14 has been proved, the latter part of the proof is omitted. □

Example 2.16. $\frac{2}{5} = 0.4, \frac{3}{7} = 0.\dot{4}2857\dot{1}, \frac{2}{15} = 0.1\dot{3}$.

Möbius function

In Section 8, we have known that Euler's function $\phi(n)$ is a multiplicative function. Because the multiplicative function $f(n)$ is not always zero, we set $f(a) \neq 0$ for $(1, a) = 1$ and $f(a)f(1) = f(a)$, $f(1) = 1$. Now we introduce two additional properties of multiplicative function.

$\mu(n)$

The first 50 values of the Möbius function are shown

Property 2.1. The product of multiplicative functions is a multiplicative function.

Property 2.2. If $f(x)$ is a multiplicative function, and $n = p_1^{\alpha_1} \dots p_k^{\alpha_k}$ is its standard factor decomposition, then

$$\sum_{d \mid n} f(d) = \prod_{i=1}^{k} (1 + f(p_i) + \dots + f(p_i^{\alpha_i})).$$

Property 2.1 is obviously true. To prove Property 2.2, we only need to note that every positive factor of n can be expressed as (Section 5)

$$n = p_1^{\beta_1} \dots p_k^{\beta_k}, \quad 0 \leq \beta_i \leq \alpha_i, 1 \leq i \leq k,$$

and $f(p_i^0) = f(1) = 1$. Then we can use the multiplicative property of $f(n)$.

Example 2.17. The arithmetical function

$$I(n) = \left[\frac{1}{n}\right] = \begin{cases} 1, n = 1 \\ 0, n = 0 \end{cases}$$

is a complete multiplicative function.

Now we introduce another important multiplicative arithmetical function — Möbius function $\mu(n)$ which was defined by the

German mathematician and astronomer August Ferdinand Möbius (1790–1868). He was a student of Gauss and was well known for the discovery of Möbius strip. The Möbius function is defined as follows:

$$\mu(n) = \begin{cases} 1, & n = 1 \\ (-1)^r, & n = p_1 \dots p_k \\ 0, & p^2 | n \end{cases}$$

where p, p_1, \dots, p_k are distinct primes.

It is obvious that this is a multiplicative function. By using Properties 2.1 and 2.2, we have

Property 2.3. If $f(x)$ is a multiplicative function and $n = p_1^{\alpha_1} \dots p_k^{\alpha_k}$ is its standard factor decomposition, then

$$\sum_{d|n} \mu(d) f(d) = \prod_{i=1}^{k} (1 - f(p_i)).$$

Take $f(x) = 1, f(x) = \frac{1}{x}$ in Property 2.3. Then by Theorem 2.9,

$$\sum_{d|n} \mu(d) = \begin{cases} 1, n = 1 \\ 0, n > 1 \end{cases}, \quad \phi(n) = \sum_{d|n} \mu(d) \frac{n}{d}.$$

The latter is the special case of the Ramanujan identity $(m = n)$

$$\frac{\mu(n/(m, n)) \phi(n)}{\phi(n/(m, n))} = \sum_{d|(m,n)} \mu(d) \frac{n}{d}.$$

Example 2.18. Let m, n be arbitrary positive integers. Then

$$(1) \sum_{d^2|n} \mu(d) = \mu^2(n); \quad (2) \sum_{d|n} |\mu(d)| = 2^k.$$

Example 2.19. The definition of the Liouville function is as follows:

$$\lambda(n) = \begin{cases} 1, & n = p_1 \dots p_{2k} \ or \ 1 \\ -1, & n = p_1 \dots p_{2k-1} \end{cases}$$

where p_i are distinct primes. We have

(1) $\lambda(n)$ is a complete multiplicative function;
(2) If

$$f(n) = \begin{cases} 1, & n \text{ is a square,} \\ 0, & \text{otherwise,} \end{cases}$$

then $f(n)$ is not a complete multiplicative function, meanwhile, $\lambda(n) = \sum_{d|n} \mu(d) f(\frac{n}{d})$.

Both Euler's function and Möbius's function have the following Dirichlet series:

$$\sum_{n=1}^{\infty} \frac{\phi(n)}{n^s} = \frac{\zeta(s)}{\zeta(s-1)}, \sum_{n=1}^{\infty} \frac{\mu(n)}{n^s} = \frac{1}{\zeta(s)}, \quad s = \sigma + it, \sigma > 1.$$

Here, $\zeta(s)$ is Riemann zeta function, whose definition will be given in Section 25. On the other hand, we know

$$\sum_{n \le x} \mu(n) = o(x).$$

This is equivalent to the prime number theorem, and the so-called Riemann hypothesis (Section 25) is equivalent to

$$\sum_{n \le x} \mu(n) = o(\sqrt{x}).$$

The number $\mu(n)$ is also the sum of the primitive n-th roots of unity, i.e.,

$$\mu(n) = \sum_{\substack{1 \le k \le n \\ (k,n)=1}} e^{2\pi i \frac{k}{n}}.$$

In addition, there is a divisor function $d(n)$, namely the number of all the positive divisors of n, which is also multiplicative. In 1849, the Germany mathematician Dirichlet (1805–1859) proved that

$$\sum_{1 \le n \le x} d(n) = x \log x + (2C - 1)x + \Delta(x).$$

Here, C is Euler constant and $\Delta(x) = O(\sqrt{x})$ is the remainder. Let $\Delta(x) = O(x^{\theta+\varepsilon})$, which is known as the Dirichlet divisor problem.

In 1915, Hardy and Landau proved that $\Delta(x) = O(x^{1/4})$ is impossible. The value of θ goes down continuously, and $\theta = \frac{1}{4}$ is considered to be possible. In 2014, Chaohua Jia and the Indian mathematician Sankaranarayanan (*Acta Arith.*, 164.2, 181–208) proved that,

$$\sum_{1 \le n \le x} d^2(n) = xP(\log x) + O(x^{1/2} \log^5 x),$$

where $P(x)$ is a polynomial with order 3.

Among non multiplicative functions, there are the famous maximum and minimum prime functions $P(n)$ and $p(n)$, the function of the number of prime factors (containing the multiplicities or not) $\Omega(n)$ and $\omega(n)$, etc. In 1984, the author (Master Thesis) used the related conclusion of the prime number theorem and arithmetical skills to get the mean value estimation of $P(n)$ (*Chinese Science Bulletin*, 24, 1481–1484),

$$\sum_{1 < n \le x} P(n) = \sum_{i=1}^{k} d_k \frac{x^2}{\ln^k x} + O\left(\frac{x^2}{\ln^{k+1} x}\right),$$

where

$$d_k = \left(\sum_{\nu=0}^{k-1} \frac{(-1)^\nu}{\nu!} \zeta^{(\nu)}(2)\right) \frac{(k-1)!}{2^k}.$$

In particular, if $k = 1$, then the main term is $\frac{\zeta(2)}{2} \frac{x^2}{\ln x} = \frac{\pi^2}{12} \frac{x^2}{\ln x}$, which is a result by Erdős in 1977. It is worth mentioning that, in the same year, the Canadian mathematician De Koninck (1948–) and the Serbia mathematician Ivić (1949–2020) also got the above mean value estimation (*Arch. Math.*, 43.1984, 37–43), but they could not give the expression for d_k. After that, the author made the mean value estimation extended to the ideal sets in an arbitrary number field (*Acta Arith.* LXX., 2.1995, 97–104).

Let $f(n), g(n)$ be two arithmetical functions. The following function

$$h(n) = \sum_{d \mid n} f(d) g\left(\frac{n}{d}\right)$$

is called the Dirichlet product, denoted as $h(n) = f(n) * g(n)$. It is easy to verify that this product satisfies the commutative law and

associative law. If $f(n)$ and $g(n)$ are multiplicative functions, then so is $h(n)$. The function $f^{-1}(n)$ is called the inverse function of $f(n)$ if it meets

$$f(n) * f^{-1}(n) = I(n).$$

Property 2.4 (Gauss' congruence). Let $(a, n) = 1$, then

$$\sum_{d|n} \mu\left(\frac{n}{d}\right) a^d \equiv 0 \ (\text{mod } n)$$

It's not difficult to prove the above congruence by the multiplicative property of Dirichlet product. In particular, when $n = p$, this is Fermat's little theorem.

Property 2.5. Set $f(n)$ a multiplicative function. It is a complete multiplicative function if and only if

$$f^{-1}(n) = \mu(n)f(n).$$

Let $\theta_k(n) = n^k, n \geq 1, k \geq 0$, and $\theta(n) = \theta_0(n)$. Then

$$I = \mu * \theta, \phi = \theta_1 * \mu.$$

According to the commutative and associative laws, $\theta_1 = \phi * \theta$, i.e.,

$$\sum_{d|n} \phi(d) = n.$$

This is an example of the so-called Möbius inversion formula, the latter (is said to have applications in condensed matter physics), which means that if $g(n)$ is a multiplicative function and

$$f(n) = \sum_{d|n} g(d),$$

then

$$g(n) = \prod_{d|n} f(d)^{\mu(n/d)}.$$

As for multiplication, there is also a Möbius inversion formula. Let $g(n)$ be a multiplicative function. If

$$f_1(n) = \prod_{d|n} g(d),$$

then

$$g(n) = \prod_{d|n} f_1(d)^{\mu(n/d)}.$$

10. Application in Cryptology

Since ancient times, in order to convey some instructions and strategies in the military, political, business and other fields, people need to make their work strictly confidential, so that only the two sides communicating know the content. So, people would set the password in the communication, which means that the two sides use a method in which they agreed in advance to translate the disclosed information (information such as ordinary text expression) into the information that only they could identify. This kind of communication is called cipher communication, such as the early written communications and wireless telegraphs later on. Now the network and satellite communications typically transmit information by converting words or letters to digital.

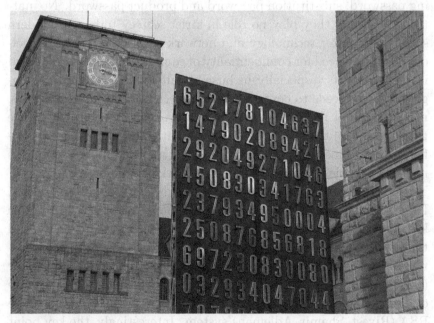

Poznan monument to Polish cryptologists who found the key to German's Enigma machine ciphers during World War II

These digitals convey information that is called the "code", which is divided into plain code and cipher. Both a telegram sent through the post office in earlier times and the international standard code with respect to personal name, which people going aboard fill in a form, are examples of plain code. Plain code is easy to be intercepted and deciphered, and so it does not have confidentiality. Therefore, the important communication needs to be translated into cipher from plain code, which requires an encryption program. After the receiving party receives the password, the password needs to be restored back into the plain code, and what this process needs is a decryption program. Implementation of encryption or decryption programs is usually with the help of appropriate tools, which are called the key or the solution key, and the above is a rough process of the password communication. Cryptology is the scientific technology of encryption and decryption. In wartime, it may help to change the course of the war.

The mathematics used in early cryptology is nothing more than shifting, substitution or both at the same time, which are called shifting password, substitution password and product password. No matter which kind, they play no role in strict secrecy today. In modern society, computer technology and network security are vastly developed, and the need for confidentiality of communication has increased in many aspects, especially in business affairs. For example, a single company needs to contact with many companies and customers, many of whom have the need for confidentiality. According to the traditional way, there will be a key and solution key agreed in advance for each customer, but that brings with it a lot of troubles. What is more, the whole society now has confidentiality requirements.

In 1976, two American mathematicians and computer experts, Diffie (1944–) and Hellman (1945–), proposed a totally new public key system, whose characteristic is the encryption procedure and the keys are open to the public, and can also be used by many customers at the same time. The following year, Rivest (1947–), Shamir (Israel, 1952–) and Adleman (1945–) of the Massachusetts Institute of Technology put forward a practical public key system called the RSA (Rivest–Shamir–Adleman) system. Interestingly, the key point

of the RSA system is the application of Qin Jiushao's *Da yan shu* (大衍术) and Euler's theorem.

RSA at MIT

Let p, q be two large prime numbers, and $N = pq$, e and d satisfy the congruence

$$ed \equiv 1 \;(\mathrm{mod}\;\; \phi(N)).$$

Here, $\phi(N)$ is the Euler function. Take the key and the solution key e, N and d, respectively. The cryptographic communication is as follows: Let a and $0 \le a \le N-1$ be the plain code. The encryption program is to translate a into b by the congruence

$$a^e \equiv b \;(\mathrm{mod}\;\; N), \quad 0 \le b \le N-1.$$

The sender sends the password b to the receiving party, and after the receiving party receives the password, the decryption program is to restore the password b back into the plain code a by the congruence

$$b^d \equiv a^{ed} \equiv a^{1+k\phi(N)} \equiv a \;(\mathrm{mod}\;\; N).$$

When $(a, N) = 1$, it can be ensured by the Euler theorem. When $(a, N) > 1$, we need to verify

$$a^{1+k\phi(N)} \equiv a \pmod{p}, \quad a^{1+k\phi(N)} \equiv a \pmod{q}$$

In the case of the first congruence, when $p|a$, it's obviously true; when $p \nmid a$, it's also right by $\phi(N) = \phi(pq) = (p-1)(q-1)$ and Fermat's little theorem.

What needs to be pointed out is that the key e, N can be open to the public, so the user only needs to keep solution key d. Anyone can send him the password by the encryption password procedures, but only he can read the information, while for other people it is almost impossible to solve it. Because if you want to get d, you must know $\phi(N)$, which requires knowing the prime factor p, q of N. When the digits of p, q are large enough, such as more than 100 digits, by the present mathematical methods, even using the most advanced computer now, no one may get the value of $\phi(N)$ in a limited time, and thus it is not possible to know d. With the establishment of the two-key password system, the single key cipher system used for thousands of years has been abandoned.

The RSA not only has a super security, but also can be used by a lot of customers. This is because if N and $\phi(N)$ are large enough, there can be a lot of pairs e_i, d_i satisfying

$$e_i d_i \equiv 1 \pmod{\phi(N)}, i = 1, 2, \cdots$$

However, with the passing of time, cryptology experts come up with new tricks to solve RSA constantly, and the way is to separate problems for different computers. The new methods include the quadratic sieve, the filed sieve, the elliptic curve algorithm, etc. The quadratic sieve is to decompose the integer with the technique of quadratic residue and continued fraction. But at present, there is no real threat to the RSA system, namely the scheme security of big number factorization. There is a reason, although the performance of computers is getting better and better, the decryption method seems to be easy. While at the same time, the number of digits of big prime numbers which can be obtained also gets higher and higher, and they can be used to set more secret keys.

Since the 1990s, American mathematicians have used the Hash function to design the Cipher system with very good results, which has a wide application in the government and financial institutions. For example, in 1991, the R of RSA — Rivest — designed the so-called MD5 algorithm, which is considered indestructible. But in 2004, this algorithm was solved by a Chinese mathematician Xiaoyun Wang (1966–). The next year, with the cooperation of the Chinese American computer theorist Qizhi Yao (1946–), she solved the international SHA-1 algorithm designed by the American National Security Agency, which caused a sensation in the world. These results indicate that the electronic signature can be forged theoretically.

The MD and SHA are, respectively, the abbreviation of Message-Digest algorithm and the Secure Hash Algorithm. In addition to applications in cryptology, number theory can also be used in the error correcting code, which is a kind of code with automatic error detection and correction of errors in the error correction procedure. For example, in each set of a recording equipment, the reason why it is able to accurately reproduce the sound lies in its error correcting code. Both the CD product and a mobile phone need to encode the voice, which benefited from the application of number theory in error correcting code.

The generalized Euler function

The French mathematician Camille Jordan (1838–1922) had a unique contribution in many areas of mathematics. For example, a Jordan matrix in linear algebra, a Jordan curve in complex analysis, a Jordan measure in real analysis. In number theory, he also defined the Jordan totient function, i.e., the number of $\{n_1, \ldots, n_k, n\}$ that satisfy $1 \leq n_1, \ldots, n_k \leq n, (n_1, \ldots, n_k, n) = 1$, where n is an arbitrary positive integer. It is not hard to get

$$J_k(n) = n^k \prod_{p|n} \left(1 - \frac{1}{p^k}\right).$$

When $k = 1$, it's the Euler totient function.

Let k be any integer, we define

$$\phi_k(n) = \sum_{\substack{1 \le i \le n \\ (i, n(n+k))=1}} 1.$$

Then $\phi_0(n) = \phi(n)$. It is easy to prove that $\phi_k(n)$ is a multiplicative function:

$$\phi_k(n) = n \prod_{p|n} \left(1 - \frac{a_p}{p}\right),$$

where

$$a_p = a_{p,k} = \begin{cases} 1, & p|k \\ 2, & p \nmid k \end{cases}.$$

When $k = 1$, we have $a_p = 2$ and

$$\phi_1(n) = n \prod_{p|n} \left(1 - \frac{2}{p}\right) = \sum_{d|n} \phi\left(\frac{n}{d}\right)(-1)^{\omega(d)}.$$

On the other hand,

$$\sum_{d|n} \phi_1(d) = \prod_{p^\alpha||n} (p-1)^\alpha.$$

We consider the distribution of twin primes on arithmetical progression. Denote $\Pi_2(x)$ the number if prime p (with $p+2$ also prime) less than or equal to x. Then the famous Hardy–Littlewood conjecture is

$$\prod_2(x) \sim 2C_2 \frac{x}{(\log x)^2} \sim 2C_2 \int_2^x \frac{dt}{(\log t)^2}.$$

Here, $C_2 = \prod_{p \ge 3} \frac{p(p-2)}{(p-1)^2} \approx 0.66016 \cdots$.

A natural conjecture is: let a and b be coprime integers and $(a+2, b) = 1$. Then

$$\sum_{\substack{p \le x \\ p \equiv a \pmod{b}}} 1 \sim \frac{\Pi_2(x)}{\phi_2(b)}.$$

For any positive integer e, we define the generalized Euler function

$$\varphi_e(n) = \sum_{\substack{1 \le i \le n/e \\ (i,n)=1}} 1 = \sum_{\substack{i=1 \\ (i,n)=1}}^{[\frac{n}{e}]} 1.$$

It is obvious that

$$\varphi_e(n) = \sum_{i=1}^{[\frac{n}{e}]} \sum_{d|(i,n)} \mu(d) = \sum_{d|n} \mu(d) \left[\frac{n}{de}\right] = \sum_{d|n} \mu\left(\frac{n}{d}\right)\left[\frac{d}{e}\right].$$

When $e = 1$, we have $\varphi_1(n) = \phi(n)$, namely the Euler phi function value. For $e > 1$, the function $\varphi_e(n)$ is not a multiplicative one, but there is

$$\varphi_e(n) = \left[\frac{n}{e}\right] - \sum_{i=1}^{k}\left[\frac{n}{ep_i}\right] + \sum_{\substack{i,j=1 \\ i \ne j}}^{k}\left[\frac{n}{ep_i p_j}\right] - \cdots + (-1)^k\left[\frac{n}{ep_1 p_2 \cdots p_k}\right].$$

In particular, when $e|\phi(n)$, we get $\varphi_e(n) = \frac{\phi(n)}{e}$.

The Euler totient function has many interesting properties. For the generalized Euler function, these properties still exist. For example,

$$\left[\frac{n}{e}\right] = \sum_{d|n} \varphi_e(d),$$

$$\Gamma_e(x) = \sum_{n=1}^{x} \varphi_3(n) = \frac{3}{e\pi^2}x^2 + O(x \log x).$$

What we are more interested in is that the Euler function has many interesting but unsolved problems. Some of them are still valid for

the generalized Euler function, and even bring some new problems. Here are three examples.

In 1907, American mathematician Carmichael (1879–1967) announced to have proved that for any positive integer n, there exists a positive integer $m \neq n$, such that $\phi(m) = \phi(n)$. Later, his proof was found to be false. In 1922, Carmichael changed his result into a conjecture, which is still an open question today. In 1999, Ford proved the conjecture proposed by the Poland mathematician Schinzel (1937–): for arbitrary $k \geq 2$, there exists m such that $\phi(x) = m$ has exactly k solutions. We guess: (1) for any even number k, there are at least 3 different solutions for $\varphi_3(n) = k$; (2) for any positive integer k, there exists m such that $\varphi_3(x) = m$ has exactly k solutions; (3) there exist infinitely many odd number k, respectively, such that. there is no solution, unique solution, or exactly two solutions for $\varphi_3(n) = k$. And what's the set density of the above 3 groups with infinitely many solutions?

In 1932, American mathematician D. H. Lehmer (1905–1991) proved that $\phi(n) \nmid n - 1$ for any arbitrary composite number n. We believe that $\varphi_2(n) \nmid n - 1$ for any composite number n but 4, 6, 9. This result has been checked for $n < 3 \times 10^9$. When $e = 3$, it's easy to see that for 3, 5, and all prime numbers p in the form of $3n + 1 : \varphi_3(p) | p - 1$; for all prime number $p > 5$ in the form of $3n + 2 : \varphi_3(p) \nmid p - 1$. And for the composite numbers, we believe that for a composite number n except for 4, 6, 8, 9, 12, 16, 21, 65, 425, 1729, 3201, 3531, 22865, we have $\varphi_3(n) \nmid n - 1$, and the result has been checked for $n < 3 \times 10^6$.

The existence of infinitely many positive integers n satisfying that $\phi(n) = \phi(n + 1)$ is a very ancient mathematics conjecture. In 1958, Schinzel even guessed that for any even number i, there exist infinitely many positive integers n satisfying $\phi(n + i) = \phi(n)$. Our problems are: (1) Except for 1, 3, is there any other odd number k such that there are solutions for $\varphi_3(n) = \varphi_3(n + 1) = k$? (2) Are there infinitely many even numbers k such that there are solutions for $\varphi_3(n) = \varphi_3(n + 1) = k$? (3) For any even number i, are there infinitely many solutions for $\varphi_3(n) = \varphi_3(n+i)$? For an arbitrary odd number $k > 3$, are there no solutions for $\varphi_3(n) = \varphi_3(n + 1) = k$?

In addition, the set A of the positive integers satisfying $\phi(n) = \phi(n + 1)$ and the set A_3 of positive integers satisfying $\varphi_3(n) = \varphi_3(n + 1)$ are almost the same. For $n \leq 10^{12}$, there are totally 11 elements that belong to A but do not belong to A_3, namely, 255, 48704, 64004, 65535, 5577825, 62603264, 1343331195, 137264655, 4293787664, 4294967295, 69494811375, 69494811376. In addition, there is also a number of 20 digits that belongs to A but does not belong to A_3, i.e., 18446743790241710144; on the other hand, 4, 5, 9, 21, 4525476564005327675 belong to A_3 but do not belong to A.

Exercises 2

1. Prove that one can choose three from any five integers such that their sum is a multiple of 3.
2. Prove that one can choose $m(\leq n)$ from any n integers such that their sum is a multiple of n.
3. A palindromic number or numeral palindrome is a number that remains the same when its digits are reversed, for example, 3, 11, 242, which are "symmetric". Show that each palindromic number with four digits is divisible by 11. Try to generalize this problem.
4. Let x be real and n a positive integer. Show that

$$[x] + \left[x + \frac{1}{n}\right] + \cdots + \left[x + \frac{n-1}{n}\right] = [nx].$$

5. Let x be real and n a positive integer. Show that $\left[\frac{[x]}{n}\right] = \left[\frac{x}{n}\right]$.
6. Let n be a positive integer. Show that $\frac{(2n)!}{n!(n+1)!}$ is an integer.
7. Show that $\phi(n)$ is 1 or an even integer.
8. Prove that

 (1) for any positive integers m and n, the number of integers $m, m + 1, \ldots, m + n - 1$ that are relatively prime to n is $\phi(n)$;
 (2) for any positive integer n we have $\phi(n^2) = n\phi(n)$; find all the positive integers such that $\phi(n)|n$.

9. Let $(m, n) = 1$. Show that $m^{\phi(m)} + n^{\phi(m)} = 1 \pmod{mn}$.

10. Prove that

 (1) if $n = p_1^{\alpha 1} \cdots p_k^{\alpha k}$, then $\sum_{d^2|n} \mu(d) = \mu^2(n)$.

 (2) $\sum_{d|n} |\mu(d)| = 2^k$.

11. Let $\{0, 1, 2\}$ and $\{0, 2, 1\}$ be two least residue systems modulo 3. The difference of the corresponding residues forms a least absolute residue system. Discuss the generalization for modulo $n (\geq 3)$.

Chapter 3

Congruences

11. Qin Jiushao's Theorem

We have defined the congruence relation and proved several famous theorems about this relation. But we have not related them to congruent equations and their solutions. The latter is an important part of elementary number theory. For simple congruent equations or small modulus, we can solve or test them by a direct calculation. For example,

$$x^3 - x + 1 \equiv 0 \ (\text{mod } 5).$$

We easily verify that $x \equiv 3 \ (\text{mod } 5)$ is the only solution. For more general congruent equations, we need to introduce the following definition.

Definition 3.1. Let $f(x)$ be a polynomial $a_n x^n + a_{n-1} x^{n-1} + \cdots + a_0$ with integral coefficients and m a positive integer. Then

$$f(x) \equiv 0 \ (\text{mod } m) \tag{3.1}$$

is called a congruence modulo m. If $m \nmid a_n$, then n is called the degree of (3.1).

By Theorem 2.1, if $f(a) \equiv 0 \ (\text{mod } m)$, then for any integer a' in the residue class R_a, the relation $f(a') \equiv 0 \ (\text{mod } m)$ holds. Therefore, we treat all the numbers that are congruent to a modulo m as one solution to (3.1).

Theorem 3.1. *The linear congruence*

$$ax \equiv b \pmod{m} \tag{3.2}$$

has a solution if and only if $(a, m)|b$, and if it holds, then the number of the solutions is $d = (a, m)$.

Proof. The necessity is obvious. We prove the sufficiency. If $(a, m)|b$, then by letting $a = a'd, b = b'd$, and $m = m'd$, we obtain $(a', m') = 1$. The congruence (3.2) is equivalent to the congruence

$$a'x \equiv b' \pmod{m'}. \tag{3.3}$$

By Theorem 2.7, we see that if x goes through the complete residue system modulo m', so does $a'x$. Hence there must be a unique solution, denoted by x'. Then all the solutions of (3.2) are $x' + km'$ with $k = 0, 1, \ldots, d - 1$. This completes the proof of Theorem 3.1. \square

In particular, if $d = 1$, then there is a unique solution for (3.2). This result was first discussed by Qin Jiushao and put in his book *Mathematical Treatise in Nine Sections*, he called it *Da Yan Shu*.

If the modulus p is a prime, then whenever a is not a multiple of p, there is a unique solution for (3.2). It means that the set $0, 1, \ldots, p-1$ forms a number field modulo p under the addition and multiplication, namely \mathbb{Z}_p.

So far what we have discussed is the congruence with one equation. Now we shall think about the congruence set with several equations. In the mathematical classic of *Sunzi Suanjing*, there is a famous question *about Objects without knowing their number*:

> There is a certain number. When divided by 3, this number has a remainder 2; when divided by 5, it has a remainder 3; and when divided by 7, it has a remainder 2. What is the number?

The author Sun Zi (Master Sun) is unknown, but the book is generally considered to be a work of the 4th century AD. In the *Eastern Jin* dynasty, which is much later than *The Art of War* by the military theorist *Sun Wu* (around 535 BC–). The question *of Objects*

without knowing their number can be expressed as follows by the congruences:

$$x \equiv 2 \pmod 3,$$
$$x \equiv 3 \pmod 5,$$
$$x \equiv 2 \pmod 7.$$

Sunzi Suanjing offered the answer 23, which is the smallest positive integer satisfying the above congruent system. The book also gave the method of solving general congruent equations modulo these 3 numbers, in which the remainder 2, 3 and 2 can be replaced by any other numbers. This is the special form of the so-called Sun Zi's theorem. Yixing (683–727), the monk and astronomer in the *Tang* dynasty, set a calendar by this, but the more general method first appeared in the masterpiece *Mathematical Treatise in Nine Sections* by Qin Jiushao in the *Song* dynasty, and he called it *The Dayan General Mathematical Method* or Da Yan Zong Shu Shu. For any positive integers m_i and b_i $(1 \le i \le k)$, Qin Jiushao considered the following congruent system,

$$x \equiv b_1 \pmod{m_1}$$
$$\vdots \tag{3.4}$$
$$x \equiv b_k \pmod{m_k}$$

and obtained the following theorem which we want to name after him.

Theorem 3.2 (Qin Jiushao' theorem). *If* $(m_i, m_j) = 1$ $(i \ne j)$, $m = m_1 \ldots m_k$, *and* $m = m_i M_i$ $(1 \le i \le k)$, *then the congruent system* (3.4) *has a unique solution*

$$x \equiv M_1' M_1 b_1 + \cdots + M_k' M_k b_k \pmod m,$$

where $M_i' M_i \equiv 1 \pmod{m_i}$, *for* $1 \le i \le k$.

Proof. By the assumption $(M_i, m_i) = 1$, there exists M_i' satisfying

$$M_i' M_i \equiv 1 \pmod{m_i}.$$

欽定四庫全書　卷一上

同而法則無二也然其極和較之用窮奇偶之

情則有為元法西法所未及者但原本法解煩

雜圖式論舛令詳加改定並釋其義俾學者易

見焉

大衍數術

大衍總數術曰置諸問數〔類名〕有四　一曰元數〔謂尾位見單〕者本門牒

舊酒悤斛艱砌帆　二曰收數〔謂尾位見分釐者假令冬〕至三百六十五日二十五

失米之顆是也

刻欲與甲子六十日為　三曰通數〔謂諸數各有分子母〕者本門問一會積年

一會而求積日之類

Da Yan Shu in *Treatise in Nine Chapters* by Qin Jiushao, the book was circulated almost 600 years before its publication in the Ming Dynasty

On the other hand, $m_j | M_i$ for $i \neq j$. Hence,

$$\sum_{j=1}^{k} M_j' M_j b_j \equiv M_i' M_i b_i \equiv b_i \pmod{m_i}$$

is a solution of (3.4).

If x_1 and x_2 are two solutions of (3.4), then

$$x_1 \equiv x_2 \pmod{m_i}, \quad 1 \leq i \leq k.$$

Note that $(m_i, m_j) = 1 (i \neq j)$, so $x_1 \equiv x_2 \pmod{m}$. This proves the uniqueness. □

Statue of Qin Jiushao, Nanjing. Photo taken by the author

Example 3.1 (Soldier counting by Han Xin). There is a band of soldiers, and there is only one soldier in the last line if they stand in five lines, five if in six lines, four if in seven lines, and ten if in eleven lines. What is the total number of soldiers?

Solution. Here $m = 5 \times 6 \times 7 \times 11 = 2310$, $M_1 = 462$, $M_2 = 385$, $M_3 = 330$, and $M_4 = 210$. Solving

$$M_i' M_i \equiv 1 \ (\text{mod } m_i), \quad 1 \le i \le 4,$$

we have $M_1' = 3$ and $M_2' = M_3' = M_4' = 1$. Then by Theorem 3.2,

$$x \equiv 3 \times 462 + 385 \times 5 + 330 \times 4 + 210 \times 10 = 6731$$

$$\equiv 2111 \ (\text{mod } 2310).$$

This example is a folk legend in Huai'an, Jiangshu Province of China, which tells us that in the late *Qin* dynasty (far before *Sun Zi*), a particular example of Qin Jiushao's theorem had already existed. As an application, we also have the following theorem.

Theorem 3.3. *Let* b_1, b_2, \ldots, b_k *run through the complete residue systems of modulo* m_1, m_2, \ldots, m_k, *respectively. Then the solution of* (3.4) *runs through a complete residue system modulo* $m = m_1 m_2 \ldots m_k$.

Usually in Chinese textbooks of number theory, Qin Jiushao's theorem is called Sun Zi's theorem, although Qin Jiushao made the greatest contribution to this theorem. Maybe the reason is that there were two writers publishing similar articles criticizing his problems in the lifestyle and moral behavior right before and after his death (actually he was a victim of the political struggle). However, the theorem is the most perfect result in the ancient Chinese history of mathematics, of which we are most proud. It appears in every elementary number theory textbooks in China and abroad, which is called the Chinese Remainder Theorem by westerners. It has corresponding extensions and wide applications in cryptography, the polynomial interpolation calculation in the numerical analysis, proof of Godel's incompleteness theorem, the theory of fast Fourier transformation, and so on, as well as the theory of abstract algebra.

The Chinese mathematician Chengdong Pan (1934–1997) once said that the reason why westerners named the above theorem in their way is that, in addition to the difficult pronunciation of a Chinese name, this is the only great theorem obtained by ancient

Chinese mathematicians. So it is a kind of prejudice. Qin Jiushao (1202–1261) had a styled name Daogu. On the Xixi road near Zhejiang University in Hangzhou, there ever existed a bridge named Daogu Bridge in memory of this scientist of the 13th century. It is said that Qin Jiushao designed and raised funds for the bridge that was then named initially by another Chinese mathematician Zhu Shijie in *Yuan* Dynasty. Unfortunately it was pulled down in the early 21st century. However, a new constructed bridge less than a hundred meters away from the old site was named Daogu Bridge in the spring of 2012, suggested by the author and supported by some local government agencies.

In 1852, Alexander Wylie (1815–1881), the English missionary and sinologist, translated "The Dayan Seeking-unity Method" and the solution method of (3.4), i.e., Theorem 3.2, to the Western. A German mathematics historian and Gauss' student Morris Cantor (1829–1920) praised Qin Jiushao as "The most fortunate genius" since Euler and Gauss had made deep researches on this subject before with theoretical proofs without being named. Lagrange (1736–1813), a French mathematician, praised Newton this way. Lagrange thought that there was only one chance to find the law of universal gravitation. It is worth mentioning that at the International Congress of Mathematicians in Paris in 1900, Hilbert delivered an invited speech on mathematics education, while Cantor delivered a plenary speech on the history of mathematics. But George Sarton (1884–1956), a Belgian-born American historian of science, who has a good reputation of "The father of science history", regarded Qin Jiushao as "one of the greatest mathematicians in his nation, in his age, and actually all ages." Strictly speaking, Sun Zi's theorem should be named as Sun Zi-Qin Jiushao's theorem, or Qin Jiushao's theorem.

Fibonacci's rabbits

About a quarter of century before the birth of Qin Jiushao, an important mathematician appeared in Europe when it still stayed in the darkness of the the Middle Ages, who is an Italian called Fibonacci (about 1170–1250). His original name is Leonardo Pisano, with the meaning of Leonardo of Pisa. When he was a child, he went with his

father Bonacci, a civil servant, to many places along the coast of the Mediterranean Sea, including today's port city of Bejaia in Algeria, where it is said that Bonacci was the consul of Pisa, and Fibonacci learned much from the masterpieces of mathematicians such as Diophantine and Khwarizmi from Greece and Arab. Fibonacci is the shorthand of Filius Bonacci, with the meaning of "son of Bonacci", which was named by an Italian historian of mathematics in 1828.

The Statue of Fibonacci in Pisa in the 19th Century

In the masterpiece *Abacus Books* (1202) of Fibonacci, it was mentioned that the "hundred fowls" problem of Zhang Qiujian and the example of Sun Zi about Qin Jiushao's theorem. The "Abacus" in

the name of the book refers to sand table for calculation, not the abacus of China. Fibonacci introduces the middle line of the fraction in the book, and he listed some questions about congruences. What is most interesting in the book is the "question of rabbits": Assuming every couple of adult rabbits can give birth to a pair of rabbits (a male and a female) every month, and the young rabbits can bear after two months. Then, beginning from a pair of young rabbits, how many rabbits are there after one year?

Fibonacci must have never thought that after more than 800 years, the rabbit's question still attracts mathematicians from one generation to another all over the world. If Qin Jiushao's theorem gives a perfect conclusion and effective calculation method, then Fibonacci sequence provides everlasting inspiration for us. Absolutely, the first 13 numbers of rabbit's sequence are: $1, 1, 2, 3, 5, 8, 13, 21, 34, 55, 89, 144, 233\ldots$ Let F_n represent the numbers of pairs of rabbits in the nth month, which is called the Fibonacci sequence or Fibonacci numbers:

$$F_0 = 0, F_1 = 1, F_n = F_{n-2} + F_{n-1} \quad (n \geq 2).$$

Graph of Fibonacci numbers

This sequence appears in many unexpected ways in nature. In plants, for example, the number of petals of many flowers just happens to be a Fibonacci number, such as 5 petals of plum blossom, 8 of Delphinium ajacis, 13 of marigold, and 21 of aster, 34, 55, and 89 of daisy. There are many interesting properties, including conclusions about division and congruence. For example, using the above

expression and by induction (for $m \geq 1$), we can obtain

$$F_{m+n} = F_m F_{n+1} + F_{m-1} F_n,$$

$$(-1)^n F_{m-n} = F_m F_{n+1} - F_{m+1} F_n.$$

The second formula was discovered by Philbert Maurice d'Ocagne (1862–1938). By the first formula, it is not difficult to prove that if n is a divisor of m, then F_n is also a divisor of F_m. Hence, for any composite number n except 4 ($F_4 = 3$), F_n is also a composite number. However, the converse conclusion is not true. For example, 53 is a prime, but $F_{53} = 953 \times 55945741$.

In 1718, an English mathematician born in France Abraham De Moivre (1667–1754) found that

$$F_n = \frac{1}{\sqrt{5}} \left\{ \left(\frac{1 + \sqrt{5}}{2} \right)^n - \left(\frac{1 - \sqrt{5}}{2} \right)^n \right\}.$$

Ten years later, this formula was proved by the Swiss mathematician Nicolas Bernoulli (1687–1759), by using the generating function of differential equations. It follows easily from this formula that

$$\frac{F_{n+1}}{F_n} \to \frac{1 + \sqrt{5}}{2} \quad (n \to \infty).$$

The value on the side is equal to $1.618\ldots$, which is the so-called golden ratio.

In 1680, Giovanni Cassini (1625–1712), the captain of the Paris Observatory found that

$$F_{n-1} F_{n+1} - F_n^2 = (-1)^n \quad (n \geq 1).$$

It is called Cassini's identity. This formula will often be used in this book, and it can be proved by a matrix method. In fact,

$$F_{n-1} F_{n+1} - F_n^2 = \begin{vmatrix} F_{n+1} & F_n \\ F_n & F_{n-1} \end{vmatrix} = \det \begin{pmatrix} 1 & 1 \\ 1 & 0 \end{pmatrix}^n = (-1)^n.$$

In 1879, the French–Belgium mathematician Eugene Charles Catalan (1814–1894) generalized it to (Catalan's identity).

$$F_n^2 - F_{n-r} F_{n+r} = (-1)^{n-r} F_r^2 \quad (n > r \geq 1).$$

Later on the Hungarian-born British mathematician Steven Vajda (1901–1995) generalized it to (Vajda's identity)

$$F_{n+i}F_{n+j} - F_nF_{n+i+j} = (-1)^n F_i F_j \quad (n, i, j \geq 0),$$

Meanwhile, the Fibonacci sequence satisfies the following identity of the determinant:

$$\begin{vmatrix} F_n & F_{n-1} & F_{n-2} \\ F_{n-1} & F_{n-2} & F_{n-3} \\ F_{n-2} & F_{n-3} & F_{n-4} \end{vmatrix} = 0.$$

In 1876, the French mathematician Francois Edouard Lucas (1842–1891) proved that for any positive integers m and n,

$$(F_m, F_n) = F_{(m,n)}.$$

This wonderful formula can also be proved by the Euclidean algorithm. From this formula we can also see that there are infinitely many primes, and any three consecutive Fibonacci numbers are coprime each other. He also defined the so-called sequence of Lucas numbers

$$L_0 = 2, \quad L_1 = 1, \quad L_n = L_{n-2} + L_{n-1} \quad (n > 1).$$

Lucas numbers have many similar and relevant properties to Fibonacci numbers, such as

$$L_n = \left(\frac{1 + \sqrt{5}}{2}\right)^n + \left(\frac{1 - \sqrt{5}}{2}\right)^n,$$

$$L_n = F_{n-1} + F_{n+1}, \quad F_n = \frac{L_{n-1} + L_{n+1}}{5},$$

$$\sum_{i=1}^{k} F_{2i-1} = F_{2k} - F_0, \quad \sum_{i=1}^{k} F_{2i} = F_{2k+1} - F_1,$$

$$\sum_{i=1}^{k} L_{2i-1} = L_{2k} - L_0, \quad \sum_{i=1}^{k} L_{2i} = L_{2k+1} - L_1.$$

From the last two formulas, we can deduce the following conclusion: for any positive integer k, $F_{2k+1} - 1 = \sum_{i=1}^{k} F_{2i} (L_{2k} - 2 = \sum_{i=1}^{k} L_{2i-1})$ is the least positive integer that needs to be decomposed into at least k Fibonacci numbers (Lucas numbers).

By using De Moivre's formula of Fibonacci number and Lucas number, we recently get the following formulas:

$$L_n^2 - L_{2n} = (-1)^n 2, \quad 5F_n^2 - L_{2n} = (-1)^{n-1} 2.$$

Meanwhile, Fibonacci numbers and Lucas numbers could be extended to the negative terms by

$$F_{-n} = (-1)^{n-1} F_n, \quad L_{-n} = (-1)^n L_n.$$

Therefore, many properties of Fibonacci numbers and Lucas numbers mentioned above are still valid for any terms of arbitrary integers.

Besides, there are the following congruences modulo odd primes or prime powers:

$$F_p \equiv 5^{\frac{p-1}{2}} \pmod{p}, \quad F_{p-\left(\frac{5}{p}\right)} \equiv 0 \pmod{p}, \quad L_p \equiv 1 \pmod{p},$$

$$F_{5n} \equiv 5F_n \pmod{5^2}, \quad L_{5n} \equiv 3^{5n-1} + 5L_n \pmod{5^2},$$

where $\left(\frac{5}{p}\right) = \pm 1$ or 0 is the Legendre symbol (cf. Section 17). From the second congruence above, we see that every prime must be a factor of some Fibonacci numbers.

Now, we will introduce the Pisano period of Fibonacci numbers, which was found by Lagrange in 1774. For any positive integer n, there exists the least positive integer $\pi(n)$, called its period, satisfying

$$F_{k+\pi(n)} \equiv F_k \pmod{n},$$

where k can be any positive integer. For example, $\pi(1) = 1, \pi(2) = 3, \pi(3) = 8, \pi(8) = 12, \pi(32) = 48$. The cycle of $\pi(8)$ is $\{011235055271\}$.

12. Wilson's Theorem

We focused on linear congruences in the previous section, and now we discuss higher degree congruences. First, we let the modulus m be an arbitrary integer, and then we take prime power for m.

Theorem 3.4. *Let the positive integers m_1, \ldots, m_k be given such that $\gcd(m_i, m_j) = 1$ for $i \neq j$. If $m = m_1 m_2 \ldots m_k$, then the system*

of congruences

$$f(x) \equiv 0 \ (\text{mod } m_i), \quad 1 \le i \le k \tag{3.5}$$

is equivalent to the congruence (3.1). *Let T_i be the number of solutions mod m_i to* (3.5), *and T the number of solutions mod m to* (3.1), *then*

$$T = T_1 T_2 \ldots T_k.$$

Proof. The equivalence is obvious. Let

$$x \equiv x_{it_i} \ (\text{mod } m_i), \quad 1 \le t_i \le T_i$$

be the T_i different solutions modulo m_i to (3.5). Then each solution to (3.5) is also a solution to the congruences:

$$x \equiv x_{1t_1} \ (\text{mod } m_1),$$
$$x \equiv x_{2t_2} \ (\text{mod } m_2), \ldots, x \equiv x_{kt_k} \ (\text{mod } m_k), \tag{3.6}$$

where $t_i = 1, 2, \ldots, T_i$ $(1 \le i \le k)$. Thus, (3.6) has the same set of solutions as (3.1). According to Qin Jiushao's theorem, there is exactly one solution to each system (3.6), so there are $T_1 T_2 \ldots T_k$ solutions modulo m to (3.6). By Theorem 3.3, these $T_1 T_2 \ldots T_k$ solutions are incongruent modulo m. Thus, the number of solutions modulo m to (3.6) is $T_1 T_2 \ldots T_k$. This completes the proof of Theorem 3.4. □

By Theorem 3.4 and the Fundamental Theorem of Arithmetic, (3.1) can be converted into congruences modulo prime powers, i.e.,

$$f(x) \equiv 0 \ (\text{mod } p^\alpha). \tag{3.7}$$

It is clear that each solution to (3.7) is also a solution of

$$f(x) \equiv 0 \ (\text{mod } p). \tag{3.8}$$

So, we consider (3.8) first.

Theorem 3.5. *Let $x \equiv x_1 \ (\text{mod } p)$, that is,*

$$x = x_1 + p t_1, \quad t_1 = 0, \pm 1, \pm 2, \ldots \tag{3.9}$$

be solution to (3.8). If $p \nmid f'(x_1)$, then they provide one solution to (3.7)

$$x = x_\alpha + p^\alpha t_\alpha, \ t_\alpha = 0, \pm 1, \pm 2, \ldots$$

that can be written as $x \equiv x_\alpha \pmod{p^\alpha}$, where $x_\alpha \equiv x_1 \pmod{p}$.

Proof. When $\alpha = 1$, Theorem 3.5 is obviously true. Suppose that Theorem 3.5 is true for $\alpha - 1$ ($\alpha \geq 2$), then (3.9) presents exactly one solution for the congruence

$$f(x) \equiv 0 \pmod{p^{\alpha-1}}.$$

Let it be

$$x = x_{\alpha-1} + p^{\alpha-1} t_{\alpha-1} (t_{\alpha-1} = 0, \pm 1, \pm 2, \ldots),$$

where $x_{\alpha-1} \equiv x_1 \pmod{p}$, put it into (3.7), then

$$f(x_{\alpha-1}) + p^{\alpha-1} t_{\alpha-1} f'(x_{\alpha-1}) \equiv 0 \pmod{p^\alpha}.$$

But since $f(x_{\alpha-1}) \equiv 0 \pmod{p^{\alpha-1}}$, hence

$$t_{\alpha-1} f'(x_{\alpha-1}) \equiv -\frac{f(x_{\alpha-1})}{p^\alpha} \pmod{p}.$$

Noting that $(f'(x_1), p) = 1$, the above congruence has exactly one solution

$$t_{\alpha-1} = t'_{\alpha-1} + p t_\alpha (t_\alpha = 0, \pm 1, \pm 2, \ldots).$$

Taking it into (3.10), we obtain a solution for (3.7):

$$x = x_{\alpha-1} + p^{\alpha-1} t'_{\alpha-1} + p^\alpha t_\alpha (t_\alpha = 0, \pm 1, \pm 2, \ldots).$$

Let $x_\alpha = x_{\alpha-1} + p^{\alpha-1} t'_{\alpha-1}$. Then

$$x = x_\alpha + p^\alpha t_\alpha (t_\alpha = 0, \pm 1, \pm 2, \ldots),$$

where $x_\alpha \equiv x_1 \pmod{p}$. This completes the proof of Theorem 3.5. \square

Corollary. *If the congruences (3.8) and $f'(x) \equiv 0 \pmod{p}$ has no common solution, then (3.7) and (3.8) have the same number of solutions.*

Now, let's talk about the number of solutions to (3.8).

Theorem 3.6. *The congruence* (3.8) *is equivalent to a congruence modulo a prime with degree no more than $p - 1$.*

Proof. By the division algorithm for polynomials (which is exactly the same as that for integers), there are polynomials with integral coefficients $q(x)$ and $r(x)$ such that

$$f(x) = (x^p - x)q(x) + r(x),$$

where the degree of $r(x)$ is no more than $p - 1$. By Fermat's Little Theorem, we have $f(x) \equiv r(x) \pmod{p}$, for any arbitrary integer x. This completes the proof of Theorem 3.6. \square

Theorem 3.7. *Let $k \leq n$, $x \equiv \alpha_i \pmod{p}$, $1 \leq i \leq k$, be k different solutions to* (3.8). *Then for any arbitrary integer x,*

$$f(x) \equiv (x - \alpha_1)(x - \alpha_2) \cdots (x - \alpha_k) f_k(x) \pmod{p},$$

where $f_k(x)$ is a polynomial of degree $n - k$ with a_n as the first term coefficient.

By Theorem 3.7, we have the following corollary and the famous Wilson's theorem, which can be easily proved by induction on k.

Corollary 3.1. *For any arbitrary integer x,*

$$x^{p-1} - 1 \equiv (x - 1)(x - 2) \cdots (x - (p - 1)) \pmod{p}.$$

Theorem 3.8 (Wilson). *If p is a prime, then*

$$(p - 1)! \equiv -1 \pmod{p}.$$

Wilson's theorem was first published by the English mathematician Edward Waring (1734–1798) in his book *Meditationes Algebraicae* (1770). In the book, he just told that this was a conjecture by his student Wilson (1741–1793), but there was no proof. One year later, Lagrange, the French mathematician, gave the first proof. Long afterwards, it was found that Leibniz, a German mathematician, was also aware of the result a century earlier, but he never published it. It is interesting that although Wilson made his imprint in the history of mathematics, he didn't engage himself in the

mathematical research later on. He became a Judge of Common Pleas, and was knighted.

Now we introduce a simple proof of Wilson's theorem.

Proof. We might assume that p is odd. For any integer a with $0 < a < p$, there exists a' with $0 < a' < p$ such that $aa' \equiv 1 \pmod{p}$. If $a = a'$, then $a^2 \equiv 1 \pmod{p}$, thus $a = 1$ or $a = p - 1$. For the rest of the numbers, $2, 3, \ldots, p - 2$, they can form $\frac{p-3}{2}$ pairs of a and a' with $aa' \equiv 1 \pmod{p}$. It follows that $2 \times 3 \times \cdots \times (p - 2) \equiv 1 \pmod{p}$, and then $(p - 1)! \equiv -1 \pmod{p}$. This completes the proof of Theorem 3.8. □

It is easy to prove that p is necessarily prime in Wilson's theorem, which makes the theorem a criterion for judging a given number is prime or not. It is said that Wilson was inspired when reading works of Alhazen (965 Basra–1040 Cairo), an Arabian physicist, astronomer, and mathematician. Alhazen, nicknamed *Ptolemy the second*, has been described as the first one since ancient Greeks who made significant contributions to the principles of optics, and the author of *Book of Optics*. He first described the vision exactly, and asserted that light came from the object that we see.

Portrait of Alhazen, the Arabian mathematician

Alhazen explored the fifth postulate in Euclid's Elements. His work had a considerable influence on its development among the later Persian mathematicians Omar Khayyám and Nasīr al-Dīn al-Tūsī, and his methodology influenced a number of western scientists including Avicenna and Roger Bacon. Alhazen solved problems involving congruences using what is now called Wilson's theorem and Qin Jiushao's theorem. Alhazen, before Euler, may have been the first to state that every even perfect number is of the form $2^{n-1}(2^n - 1)$ where n and $(2^n - 1)$ are primes, but he was not able to prove this result successfully.

Example 3.2. Let p be a prime number. If $p \equiv 1 \pmod 4$, then

$$x \equiv \pm \left(\frac{p-1}{2} \right)! \pmod p$$

are solutions to the congruence

$$x^2 \equiv -1 \pmod p.$$

If $p \equiv 3 \pmod 4$, then the above congruence has no solution. Otherwise,

$$x^{p-1} = (x^2)^{\frac{p-1}{2}} \equiv (-1)^{\frac{p-1}{2}} = -1 \pmod p,$$

which contradicts Fermat's Little Theorem.

Theorem 3.9 (Lagrange). *Let p be a prime number and $p \nmid a_n$. Then the number of solutions (including repeated ones) to the congruence*

$$a_n x^n + a_{n-1} x^{n-1} + \cdots + a_0 \equiv o \pmod p$$

does not exceed n.

Proof. Suppose the above formula has $n + 1$ solutions $x \equiv \alpha_i \pmod p$ $1 \leq i \leq n + 1$. Then it follows from Theorem 3.7 that

$$f(x) \equiv a_n (x - \alpha_1)(x - \alpha_2) \cdots (x - \alpha_n) \pmod p.$$

Since $f(\alpha_{n+1}) \equiv 0 \pmod{p}$,

$$\alpha_n(\alpha_n - \alpha_n)(\alpha_n - \alpha_n) \cdots (\alpha_n - \alpha_n) \equiv 0 \pmod{p}.$$

which is a contradiction since $p \nmid a_n$, α_i $(1 \leq i \leq n+1)$ are incongruent modulo p. This completes the proof of Theorem 3.9.

Theorem 3.9 may not hold if the modulus is a composite number. For example, there are six solutions of $x^3 - x \equiv 0 \pmod{6}$.

In the end, we discuss the sufficient and necessary conditions for the nth-degree congruence (3.8) to have n solutions. Without loss of generality, we can assume that $a_n = 1$. \square

Theorem 3.10. *Let $n \leq p$. There are n solutions to*

$$f(x) = x^n + a_{n-1}x^{n-1} + \cdots + a_0 \equiv 0 \pmod{p} \tag{3.9}$$

if and only if p divides all the coefficients of the remainder when $f(x)$ divides $x^p - x$.

Proof. Since the leading coefficient of $f(x)$ is 1, it follows from the division algorithm that there exist integral coefficient polynomials $q(x)$ and $r(x)$ satisfying

$$x^p - x = f(x)q(x) + r(x), \tag{3.10}$$

where the degree of $r(x)$ is less than n and the degree of $q(x)$ is $p-n$. If (3.9) has n solutions, then by Fermat's Little Theorem, they must be solutions to $x^p - x \equiv 0 \pmod{p}$. It follows from (3.10) that they are solutions to $r(x) \equiv 0 \pmod{p}$. But the degree of $r(x)$ is less than n, so by Lagrange's theorem, p divides all the coefficients of $r(x)$.

Conversely, if all the coefficients of $r(x)$ can be divided by p, then by Fermat's little theorem and (3.10), the congruence

$$f(x)q(x) \equiv 0 \pmod{p}$$

holds for any integer x. By Lagrange's theorem, the number of solutions of (3.9) is less than or equal to n, and the number of solutions to $q(x) \equiv 0 \pmod{p}$ is less than or equal to $p - n$. Thus, the

number of solutions to (3.9) must be n. This completes the proof of Theorem 3.10. \square

Example 3.3. Let p be a prime and $d|p-1$. Then the congruence

$$x^d \equiv 1 \pmod{p}$$

has d solutions.

In the end, the author wants to point out that, since Wilson's theorem is a criterion for primes, we can find a criterion for the kth-power twin prime pairs (cf. Conjecture A in Section 5).

Let k be an arbitrary positive integer. Then $\{n, n^k - 2\}$ is a prime pair if and only if

$$4\{(n-3)! + 1\} \equiv -(n-2) \pmod{n(n-2)}, \quad \text{for } k = 1,$$

$$4\{(n^k - 3)! + 1\} \equiv 2(n-1)!(n^k - 2) \pmod{n(n^k - 2)}, \quad \text{for } k > 1.$$

If k is an odd integer, then $\{m, m^k + 2\}$ is twin prime if and only if

$$4\{(m-1)! + 1\} \equiv -m \pmod{m(m+2)}, \quad \text{for } k = 1,$$

$$4\{(m^k + 1)! + 1\} \equiv -2(m-1)!(m^k + 2) \pmod{m(m^k + 2)},$$

$$\text{for } k > 1.$$

Remark. Double factorial is defined as follows, if n is positive odd integer, then $n!! = n \times (n-2) \times \cdots \times 3 \times 1$, and if n is positive even integer, then $n!! = n \times (n-2) \times \cdots \times 4 \times 2$. It is well known that

$$(p-1)!! \equiv \begin{cases} (-1)^{\frac{\alpha+1}{2}} \pmod{p} & \text{if } p \equiv 3 \pmod{4}, \\ (-1)^{\frac{\alpha+1}{2}} i_p \pmod{p} & \text{if } p \equiv 1 \pmod{4}, \end{cases}$$

where μ is the number of j satisfying $1 \leq j < \frac{p}{2}$, $1 \leq j^{-1} < \frac{p}{2}$, i_p is the unique solution of $j_p^2 \equiv -1 \pmod{p}$, $1 \leq i_p < \frac{p}{2}$, Meanwhile, by Wilson's theorem, $(p-2)!! \equiv -(p-1)!! \pmod{p}$.

We will show that the coefficients and their special values of the series expression of elliptic integral are closely related to the double factorials. The so-called elliptic integral is a concept in analysis.

It first appeared in some problems related to the calculation of length for elliptic arc, which cannot be expressed by basic functions. However, by simplifying the formula appropriately, each elliptic integral can be transformed into an integral involving only rational functions and three classical forms. That is to say, the first, second and third type of elliptic integrals. Meanwhile, each type of elliptic integral is divided into complete integrals and incomplete integrals. We find that in each of the three kinds of complete elliptic integrals, the coefficients and their special values of the series expressions are related to the double factorial, and each of them satisfies some congruence. Now, we give the definitions of the first two kinds of complete elliptic integrals and their congruences.

The First Kind of Complete Elliptic Integrals

$$K(k) = \int_0^{\frac{\pi}{2}} \frac{d\theta}{\sqrt{1 - k^2 \sin^2 \theta}} = \int_0^1 \frac{dt}{\sqrt{(1 - t^2)(1 - k^2 t^2)}}$$

$$= \frac{\pi}{2} \sum_{n=0}^{\infty} (P_{2n}(0))^2 k^{2n}$$

$$= \frac{\pi}{2} \left\{ 1 + \left(\frac{1}{2}\right)^2 k^2 + \left(\frac{(2n-1)!!}{(2n)!!}\right)^2 k^{2n} + \cdots \right\}.$$

Let the sum of the first n turns in above series be $K_n(k) = \frac{\pi}{2} L_n(k)$, then $L_n(1) = \sum_{k=0}^{n} \left(\frac{(2k-1)!!}{(2k)!!}\right)^2$. Taking $n = \frac{p-2}{2}$, by the properties of double factorials, we have $L_{\frac{p-1}{2}}(1) \equiv (-1)^{\frac{p-1}{2}} \pmod{p^2}$.

The second kind of Complete Elliptic Integrals

$$E(k) = \int_0^{\frac{\pi}{2}} \sqrt{1 - k^2 \sin^2 \theta} \, d\theta = \int_0^1 \frac{\sqrt{1 - k^2 t^2}}{\sqrt{1 - t^2}} dt$$

$$= \frac{\pi}{2} \sum_{n=0}^{\infty} \left(\frac{(2n)!!}{2^{2n}(n!)^2}\right)^2 \frac{k^{2n}}{1 - 2n}$$

$$= \frac{\pi}{2} \left\{ 1 + \left(\frac{1}{2}\right)^2 \frac{k^2}{1} - \cdots - \left(\frac{(2n-1)!!}{(2n)!!}\right)^2 \frac{k^{2n}}{1 - 2n} - \cdots \right\}.$$

Let the sum of the first n turns in above series be $E_n(k) = \frac{\pi}{2}F_n(k)$, then by induction we deduce that

$$F_n(1) = 1 - \left(\frac{1}{2}\right)^2 \frac{1^2}{1} - \cdots - \left(\frac{(2n-1)!!}{(2n)!!}\right)^2 \frac{1^{2n}}{1-2n}$$

$$= \left(\frac{(2n-1)!!}{(2n)!!}\right)^2 (2n+1).$$

Let $n = \frac{p-1}{2}$. By the properties of double factorials, we have

$$F_{\frac{p-1}{2}}(1) \equiv p \pmod{p^2}$$

The third kind of Complete Elliptic Integrals

$$\prod(l,k) = \int_0^{\frac{\pi}{2}} \frac{d\theta}{(1 - l\sin^2\theta)\sqrt{1 - k\sin^2\theta}}$$

$$= \int_0^1 \frac{dt}{(1 - lt^2)\sqrt{(1-t^2)(1-kt^2)}}.$$

This integral is a double series. When one variable is fixed, the coefficients and their special values of the other variable satisfy some congruence, which is similar to the ones of the first and second complete elliptic integrals.

A theorem that Gauss did not prove

In his *Disquisitiones Arithmeticae* published in 1801, Gauss generalized Wilson's theorem from prime number modulus to composite number modulus, which is another significant work since Euler generalized Fermat's Little Theorem from prime number modulus to composite number modulus. Gauss concluded:

Theorem (Gauss.) *Let $m > 1$ be an arbitrary integer. Then*

$$\prod_{\substack{1 \le i \le m \\ (i,m)=1}} i \equiv \begin{cases} -1 \pmod{m} & \text{if } m = 2, 4, p^\alpha, 2p^\alpha \\ & \quad (p \text{ is an odd prime number}), \\ 1 \pmod{m} & \text{otherwise.} \end{cases}$$

It is a pity that Gauss himself didn't prove it. According to L. E. Dickson's *History of the Theory of Numbers*, it was German

mathematician A. L. Crelle (1780–1855) who first proved it in 1832. He founded the well-known Journal of Crelle (*Journal für die reine und angewandte Mathematik*), and published in the first issue Abel's detailed proof of no solution in roots for the quintic equation. So far, the simplest proof is drawn from the lemma below which is a special case of Theorem 3.7 in Section 20.

Lemma. *There must be solutions to the congruence*

$$x^2 \equiv 1 \pmod{m}, \quad m = 2^\alpha p_1^{\alpha_1} \cdots p_k^{\alpha_k}, \quad k \geq 1,$$

and the number of solutions is 2^k *if* $\alpha = 0$ *or* 1; 2^{k+1}, *if* $\alpha = 2$; 2^{k+2}, *if* $\alpha \geq 3$.

Proof of Gauss's theorem. If $m = 2$ or 4, we can verify it directly. In fact, if $m = 2$, then both congruences hold. For the rest of m, it follows from the proof of Wilson's theorem.

For any $i, 1 \leq i \leq m, (i, m) = 1$, there exists a unique i', with $1 \leq i' \leq m, (i', m) = 1$ such that

$$ii' \equiv 1 \pmod{m}.$$

They are all in pairs. Now it is enough to consider solutions to the congruence

$$x^2 \equiv 1 \pmod{m}.$$

If i is a solution to the above congruence, then $-i$ must be a solution, too. Obviously, they are different solutions modulo m, and the contribution of each pair to the product is

$$i(-i) = -i^2 \equiv -1 \pmod{m}.$$

Then, it follows from the above lemma that the number of pairs like this is odd if and only if $m = p^\alpha$ or $m = 2p^\alpha$. The number is even for the rest of m. Thus, Gauss's theorem holds.

In 1938, the Russian woman mathematician E. Lehmer (D. H. Lehmer's wife) defined the Wilson prime as a prime number p such that

$$p^2 | (p-1)! + 1,$$

and she conjectured that there are infinitely many Wilson primes. However, so far the only known Wilson's primes are 5, 13, and 563.

Let m, k be positive integers, p an odd prime number, and $\phi(m) > k$. As a student the author generalized Gauss's theorem to

$$\sum_{\substack{1 \le i_1 < i_2 \le m \\ (t_1 t_2 \ldots t_k, m)=1}} i_1 i_2 \ldots i_k$$

$$\equiv \begin{cases} (-1) \dbinom{\phi(m) - 1}{k - 1} \pmod{m}, & \text{if } m = 2, 4, p^\alpha, 2p^\alpha, \\ 1 \pmod{m}, & \text{otherwise.} \end{cases}$$

When $k = 1$, this is Gauss's theorem. In 2005, the conclusion was generalized to the group of ideal classes modulo m (Yu Chengyan, Zhou Xia, *Generalization of Gauss' theorem on the groups of ideal classes modulo m*, *Journal of Zhejiang University*, **32**(3), 251–252). It is worth mentioning that when all unequal signs in the left sum above are changed to be less than or equal signs, we can get another generalization of Gauss' theorem, only if the index on the right side of the congruence be changed to $\dbinom{\phi(m) + k - 1}{k - 1}$.

Finally, we introduce the Carmichael function and Carmichael's theorem, the latter is a generalization of Euler's theorem, with the form similar to Gauss' theorem. The American mathematician Robert Carmichael (1879–1967) defined the following arithmetic function which was later named after him. Let $n > 1$ be an integer, and let $\lambda(n)$ be the least positive integer m satisfying

$$a^m \equiv 1 \pmod{n}$$

for any integer a coprime to n. By using the properties in Chapter V and Qin Jiushao's theorem, one can deduce the following Carmichael theorem.

Theorem (Carmichael).

$$\lambda(n) = \begin{cases} \phi(n), & \text{if } n = 2, 4, p^\alpha, 2p^\alpha, \\ \dfrac{\phi(n)}{2}, & \text{if } n = 8, 16, 32, \ldots, \end{cases}$$

and if $n = p_1^{\alpha_1} \ldots p_k^{\alpha_k}$ is the standard factorization of n, then

$$\lambda(n) = [\lambda(p_1^{\alpha_1}), \ldots, \lambda(p_k^{\alpha_k})].$$

The Göttingen Observatory and former residence of Gauss and Riemann, photo taken by the author.

13. Diophantine Equation

One thousand five hundred years ago, in the South and North dynasties (368–534), Zhang Qiujian, a Chinese mathematician in the Northern Wei dynasty, solved the following problem:

> *If a rooster worths five coins, a hen three coins, and three chickens together worth one coin, how many roosters, hens, and chickens total 100 can be bought for 100 coins?*

This "hundred fowls" problem originally appeared in *Zhang Qiujian Suanjing* (*Zhang Qiujian's Mathematical Manual*). The book, although its writing time is obscure, had been widely spread in Sui Dynasty, and listed as one of the Suanjing Shishu (*Ten Mathematical Manuals*) (*Sunzi Suanjing* among them) in Tang dynasty. Besides, in folklore, *Zhang Qiujian* was a child prodigy, and this question was asked by prime minister to test him.

Let x, y, z be the numbers of roosters, hens and chicks bought, respectively. Then the problem can be stated in two equations in three unknowns as follows:

$$\begin{cases} 5x + 3y + \dfrac{1}{3}z = 100, \\ x + y + z = 100. \end{cases}$$

Let's eliminate z. Through proper simplification, we have

$$7x + 4y = 100.$$

Then, to solve the problem, we only need to get the non-negative integer solutions to the above equation. This equation is an example of indeterminate linear equation in two variables, whose general form is

$$ax + by = c, \tag{3.11}$$

where a, b, c are fixed integers, and a, b are not both zero.

The indeterminate equation is one of the oldest branches in number theory. Diophantus (246–330), the last notable mathematician in ancient Greek, made a systematic study of such equations, so they are also known as Diophantus equations. In general, the number of unknown variables must be more than that of equations, and only the integer or positive integer solutions are searched or studied. There are three questions asked in the study of Diophantus equation: Are there any solutions? How many if there exist solution? How to find all solutions?

By the definition of congruence, (3.11) has solutions if and only if the congruence (we may assume that $b > 0$)

$$ax \equiv c \pmod{b}$$

has solutions. Thus, as with division, linear Diophantus equations can be defined by congruence.

Now we give a necessary and sufficient condition for the existence of solutions to (3.11), and the method to obtain the general solution if there exists one.

DIOPHANTI
ALEXANDRINI
ARITHMETICORVM
LIBRI SEX,
ET DE NVMERIS MVLTANGVLIS
LIBER VNVS.

Nunc primùm Græcè & Latinè editi, atque absolutissimis
Commentariis illustrati.

AVCTORE CLAVDIO GASPARE BACHETO
MEZIRIACO SEBVSIANO, V.C.

LVTETIAE PARISIORVM,
Sumptibus Sebastiani Cramoisy, via
Iacobæa, sub Ciconiis.
M. DC. XXI.
CVM PRIVILEGIO REGIS.

Title page of the 1621 edition of Diophantus' *Arithmetic* in Latin version

The Latin translator, French mathematician and poet Bachet

Theorem 3.11. *Equation* (3.11) *has a solution if and only if* $(a, b)|c$.

Proof. The necessity is clear. We prove the sufficiency.

Suppose $(a, b)|c$. Assume that $c = c_1(a, b)$. By Theorem 1.6, there exist integers s and t such that

$$as + bt = (a, b).$$

Take $x_0 = sc_1$ and $y_0 = tc_1$. We have $ax_0 + by_0 = c$. It follows that (3.11) has the integer solution (x_0, y_0). This completes the proof of Theorem 3.11. □

Theorem 3.12. *Suppose* (3.11) *has solution* (x_0, y_0). *Assume* $(a, b) = d, a = a_1 d,$ *and* $b = b_1 d$. *Then all the solutions to* (3.11) *are*

$$x = x_0 - b_1 t, y = y_0 + a_1 t, \quad t = 0, \pm 1, \pm 2, \ldots. \tag{3.12}$$

Proof. It is easy to verify that for any integer t, each pair (x, y) satisfies (3.11). Suppose (x, y) is a solution to (3.11). Since (x_0, y_0) is also a solution, we have

$$a(x - x_0) + b(y - y_0) = 0.$$

It follows from the assumption that

$$a_1(x - x_0) + b_1(y - y_0) = 0,$$

where $(a_1, b_1) = 1$. By Lemma 3.2, $y - y_0 = a_1 t$, i.e., $y = y_0 + a_1 t$. Substitute this into the above equation, we obtain $x = x_0 - b_1 t$. This completes the proof of Theorem 3.12. □

Now we want to discuss how to solve (3.11). If the solution exists, then by the proof of Theorem 3.11, we might first consider the equation:

$$ax + by = (a, b). \tag{3.13}$$

It is equivalent to

$$\frac{a}{(a, b)}x + \frac{b}{(a, b)}y = 1.$$

That is to say, we need get an integer solution to the equation:

$$ax + by = 1, \quad (a, b) = 1. \tag{3.14}$$

We can assume that a and b are both positive integers. Using the Euclidean algorithm for a and b, if r_n is the last non-zero remainder, we have $r_n = 1$. By working backward through the equations, 1 can be represented as a linear combination of a and b. Thus, (x_0, y_0) is a solution to (3.14). It follows that (dx_0, dy_0) is a solution to (3.13).

Example 3.4 (The "hundred fowls" problem). Find all the non-negative solutions to $7x + 4y = 100$.

We first solve $7x + 4y = 1$. By the division algorithm, we have

$$7 = 4 \times 1 + 3, \quad 4 = 3 \times 1 + 1.$$

Evidently, $x = -1, y = 2$. Thus, $x = -100, y = 200$ is a solution to the original equation. It follows from Theorem 3.12 that all the solutions to the original equation are

$$x = -100 - 4t, y = 200 + 7t, \quad t = 0, \pm 1, \pm 2, \ldots.$$

To solve the "hundred fowls" problem, it is necessary that x, y and z are all non-negative integers, so we take $t = -28, -27, -26, -25$. Thus, we get all the solutions to the "hundred fowls" problem

$$(x, y, z) = (12, 4, 84), (8, 11, 81), (4, 18, 78), (0, 25, 75).$$

Example 3.5. There are $[\frac{N}{ab}]$ or $[\frac{N}{ab}] + 1$ non-negative solutions to the equation

$$ax + by = N, \quad a > 0, b > 0, (a, b) = 1.$$

Now we will discuss how to solve linear equation of n unknowns. Let $n \geq 2$ and a_1, a_2, \ldots, a_n be not all zeros, we have the following result.

Theorem 3.13. *The equation*

$$a_1 x_1 + a_2 x_2 + \cdots + a_n x_n = N \tag{3.15}$$

has a solution if and only if $(a_1, a_2, \ldots, a_n) | N$.

Proof. The necessity is clear. We prove the sufficiency.

Let $d = (a_1, a_2, \ldots, a_n)$ and $d | N$. We will prove it by induction. If $n = 2$, then there is one solution to (3.15) by Theorem 3.11. Suppose it is true for $n - 1$ unknowns, and consider the case of n unknowns.

Let $d_2 = (a_1, a_2)$. Then $(d_2, a_3, \ldots, a_n) = d|N$. By the inductive assumption, we see that the equation

$$d_2 t_2 + a_3 x_3 + \cdots + a_n x_n = N$$

has a solution, written as t_2', x_3', \ldots, x_n'. Since $a_1 x_1 + a_2 x_2 = d_2 t_2'$, if x_1', x_2' is a solution, then

$$a_1 x_1' + a_2 x_2' + \cdots + a_n x_n' = N.$$

With this we have proved Theorem 3.13. □

A method to solve (3.15) is obtained from the proof of Theorem 3.13.

Example 3.6. Find all the solutions to $15x + 10y + 6z = 61$.

Example 3.7. Express $\frac{19}{60}$ as the sum of three irreducible fractions.

Let $0 < a_1 < a_2 < \cdots < a_n$ and $(a_1, a_2, \ldots, a_n) = 1$. Consider the non-negative solution to (3.15). It is known that there must exist a solution to (3.15), if N is sufficiently large. Denote $G(a_1, a_2, \ldots, a_n)$ the largest positive integer N such that there is no non-negative solution to (3.15). Questions about how to find and estimate $G(a_1, a_2, \ldots, a_n)$ is called Frobenius coin problem, and G is called Frobenius number. For example, $G(2, 3) = 1$ and $G(3, 4) = 5$. Frobenius (1849–1917) was a German mathematician specializing in differential equations and group theory, and he was an amateur in number theory. In 1884, Sylvester proved that

$$G(a_1, a_2) = (a_1 - 1)(a_2 - 1) - 1,$$

and he also asserted that the number of positive integers that cannot be expressed as (3.15) is $(a_1 - 1)(a_2 - 1)/2$.

If $n = 3$, the conclusion is also known, but in the way of iterative computation without a clear expression. If $n \geq 4$, then it remains an open question.

If $n = 2$ and each unknown in (3.15) does not take value 1, then the largest number for which there is no solution is $(a_1 + 1)(a_2 + 1) - 1$, and the number of positive integers that cannot be expressed is $(a_1 + 1)(a_2 + 1)/2$. The second largest number such that there is no

solution to (3.15) is $(a_1 - 1)(a_2 - 2)$ or $(a_1 + 1)(a_2 - 1) + 1$ under the condition without limit and with limit, respectively.

Remark. Denote $G_k(a_1, a_2)(k \geq 1)$ the largest positive integer N such that there doesn't exist k or more solutions to (3.15), then we have

$$G_k(a_1, a_2) = ka_1a_2 - a_1 - a_2,$$

and the number of positive integers that (3.15) hasn't got k or more solutions is

$$(k^2 - k + 1)a_1a_2 - a_1 - a_2 - 2k + 3.$$

The Pythagorean triple

The earliest Diophantine equation might be the following quadratic equation with two unknowns

$$x^2 + y^2 = z^2. \tag{P}$$

Any solution to (P) is called a Pythagorean triple. As mentioned at the beginning of this book, Chinese have long known that $(3, 4, 5)$ is the minimum positive integer solution. But the clay tablets left by Babylonians showed that they might know the existence of the triple and Pythagorean theorem even earlier, and they also calculated the approximation of $\sqrt{2}$. However, the ancient Greeks first found infinitely many integer solutions to (P), namely,

$$x = 2n + 1, \quad y = 2n^2 + 2n, \quad z = 2n^2 + 2n + 1.$$

According to Proclus (c. 410–485), the last main philosopher in ancient Greece, the above answer due to Pythagoras. He was particularly interested in the right triangle of which hypotenuse is 1 longer than one leg. For example, take 1, 2 and 3 for n, we have three solutions (3, 4, 5), (5, 12, 13) and (7, 24, 25). Another positive integer solution which is the third least is (8, 15, 17), and Proclus believed that it was Plato who found it. Or rather, what he studied is the following array:

$$x = 2n, \quad y = n^2 - 1, \quad z = n^2 + 1.$$

If n is positive and even, then each set is an irreducible solution to (P). Obviously, these solutions correspond to the right triangle of which the hypotenuse is 2 longer than one leg.

Euclid provided all the irreducible solutions to the equation in *Elements of Geometry*, but did not indicate its provenance. He pointed out in the book that all the solutions to (P) satisfying $(x, y) = 1$ are

$$x = 2ab, y = a^2 - b^2, z = a^2 + b^2,$$

where $a > b > 0, (a, b) = 1, a + b \equiv 1 \pmod 2$.

Clay tablet Plimpton 322, collected in the library of Columbia University, on which there are examples of Pythagorean triples in sexagesimal

This is called Euclid's formula. The proof depends on the following conclusion: If a square number is equal to the product of two coprime positive integers, then these two positive integers are both square numbers. Reforming (P) into

$$\left(\frac{x}{2}\right)^2 = \frac{z+y}{2}\frac{z-y}{2}.$$

Note that $(\frac{z+y}{2}, \frac{z-y}{2}) = 1$. We take $\frac{z+y}{2} = a^2$ and $\frac{z-y}{2} = b^2$. Thus obtaining all the irreducible solutions to (P). In particular, taking

$a = n$, $b = 1$ and $a = n + 1$, $b = n$ in (P), we get the above results of Plato and Pythagoras, respectively. Moreover, for any prime $p \equiv \pm 1 \pmod 8$, we can show that there are infinitely many Pythagoras triples with p as the difference of two legs.

With Euclid's formula, we can get all the positive integer solutions to (P), by multiplying each of the above solutions by any positive integer.

After more than 2000 years, one day in 1637, in Toulouse, southern France, when Fermat, the king of amateur mathematicians, was reading the Latin edition of Diophantus's *Arithmetica*, in the margin of the page where there is Pythagorean theorem (with the number Problem 8), he wrote down the following words:

> *It is impossible to separate a cube into two cubes, or a fourth power into two fourth powers, or in general, any power higher than the second, into two like powers.*

This is the well-known Fermat's Last Theorem. Next to this note, Fermat added another note: "*I have discovered a truly marvelous proof of this, which this margin is too narrow to contain.*"

However, by using the method of descent, Fermat gave the proof for the $n = 4$, it needs to use Euclid's formula twice. In fact, what Fermat proved is a stronger result, he showed that the equation

$$x^4 + y^4 = z^2, \quad (x, y) = 1$$

has no positive integer solution. Suppose the above equation has a solution, then by Euclid's formula,

$$x^2 = 2st, \quad y^2 = s^2 - t^2, \quad z = s^2 + t^2, \quad s > t, \quad (s, t) = 1.$$

Hence,

$$t^2 + y^2 = s^2, \quad (t, y) = 1,$$

where both y and s are odd, t is even, $(s, 2t) = 1$. From $x^2 = s(2t)$, we have $s = u^2$, $2t = v^2$. Since

$$t = 2ST, \quad y = S^2 - T^2, \quad s = S^2 + T^2, (S, T) = 1,$$

we have

$$\left(\frac{v}{2}\right)^2 = ST, \quad S = X^2, \quad T = Y^2.$$

Therefore,

$$X^4 + Y^4 = S^2 + T^2 = s = u^2, \quad (X, Y) = 1.$$

Taking $Z = u$, then $Z = u \le u^2 = s < s^2 + t^2 = z$. A contradiction by descent!

It is worth mentioning that like *Elements*, *Arithmetica* consists of 13 books. In 1464, the Greek version of the first six books were found in Venice, Italy, and late, the Arabic version of the four books were found in Mashhad, Iran. Meanwhile, an interesting generalization of Fermat's Last Theorem will be discussed in Section 32.

Remark. Let n be a positive integer n, for any four adjacent Fibonacci numbers or Lucas numbers, we can construct Pythagoras triples by using their recurrence, i.e., $\{F_n F_{n+3}, 2F_{n+1} F_{n+2}, F_{n+1}^2 + F_{n+2}^2\}$, $\{L_n L_{n+3}, 2L_{n+1} L_{n+2}, L_{n+1}^2 + L_{n+2}^2\}$.

14. Lucas' Congruence

Let n be a natural number, and $0 \le k \le n$. A binomial coefficient or combinatorial number $\binom{n}{k}$ is the coefficient of the x^k term in the polynomial expansion of the nth-degree polynomial $(1 + x)^n$. It follows from the definition that a binomial coefficient must be a positive integer. In Section 5, the values of the binomial coefficients are given by the expression

$$\binom{n}{k} = \frac{n!}{k!(n-k)!}, \quad 0 \le k \le n.$$

The notation $\binom{n}{k}$ was introduced in 1826 by Ettingshausen (1796–1878), a German mathematician and physicist, but in fact, this coefficient is the $(k+1)$th number from left in the $(n+1)$th row of Pascal's triangle. The first seven rows of the triangle were presented early in the 13th century in *Xiangjie Jiuzhang Suanfa* by Yang Hui (active in academy from 1261 to 1275), a Chinese mathematician in the Southern Song Dynasty, known for his study of magic squares. Yang Hui mentioned in his book that the triangle was quoted from *Shi Suo Suan Shu* (now lost unfortunately) by a Chinese mathematician of the Northern Song Dynasty, Jia Xian

The Jia Xian-Yang Hui triangle in *Yongle Canon*

(active in academy about 1050), in the 11th century. Hence it is sometimes called Jia Xian's triangle or Yang Hui's triangle. Before or after them, an Indian, a Persian, and an Italian found binomial coefficients respectively, but no one seems to draw a triangle graph for them as Jia Xian or Yang Hui did in their books.

Binomial coefficients are of importance in combinatorics. There are also many beautiful congruences related to them in number theory. In 1852, Ernst Kummer proved that for a prime number p, the

maximum integer a such that $p^a | \binom{n}{k}$ is equal to the number of carries when k is added to $n - k$ in base p. In 1879, Lucas obtained the following theorem.

Theorem 3.14 (Lucas). *For any prime p and $0 \le s \le p - 1$,*

$$\binom{p-1}{s} \equiv (-1)^s \pmod{p}.$$

Proof. If $s = 0$, then the theorem is obvious. Now we assume $s \ge 1$. Then

$$(p - 1)(p - 2) \cdots (p - s) \equiv (-1)(-2) \cdots (-s) = (-1)^s s! \pmod{p}.$$

Let $(p - 1)(p - 2) \cdots (p - s) = (-1)^s s! + pt$. Then

$$\binom{p-1}{s} = \frac{(p-1)(p-2)\cdots(p-s)}{s(s-1)\cdots 1} = \frac{(-1)^s s! + pt}{s!} = (-1)^s + \frac{pt}{s!}.$$

Since $(p, s!) = 1$, we have $s! | t$. It follows that

$$\binom{p-1}{s} \equiv (-1)^s \pmod{p}.$$

This completes the proof of Theorem 3.14. □

The portrait of Lucas, who was killed by a piece of broken tile

Remark 3.1. If we replace p by a composite number n and take the smallest prime factor (*denoted* p) for s, then Lucas' congruence is no longer true, but it is still true if $0 \leq s < p$. If n is odd and $n - p \leq s \leq n - 1$, then the congruence holds, too. These are all trivial solutions. The study about non-trivial solutions is more interesting.

If n is a power of 2 or the square of an odd prime number, then there is no non-trivial solution. If $n = pk$ and each prime factor of k is greater than p^2, then $s = p^2$ is a non-trivial solution if and only if $\binom{k-1}{p} \equiv (-1)^p \pmod{p}$. In particular, if $p = 2$, then $k \equiv 3 \pmod 4$, and if $p = 3$, then $k \equiv 7, 17 \pmod{18}$.

Remark 3.2. For any prime p, the congruence

$$\binom{p-1}{s} \equiv (-1)^s \pmod{p^2}$$

has a solution $1 < s < \frac{p-1}{2}$, if and only if

$$p \mid s! \left(1 + \frac{1}{2} + \cdots + \frac{1}{s} \right).$$

The set of p less than 100 which satisfy the above condition is $\{11, 29, 37, 43, 53, 61, 97\}$. This is also the set of primes that divide the numerator of the sth harmonic number $H(s) = \sum_{i=1}^{s} \frac{1}{i} (s \geq 1)$ with $s \geq p - 1$. (cf. A 001008 in OEIS). The density of these primes is about 0.4, but we cannot judge whether there are infinitely many.

Meanwhile, we want to ask if there are distinct odd primes p and q such that

$$\binom{p-1}{s} \equiv (-1)^s \pmod{q^2},$$

$$\binom{q-1}{t} \equiv (-1)^t \pmod{p^2},$$

where $1 < s < \frac{p-1}{2}$, $1 < t < \frac{q-1}{2}$. We have searched and found no solution for $p, q < 65000$.

In 1878, one year before the proof of Theorem 3.14, Lucas obtained another more famous congruence, usually called Lucas' theorem. Let's first assume that $\binom{0}{0} = 1$, $\binom{r}{s} = 0$ if $r < s$.

Theorem 3.15 (Lucas). *For any positive integers m and n, if $n = n_0 + n_1 p + \cdots + n_d p^d$ and $m = m_0 + m_1 p + \cdots + m_d p^d$ are the base p expansions of m and n, respectively, then the following congruence relation holds:*

$$\binom{n}{m} \equiv \binom{n_0}{m_0} \binom{n_1}{m_1} \cdots \binom{n_d}{m_d} \pmod{p}.$$

Corollary 3.2. *Under the notation of Theorem 3.15, $p \mid \binom{n}{m}$ if and only if there exists some i satisfying $n_i < m_i$.*

There are several ways to prove Theorem 3.15. We will give a proof based on generating functions. Before that we introduce a lemma which is a special case of the previous corollary.

Lemma 3.1. *For any prime p and positive integer j, if $0 < i < p^j$, then $p \mid \binom{p^j}{i}$.*

Proof. Suppose the standard factorization of $\binom{p^j}{i}$ is

$$\binom{p^j}{i} = \prod_q q^{\alpha_q}.$$

We only need to prove that $\alpha_p \geq 1$. In fact,

$$\alpha_p = \sum_{k=0}^{j} \left\{ \left[\frac{p^j}{p^k} \right] - \left[\frac{i}{p^k} \right] - \left[\frac{p^j - i}{p^k} \right] \right\}.$$

By the property of $[x]$, each term in the sum is non-negative, and the last term is 1, so we have $\alpha_p \geq 1$. The lemma is proved. $\qquad\square$

More generally, from Kummer's conclusion we can easily obtain that $\frac{n}{(n,k)} \mid \binom{n}{k}$.

Proof of Theorem 3.15. In terms of binomial expansion, Lemma 3.1 means that

$$(1+x)^{p^j} \equiv 1 + x^{p^j} \pmod{p}.$$

Thus,

$$\sum_{m=0}^{n} \binom{n}{m} x^m = (1+x)^n = \prod_{j=0}^{d} ((1+x)^{p^j})^{n_j}$$

$$\equiv \prod_{j=0}^{d} (1+x^{p^j})^{n_j} = \prod_{j=0}^{d} \left(\sum_{m_j=0}^{n_j} \binom{n_j}{m_j} x^{m_j p^j} \right)$$

$$\equiv \sum_{m=0}^{n} \left(\prod_{j=0}^{d} \binom{n_j}{m_j} \right) x^m \pmod{p}.$$

Then Theorem 3.14 follows from the property of generating functions.

There is an obvious corollary or equal form of Theorem 3.15.

Theorem 3.16. *For any positive integers n and m, if n_0 and m_0 are the least non-negative residues of n and m modulo p, respectively, then*

$$\binom{n}{m} \equiv \binom{\left[\frac{n}{p}\right]}{\left[\frac{m}{p}\right]} \binom{n_0}{m_0} \pmod{p}.$$

Corollary 3.3. *If p and q are distinct odd primes, then*

$$\binom{pq-1}{\dfrac{pq-1}{2}} \equiv \binom{p-1}{\dfrac{p-1}{2}} \binom{q-1}{\dfrac{q-1}{2}} \pmod{pq}.$$

The last congruence was obtained by the author in 2002 (Acta. Math. Sinica **18**. *277–288). Indeed, we have a better result, that is, if p, q are distinct odd primes, then*

$$\binom{pq-1}{\left[\dfrac{pq-1}{e}\right]} \equiv \binom{p-1}{\left[\dfrac{p-1}{e}\right]} \binom{q-1}{\left[\dfrac{q-1}{e}\right]} \pmod{pq},$$

where e = 2, 3, 4 or 6. In Section 29, we will show more stronger congruences.

Meanwhile, we have the following result.

Theorem 3.17. *If $p \geq 5$ is a prime and n is a positive integer, then*

$$\sum_{s=0}^{p-1} (-1)^s \binom{p-1}{s}^n \equiv \begin{cases} \binom{np-2}{p-1} \pmod{p^4} & \text{if } n \text{ is odd,} \\ 2^{n(p-1)} \pmod{p^3} & \text{if } n \text{ is even.} \end{cases}$$

$$\sum_{s=0}^{p-1} \binom{p-1}{s}^n \equiv \begin{cases} \binom{np-2}{p-1} \pmod{p^4} & \text{if } n \text{ is even,} \\ 2^{n(p-1)} \pmod{p^3} & \text{if } n \text{ is odd.} \end{cases}$$

In 2017, we (Zhong Hao and Tianxin Cai, *Int. J. Number Theory* **13**(6), 1617–1625) showed that both Fibonacci numbers and Lucas number have some properties as Lucas' theorem. Let $n = n_0 + n_1 p + \cdots + n_d p^d$ with $0 \leq n_i \leq p-1 (0 \leq i \leq d)$. Then

$$L_{n+1} \equiv \prod_{i=0}^{d} L_{n_i+1} \pmod{p}, \quad F_{5n+1} \equiv \prod_{i=0}^{d} F_{5n_i+1} \pmod{p}$$

holds for all n if and only if $p = 5$.

Lucas died in an unusual circumstance. At the banquet of the annual congress of the *Association française pour l'avancement des sciences* a waiter dropped some crockery, and a piece of broken plate cut Lucas on the cheek. He died a few days later of a severe skin inflammation probably caused by septicemia. Lucas was only 49 years old when he died.

It's worth mentioning that in 2012, a French mathematician Ayad and a Canadian mathematician Kihel (*Int. J. Number Theory*, **8**, 2045–2057) studied odd composite number n satisfying

$$\binom{n-1}{\frac{n-1}{2}} \equiv (-1)^{\frac{n-1}{2}} \pmod{n}.$$

They found the minimum $n = 5907 = 3 \times 11 \times 179$ and conjectured that the above congruence has no solution if n is a product of two distinct odd primes.

Covering system

The notion of covering systems was introduced by Paul Erdős in the early 1930s. A covering system is a collection of finitely many congruences

$$\{a_1 \ (\text{mod } n_1), a_2 \ (\text{mod } n_2), \ldots, a_k \ (\text{mod } n_k)\}$$

that contains every integer. The following are examples of covering systems:

$\{0 \ (\text{mod } 3), \ 1 \ (\text{mod } 3), \ 2 \ (\text{mod } 3)\}$,

$\{1 \ (\text{mod } 2), \ 2 \ (\text{mod } 4), \ 0 \ (\text{mod } 8), \ 4 \ (\text{mod } 8)\}$,

$\{0 \ (\text{mod } 2), \ 0 \ (\text{mod } 3), \ 1 \ (\text{mod } 4), \ 5 \ (\text{mod } 6), \ 7 \ (\text{mod } 12)\}$.

A covering system is called disjoint if no two members overlap. The first two examples are disjoint, and the first is also a complete residue system. A covering system is called distinct (or incongruent) if all the moduli are different (and greater than 1). The third example is distinct.

Let A be such a subset of $\{0, 1, \ldots, n-1\}$ that for any remainder residue $N_i (0 \leq i < n)$ modulo n, there exist $b, c \in A$, $b + c \equiv i \ (\text{mod } n)$, i.e., $b + c \in N_i$. Let A_n be such a subset with the least elements and denote $a_n = |A_n|$. Then the first 27 of a_ns are $\{1, 2, 2, 3, 3, 4, 4, 4, 4, 5, 5, 5, 5, 6, 6, 6, 6, 7, 7, 7, 7, 8, 8, 8, 8, 8, 8\}$. For example, $A_1 = \{0\}, A_2 = A_3 = \{0, 1\}, A_4 = A_5 = \{0, 1, 2\}$, $A_6 = A_7 = A_8 = A_9 = \{0, 1, 3, 4\}, A_{10} = \cdots = A_{13} = \{0, 1, 3, 5, 6\}$, $A_{14} = \cdots = A_{17} = \{0, 1, 3, 5, 7, 8\}, A_{18} = \cdots = A_{21} = \{0, 1, 3, 5, 7, 9, 10\}, A_{22} = \cdots = A_{27} = \{0, 1, 3, 4, 9, 10, 12, 13\}$.

However, what is the general expression of A_n or a_n?

Let c be the minimum of modulus of an incongruent covering system. Then $c = 2$ in Example 3, abbreviated to $\{0(2), 0(3), 1(4), 5(6), 7(12)\}$. This was discovered by Erdős, which is the first incongruent covering system. But it is not unique, for $\{0(2), 0(3), 1(4), 3(8), 7(12), 23(24)\}$ is another example. For $c = 3$ Erdős also provided an example, that is $\{0(3), 0(4), 0(5), 1(6), 1(8), 2(10), 11(12), 1(15), 14(20), 5(24), 8(30), 6(40), 58(60), 26(120)\}$, where the

modulus are in the set of the divisors greater than or equal to 3 of 120. If $c = 4$, then the modulus are in the set of the divisors greater than or equal to 4 of 2880. As for $c = 40$, such a covering system consists of more than 10^{50} congruences.

There are two notable conclusions about covering systems, and the method used in the proof is elementary.

Theorem. *For any incongruent covering system,*

$$\sum_{j=1}^{k} \frac{1}{n_j} > 1,$$

and for any disjoint covering system, $(n_i, n_j) \neq 1$ $(1 \leq i < j \leq k)$ and

$$\sum_{j=1}^{k} \frac{1}{n_j} = 1.$$

It is obvious that a disjoint covering system is not incongruent, and vice versa.

About incongruent covering systems, Erdős had two well-known conjectures.

Conjecture 1. *For any positive integer $c > 1$, there exists an incongruent covering system with the minimum modulus c.*

Conjecture 2. *An incongruent covering, whose modulus are all odd, does not exist.*

Recently, Bob Hough of the University of Oxford denied Conjecture 1 on internet (arXiv:1307.0874), he showed that the least modulus of a distinct covering system of congruences is at most 10^{18}.

Finally, we give a conclusion about covering system, which was proved by Indian–British–American mathematician Sarvadaman Chowla (1907–1995) in 1948.

Theorem (Chowla). *Let* $n \geq 3$, *and let* a_1, a_2, \ldots, a_n *and* b_1, b_2, \ldots, b_n *are two arbitrarily complete residue systems modulo* n. *Then* $a_1 b_1, a_2 b_2, \ldots, a_n b_n$ *cannot form a complete residue system. That means*

$$\{a_1 b_1 \ (\text{mod } n), a_2 b_2 \ (\text{mod } n), \ldots, a_n b_n \ (\text{mod } n)\}$$

cannot form a covering system.

Proof. First of all, suppose $4 | n$, if $a_1 b_1, a_2 b_2, \ldots, a_n b_n$ forms a complete residue system modulo n, both odd and even numbers are half of each. Without losing its generality, let $a_1 b_1, a_2 b_2, \ldots, a_{n/2} b_{n/2}$ be odd, then $a_1, a_2, \ldots, a_{n/2}$ and $b_1, b_2, \ldots, b_{n/2}$ are also odd. From the definition of complete residue system, there exists $\frac{n}{2} < j \leq n$, satisfying $a_j b_j \equiv 2 \ (\text{mod } 4)$. While $a_j \equiv b_j \equiv 0 \ (\text{mod } 2)$, a contradiction!

Now suppose $4 \nmid n$, then $n = qm$, where m is odd, $q = p$ or $2p$, p is odd prime. By Gauss' theorem,

$$\prod_{\substack{j=1 \\ (j,q)=1}}^{q} j \equiv -1 \ (\text{mod } q)$$

hence

$$\prod_{\substack{j=1 \\ (j,q)=1}}^{n} j \equiv \prod_{\substack{j=1 \\ (j,q)=1}}^{q-1} j \prod_{\substack{j=q+1 \\ (j,q)=1}}^{2q-1} j \cdots \prod_{\substack{j=(m-1)q+1 \\ (j,q)=1}}^{n=1} j \equiv (-1)^m = -1 \ (\text{mod } q).$$

If $n > 2$, $a_1 b_1, a_2 b_2, \ldots, a_n b_n$ is a complete residue system modulo n, then

$$-1 \equiv \prod_{\substack{j=1 \\ (a_j b_j, q)=1}}^{n} a_j b_j \equiv \prod_{\substack{j=1 \\ (a_j, q)=1}}^{n} a_j \prod_{\substack{j=1 \\ (b_j, q)=1}}^{n} b_j \equiv (-1)^2 = 1 \ (\text{mod } q).$$

A contradiction! This completes the proof of the theorem. \square

It is easy to see that except $\{3, 5\}$, all the other twin primes can be written as $\{6n \pm 1\}$, where the set for all n is $\{1, 2, 3, 5, 7, 10, 12, 17, 18, 23, \ldots\}$. In 2002, Jon Perry discovered that the elements in the set are those n such that the equation

$$n = 6ab \pm a \pm b$$

has no positive integer solution. It is easy to prove it by contradiction. Hence, the twin prime conjecture is true if and only if there exists infinitely many n so that the above equation has no solution. Or for any integer x, the congruences $\{\pm a \pmod{6a \pm 1}, \ a \geq 1\}$ cannot contain all the integers greater than x, i.e.,

$$\mathbb{Z} \cup (x, \infty) \subsetneq \{\pm a \pmod{6a \pm 1}, a \geq 1\}.$$

And we observe that all the positive integer n such that $\{4n \pm 1\}$ are twin primes are those n such that the equation

$$n = 4ab \pm a \pm b$$

has no positive integer solution.

15. The Truth of Primes

Lagrange not only gave a proof of Wilson's theorem for the first time, but also stated that Wilson's theorem can be used as a criterion for a prime, that is, $n > 1$ is a prime if and only if

$$(n - 1)! \equiv -1 \pmod{n}.$$

The sufficiency follows from Wilson's theorem, and the necessity can be proved as below. Note that any real factor d of n can divide $(n-1)!$. Then $d|1$, so n must be a prime. However, the criterion has little practical significance, for there is no efficient method to quickly calculate $n!$

Now we will prove that, the Lucas congruence of Theorem 3.14 also can be used as a criterion for primes, which means that $n > 1$

Portrait of French mathematician Lagrange

is a prime if and only if for any integer $0 \leq s \leq n-1$,

$$\binom{n-1}{s} \equiv (-1)^s \pmod{n}.$$

Only the necessity needs to be proved. Suppose n is a composite number and p is the least prime factor of n with $n = pk, p \leq k$. Take $s = p$. Then

$$\binom{n-1}{s} = \frac{(pk-1)\cdots(pk-(p-1))}{(p-1)!}(k-1).$$

Note that $((p-1)!, pk) = 1$, hence

$$\binom{n-1}{s} \equiv (-1)^{p-1}(k-1) \pmod{pk}.$$

Since $pk \nmid k$, it follows from the above two formulas that

$$\binom{pk-1}{p} \equiv (-1)^p \pmod{pk}.$$

does not hold.

Further, from the above discussion and Remark 3.1 in Section 14, we have the following criterion.

Criterion 1. An integer $n \geq 2$ is a prime if and only if

$$\binom{n-1}{s} \equiv (-1)^s \pmod{n}$$

holds for any integer s satisfying $0 \leq s \leq \sqrt{n}$.

This criterion is easier to check than Wilson's theorem. Next we introduce two criteria relevant to binomial coefficients and congruences without proof.

Criterion 2 (Babbage, 1819). An integer $n \geq 2$ is a prime if and only if

$$\binom{n+m}{n} \equiv 1 \pmod{n}$$

holds for any integer m satisfying $0 \leq m \leq n-1$.

The computer pioneer Babbage on a stamp of UK

Criterion 3 (Mans-Shanks, 1972). An integer $n \geq 3$ is a prime if and only if

$$\binom{m}{n-2m} \equiv 1 \pmod{m}$$

holds for any integer m satisfying $0 \leq 2m \leq n$.

But, is the inverse proposition of Fermat's Little Theorem true? The answer is negative.

We want to talk about an anecdote here. In 1898, an English physicist wrote that, the sinologist Sir Wade who had been in China for more than 40 years brought home some Chinese classics, and there was a mathematical paper of Confucius's time proving that n is a prime if and only if

$$2^n \equiv 2 \pmod{n}.$$

This conclusion was spread as Chinese hypothesis. In fact, it is relevant to mathematician Li Shanlan (1811–1882) in *Qing* dynasty. Li Shanlan was the first man publishing paper about prime numbers in China. Once, he thought he had gotten the above results, then he told his friend Hua Hengfang (1833–1902). Later Li Shanlan found his proof wrong, so he did not publish it. After his death, Hua announced Li's result, and was falsely reported to Europe.

As a matter of fact, as early as 1819, the French mathematician Pierre Frederic Sarrus (1798–1861) had found that,

$$2^{341} \equiv 2 \pmod{341},$$

but $341 = 11 \times 31$ is a composite number, and this is the least composite number satisfying the above congruence. Some other examples are

$$3^{91} \equiv 3 \pmod{91}, 91 = 7 \times 13; \quad 4^{15} \equiv 4 \pmod{15}, \ 15 = 3 \times 5.$$

Actually, for any integer a, the number of counter examples like this is infinite. Now let's define the pseudoprimes.

Definition 3.2. Let $a > 1$. If there is a composite number n satisfying

$$a^n \equiv a \pmod{n},$$

then we call n a pseudoprime to base a.

Theorem 3.18. *If n is a pseudoprime to base 2, then $2^n - 1$ is also a pseudo-prime to base 2.*

Proof. Let $l = 2^n - 1$. Then $n|(l-1)$. Writing $l - 1 = nk$, we obtain that

$$2^{l-1} - 1 = 2^{nk} - 1 = (2^n - 1)(2^{n(k-1)} + \cdots + 2^n + 1)$$
$$= l(2^{n(k-1)} + \cdots + 2^n + 1).$$

Thus, $l|(2^{l-1} - 1)$, that is, $2^l \equiv 2 \pmod{l}$. And for n is a composite number, l is a composite number, too. It follows that l is a pseudoprime to base 2. Theorem 3.18 is proved. □

From Theorem 3.18, we can draw a conclusion that there exist infinitely many pseudoprimes to base 2. This can be generalized as follows.

Theorem 3.19. *For any integer $a > 1$, there exist infinitely many pseudoprimes to base a.*

Proof. We assume that an odd prime $p \nmid a(a^2 - 1)$. Take $n = \frac{a^{2p}-1}{a^2-1} = \frac{a^p-1}{a-1} \frac{a^p+1}{a+1}$. Then n is a composite number. It is easy to see that

$$(a^2 - 1)(n - 1) = a^{2p} - a^2 = a(a^{p-1} - 1)(a^p + a). \qquad (3.16)$$

Since $p|a^{p-1} - 1, a^2 - 1|a^{p-1} - 1$ and $(p, a^2 - 1) = 1$, we have $p(a^2 - 1)|a^{p-1} - 1$. Note that $2|a^p + a$, hence by (3.16), $2p(a^2 - 1)|(a^2 - 1)(n - 1)$. It follows that $2p|(n - 1)$.

Let $n = 1 + 2pm$. By $a^{2p} = n(a^2 - 1) + 1 \equiv 1 \pmod{n}$, we have $a^{n-1} = a^{2pm} \equiv 1 \pmod{n}$, i.e., $a^n \equiv a \pmod{n}$. Hence n is a pseudoprime to base a. Since $a > 1$, the number of odd primes p satisfying $p \nmid a(a^2 - 1)$ is infinity.

Finally we introduce Carmichael's numbers. This is a kind of special pseudoprime, which was raised and studied first by Carmichael. □

Definition 3.3. A Carmichael number is a composite number n that is the pseudoprime to base a for any integer $a > 1$ with $(a, n) = 1$.

Example 3.8. The integer 561 is the least Carmichael number.

Here we only prove that 561 is a Carmichael number. Since $561 = 3 \times 11 \times 17$, if $(a, 561) = 1$, then $(a, 3) = 1$, $(a, 11) = 1$, and $(a, 17) = 1$.

It follows from Fermat's Little Theorem that $a^2 \equiv 1 \pmod 3$, $a^{10} \equiv 1 \pmod{11}$, and $a^{16} \equiv 1 \pmod{17}$. Note that $[2, 10, 16] = 80$, so $a^{80} \equiv 1 \pmod{561}$. Since 560 is the multiple of 80, we see that 561 is a pseudoprime to base a, and it follows that 561 is a Carmichael number.

Example 3.9. Let $n = p_1 p_2 \ldots p_k$, where p_i are different primes. If $k > 2$ and $p_i - 1 | n - 1$ for any i, then n is a Carmichael number.

Some other small Carmichael numbers are $1105 = 5 \times 13 \times 17, 1729 = 7 \times 13 \times 19$, $2465 = 5 \times 17 \times 29, 2821 = 7 \times 13 \times 31$, where 1729 is also called Hardy–Ramanujan number. As the English mathematician Hardy recalled, once he went to visit the sick Indian mathematician Ramanujan by taxi, the number of the taxi is 1729. Ramanujan told Hardy that it is the least positive integer which can be expressed as a sum of two cubes in two ways, i.e., $1729 = 1^3 + 12^3 = 9^3 + 10^3$.

In 1939, J. Chernick found that, all the numbers like $(6k+1)(12k+1)(18k+1)$ were Carmichael numbers if these three factors were all primes. This conclusion can be easily deduced from the so-called Korselt criterion (1899): a positive integer n is Carmichael number if and only if n has no square factor, and $p - 1 | n - 1$ for any prime factor p of n. Obviously, a number satisfying this criterion must be an odd number with at least three different prime factors. In 1915, the Chinese amateur Jianchun Yu who was then a packer in a bookstore warehouse in Hangzhou found that all the members like $(6k+1)(12k+1)(24k^2+6k+1)$ are Carmichael numbers if these three factors are all primes. Erdős conjectured that there exist infinitely many Carmichael numbers. This was proved by Alford, Granville and Pomerance in 1994 (*Ann. of Math.*, **139**, 703–722).

The pseudoprimes we mentioned above are all odd. As for even pseudoprimes, there are also infinitely many, and the least one was found by D. H. Lehmer, which is $161038 = 2 \times 73 \times 1103$. We will give two large even pseudoprimes below.

In 1972, Rotkiewicz asked if there is a pseudoprime to base 2 like $2^N - 2$. In 1989, McDaniel found a number $N = 465794$ such that $2^N - 2$ is a pseudoprime. In 2007, we (Tianxin Cai, Feifeng Wang,

Xia Zhou, *J. Algebra, Number Theory Appl.* **9**, 237–239) found a smaller one $N = 111482$ such that $2^N - 2$ is a pseudoprime to base 2.

Note that $N-1 = 111481 = 23 \times 37 \times 131$. It follows from Fermat's Little Theorem that

$$2^N \equiv 2^8 \equiv 3 \pmod{23},$$

$$2^N \equiv 2^{26} \equiv 3 \pmod{37},$$

$$2^N \equiv 2^{72} \equiv 3 \pmod{131}.$$

Thus, $2^N \equiv 3 \pmod{n}$, which means that $N - 1|2^N - 3$. Hence, $2^{N-1} - 1|2^{2^N-3} - 1$. Let $t = 2^N - 2$. We have

$$2^t \equiv 2 \pmod{t}.$$

So t is the pseudoprime to base 2 as we wanted. With computer's help, we can verify that $N = 111482$ is the least number such that $2^N - 2$ is a pseudoprime to base 2.

The chain of primes or composite numbers

Now, we want to introduce a new concept, from a manuscript of the author (Chains of primes or composites on $\frac{ax+by}{(ax,by)}$, *Research Reports, Institute of Mathematics, Zhejiang University* **7**(2) (1998) 1–13).

Definition. A sequence of n lattice points (the abscissa and ordinate are both integers) $\{(x_i, y_i)\}(1 \leq i \leq n)$ on the plane is called a chain if it satisfies

$$x_1 = y_n = 1, |x_i - x_{i+1}| \leq 1, |y_i - y_{i+1}| \leq 1|, |x_i - x_{i+1}| + |y_i - y_{i+1}| \geq 1,$$
$$1 \leq i < n.$$

Let $f(x, y)$ be a number theory function with two variables. The so-called prime numbers chain (or composite numbers chain) mean that for any $1 \leq i \leq n$, $\{(x_i, y_i)\}$ is always a pair of prime (or composite number). If this chain exists, then n is called the length of the chain. Two chains are disjoint, if they have no common lattice point. The chain with shortest length is called the shortest chain. It is not unique sometimes.

$\frac{x+y}{(x,y)}$	y=1	2	3	4	5	6
x=1	2	3	♡	5	♠	
2	3	2	5	3	7	♠
3	♡	5	2	7	♠	
4	5	3	7	2	♠	
5	♠	7	♠	♠		
6		♠				

Chains of composite numbers

$\frac{2x+y}{(2x,y)}$	y=1	2	3	4	5	6	7	8
x=1	3	2	5	3	7	♡	♡	5
2	5	3	7	2	♡	5	11	
3	7	♡	3	5	11			
4	♡	5	11					
5	11							

Chains of primes

Example. $f(x,y) = \frac{x+y}{(x,y)}$. One of the shortest composite number chains is

$$\{(5,1),(6,2),(5,3),(5,4),(4,5),(3,5),(2,6),(1,5)\}. \qquad \text{(C)}$$

The length is 8. We make the chart as below, where the heart and spade both represent composite numbers. The spades form the shortest chain of composites, that surround the primes, which seems like the fight in the game of Go (Weiqi).

For any number theory function with two variables $f(x,y)$, a natural problem is: does it have a prime chain or a composite number chain? Further, whether is the number of disjoint chains

infinity, and how to construct them? Now we consider the function $f(x, y) = \frac{ax+by}{(ax,by)}$, where a and b are coprime positive integers. When $a = b = 1$, it is easy to know that for all primes p,

$$\{(p-1, 1), (p-2, 2), \ldots, (1, p-1)\}$$

are prime chains, which are disjoint with respect to different p, and the shortest chain contains only one lattice point (1,1), which means that the length is 1. If $(a, b) \neq (1, 1)$, then we need the following (cf. Halberstam and Richert, *Sieve Methods*, Academic Press, 1976).

k-Tuple primes conjecture. *Suppose a_i and b_i are integers, $a_i > 0$, and $(a_i, b_i) = 1 (1 \leq i \leq k)$. If the polynomial*

$$\prod_{1 \leq i \leq k} (a_i x + b_i) \tag{K}$$

has no fixed prime factor, then there are infinitely many n such that $a_i n + b_i (1 \leq i \leq n)$ are all primes.

When $k = 2, a_1 = a_2 = 1, b_1 = 0, b_2 = 2$. This is the well-known twin prime conjecture.

Theorem A. *Let a and b be coprime positive integers with $2 < a + b \leq 5$. If the k-tuple prime conjecture is true, then the number theory function $\frac{ax+by}{(ax,by)}$ has infinitely many disjoint prime chains.*

Theorem B. *Let a and b be coprime odd positive numbers. Then the number theory function $\frac{ax+by}{(ax,by)}$ has infinitely many disjoint composite number chains.*

We only give the proof of Theorem A under the condition that $(a, b) = (2, 1)$. using Theorem 3.9 (Lagrange). Denote the number of solutions to the congruence in Theorem 3.9 as $\rho(f, p)$. For any prime p, if p is not the fixed factor of $f(t)$, then $\rho(f, p) < p$. And if $f(t)$ has no fixed factors, then $\rho(f, p) < p$ for any prime p.

In (K), we take $(a_1, b_1) = (2, 1), (a_2, b_2) = (1, 0)$, and $f(t) = (2t + 1)t$. If $p = 2$, then $2 | f(t)$ if and only if $t \equiv 0 \pmod{2}$, i.e., $\rho(f, 2) = 1 < 2$. If $p \geq 3$, then p cannot divide the first coefficient of $f(t)$. According to Lagrange's theorem, $\rho(f, p) < p$. Thus, if the k-tuple primes conjecture is true, then there exist infinitely many

primes t such that $2t + 1$ are primes. Construct a chain as follows:

$$\left\{ \left(t - \left[\frac{k}{2} \right], k \right) \middle| 1 \le k \le 2t - 2 \right\}.$$

By periodicity, it is not difficult to prove that the above formula forms prime chains of $\frac{2x+y}{(2x,y)}$. Obviously they are disjoint. This completes the proof of Theorem A.

Conjecture. For any pair of co-prime positive integers a and b, if the k-tuple prime conjecture is true, then $\frac{ax+by}{(ax,by)}$ has infinitely many prime chains.

If a and b are one odd and one even, what about the composite number chain? What is the length of the shortest chain? Note that chain C contains four consecutive prime numbers 2, 3, 5, and 7, so a new question is whether all the composite number chains contain consecutive prime numbers.

Exercises 3

1. Find the smallest positive integer $n > 2$ such that $2|n$, $3|n+1$, $4|n+2$, $5|n+3$, $6|n+4$.

2. Prove that

 (1) for arbitrary distinct prime numbers p_1, p_2, \ldots, p_n, there exist n consecutive integers such that the kth one is divisible by p_k for $k = 1, \ldots, n$.

 (2) there exist n consecutive integers such that each one of them has a square factor.

3. Let F_n be the nth Fibonacci number. Show that for any integer $n > 0$,

$$F_{n-1}F_{n+1} - F_n^2 = (-1)^n.$$

4. Show that the system

$$\begin{cases} x \equiv a \pmod{m} \\ x \equiv b \pmod{n} \end{cases}$$

 has a solution if and only if $a \equiv b \pmod{(m,n)}$. If a solution exists, it is unique modulo $[m, n]$.

5. Prove that for any odd prime p,

 (1) $1^2 3^2 \cdots (p-2)^2 \equiv (-1)^{\frac{p+1}{2}} \pmod{p}$;

 (2) $2^2 4^2 \cdots (p-1)^2 \equiv (-1)^{\frac{p+1}{2}} \pmod{p}$.

6. Prove that for any positive integer m and prime $p > 5$, the Diophantine equation $(p-1)! + 1 = p^m$ has no solution.

7. If p is prime, show that $\left(\left(\frac{p-1}{2} \right)! \right)^2 + (-1)^{\frac{p-1}{2}} \equiv 0 \pmod{p}$.

8. If p is prime, $n = 1 + 2 + \cdots + (p-1)$. Show that $(p-1)! \equiv p - 1 \pmod{n}$.

9. Let $x_1 = 3$, $y_1 = 4$, $z_1 = 5$,

$$\begin{cases} x_{n+1} = 3x_n + 2z_n + 1, \\ y_{n+1} = 3x_n + 2z_n + 2, \\ z_{n+1} = 4x_n + 3z_n + 2. \end{cases}$$

Show that for any positive integer n, $\{x_n, y_n, z_n\}$ is Pythagoras triple, i.e., $x_n^2 + y_n^2 = z_n^2$.

10. If p is prime, $0 < a < p$, show that the congruence $ax \equiv b \pmod{p}$ has a solution

$$x \equiv b(-1)^{a-1} \frac{(p-1) \cdots (p-a+1))}{a!} \pmod{p}.$$

11. Prove that the Diophantine equation $x^3 + y^3 + z^3 = x + y + z = 3$ has exactly 4 solutions: $(x, y, z) = (1, 1, 1), (4, 4, -5), (4, -5, 4), (-5, 4, 4)$.

12. If $n = p_1 p_2 \ldots p_k$, with prime p_i distinct, $k > 2$, and for any i, $p_i - 1 | n - 1$, show that n is Carmichael number.

Chapter 4

Quadratic Residue

16. Quadratic Congruences

We have discussed linear congruences in the previous chapters. This kind of problem is simple and useful. Now, we are going to have a systematic study of quadratic congruences, the character of which is incisive and elegant. Of course, the quadratic congruence is not only beautiful, it also has applications to combinatorial and geometrical structures. Although Fermat, Euler, Lagrange, Legendre and other number theorists of the 17th and 18th centuries proved some theorems and made a variety of conjectures about quadratic residues, Gauss was the first one to give a systematic treatment. He also introduced the terminology "quadratic residue" and "quadratic non-residue."

Portrait of Gauss, at 26, the stamp was issued by the former Democratic Republic of Germany

The general form of a quadratic congruence is

$$ax^2 + bx + c \equiv 0 \pmod{m}, (a, m) = 1.$$

It may have solutions in some cases, and no solution in other cases, just like linear congruences. For instance, congruence $x^2 - 2 \equiv 0 \pmod 3$ has no solution. Thus, the first thing to discuss is when the equation has a solution. According to the fundamental theorem of arithmetic and other conclusions that we discussed previously, we only need to consider the case where the modulus is a prime power. Moreover, we only need to consider quadratic congruences of the form

$$x^2 \equiv a \pmod{p^\alpha}, \quad (a, p) = 1$$

or more generally, the following one

$$x^2 \equiv a \pmod{m}, \quad (a, m) = 1. \tag{4.1}$$

Definition 4.1. If (4.1) has solutions, a is called a quadratic residue modulo m. Otherwise, a is called a quadratic non-residue modulo m.

First, we consider the case where $m = p$ is an odd prime, namely, the congruence

$$x^2 \equiv a \pmod{p}, \quad (a, p) = 1. \tag{4.2}$$

Theorem 4.1. (Euler). *If $(a, p) = 1$, then a is a quadratic residue modulo p if and only if*

$$a^{\frac{p-1}{2}} \equiv 1 \pmod{p}, \tag{4.3}$$

and a is a quadratic nonresidue modulo p if and only if

$$a^{\frac{p-1}{2}} \equiv -1 \pmod{p}. \tag{4.4}$$

Moreover, if a is a quadratic residue modulo p, then (4.2) has exact two solutions.

Proof. Since $(x^2 - a)|(x^{p-1} - a^{\frac{p-1}{2}})$, namely, there is a polynomial $q(x)$ with integer coefficients such that

$$x^{p-1} - a^{\frac{p-1}{2}} = (x^2 - a)q(x),$$

there holds that,

$$x^p - x = x(x^{p-1} - a^{\frac{p-1}{2}}) + (a^{\frac{p-1}{2}} - 1)x$$

$$= (x^2 - a)xq(x) + (a^{\frac{p-1}{2}} - 1)x.$$

If a is a quadratic residue modulo p, according to Fermat's Little Theorem, (4.3) satisfied. Then by Theorem 3.10, equation (4.2) has two solutions. Conversely, if (4.3) holds, using the same theorem, we deduce that a is a quadratic residue modulo p.

On the other hand, according to Fermat's Little Theorem, if $(a, p) = 1$, $a^{p-1} \equiv 1 \pmod{p}$, thus,

$$(a^{\frac{p-1}{2}} - 1)(a^{\frac{p-1}{2}} + 1) \equiv 0 \pmod{p}.$$

Since $p > 2$, only one of (4.3) and (4.4) holds. Hence, a is a quadratic non-residue modulo p if and only if

$$a^{\frac{p-1}{2}} \not\equiv 1 \pmod{p}.$$

Therefore, (4.3) is a sufficient and necessary condition for a to be a quadratic residue modulo p. □

Theorem 4.2. *In a reduced residue system modulo p, the number of quadratic residues and the number of quadratic non-residues are both $\frac{p-1}{2}$. Moreover, the $\frac{p-1}{2}$ quadratic residues have a one-to-one map modulo p to the sequence*

$$1^2, 2^2, \ldots, \left(\frac{p-1}{2}\right)^2. \tag{4.5}$$

Proof. From Theorem 4.1, the number of quadratic residues modulo p equals the number of solutions to congruence $x^{\frac{p-1}{2}} \equiv 1 \pmod{p}$. Since $x^{\frac{p-1}{2}} - 1 | x^p - x$, according to Theorem 3.10, the number of quadratic residues is $\frac{p-1}{2}$, therefore the number of quadratic non-residues is $p - 1 - \frac{p-1}{2} = \frac{p-1}{2}$.

Now we are going to prove the second half of the theorem. Obviously, each number in (4.5) is a quadratic residue modulo n, and they are incongruent each other, since if $k^2 \equiv l^2 \pmod p$ with $1 \le k$, $l \le \frac{p-1}{2}$ would imply that $k = l$ holds, which is impossible. □

Example 4.1. Compute all the quadratic residues and non-residues modulo 23 and 27, respectively.

Example 4.2. If $p = 4m + 1$ is a prime, then $x \equiv \pm(2m)! \pmod p$ are two solutions of the following congruence:

$$x^2 + 1 \equiv 0 \pmod p.$$

Finally, we present an interesting conclusion.

Proposition 4.1 (Gauss). *Let p be an odd prime. Then for any positive integer n, the following binary quadratic congruence*

$$n \equiv x^2 + y^2 \pmod{2p}$$

always has solutions.

Proof. The proposition is obvious if n can be represented as the sum of two squares. Now we assume that n cannot be represented as the sum of two squares. We consider the following $2p + 1$ numbers:

$$1^2, 2^2, 3^2, \ldots, p^2; \quad n, n - 1^2, n - 2^2, n - 3^2, \ldots, n - p^2.$$

Since n cannot be represented as the sum of two squares, they are distinct to each other. However, there are only $2p$ residues modulo $2p$, thus, at least two of them equals to each other modulo $2p$. If both of them are in $\{1^2, 2^2, 3^2, \ldots, p^2\}$, then there exist $1 \le i < j \le p$, such that $i^2 \equiv j^2 \pmod{2p}$, which is impossible. Similarly, we can prove that they are not in $\{n, n - 1^2, n - 2^2, n - 3^2, \ldots, n - p^2\}$ at the same time. Thus, there exist $1 \le x \le p - 1$ and $0 \le y \le p - 1$ such that $x^2 \equiv n - y^2 \pmod{2p}$, namely, $n \equiv x^2 + y^2 \pmod{2p}$. □

Remark 4.1. In fact, what Gauss got is more general result: Let m be any positive integer, $p^2 \nmid m$ for $p = 2$ and all the primes $p \equiv 3 \pmod 4$. Then for any positive integer n, the following binary quadratic congruence

$$n \equiv x^2 + y^2 \pmod m \tag{G}$$

always has solutions.

On the other hand, we observe that if m is positive integer, then for any integer n, the congruence

$$n \equiv \binom{x}{2} + \binom{y}{2} \pmod{m}$$

always has solutions.

This conclusion follows from a new result by Quanhui Yang and Min Tang (*J. Number Theory* **155** (2015) 1–12). They showed by the circle method that, when $8 \mid m$, (G) has solutions if and only if $n \equiv 2 \pmod 8$. In fact, the solution x and y must be odd. Denote them by $2x_1 - 1$ and $2y_1 - 1$, and set $m = 8m_1$, $n = 8n_1 + 2$, and put them into (G). Then we can deduce that $n_1 \equiv \binom{x_1}{2} + \binom{y_1}{2} \pmod{m_1}$. However, we believe that there exists a elementary proof.

Meanwhile, we also want to know whether the following congruence is solvable or not:

$$n \equiv \binom{x}{3} + \binom{y}{3} \pmod{m}.$$

And if $p \neq 5$ is a prime, then for any positive integer n, whether the following congruence always has solutions:

$$n \equiv x^2 + y^4 \pmod{2p}.$$

Integers in the Gaussian ring

Let $i = \sqrt{-1}$ and $A = Z[i] = \{a + bi \mid a, b \in Z\}$. It is easy to see that $Z[i]$ is closed under addition and multiplication, and each element in $Z[i]$ corresponds to an integral point(lattice) in the complex plane. Gauss was the first mathematician to study this kind of number, hence we call $Z[i]$ the Gaussian integer ring or the Gaussian ring, and each element in the set a Gaussian integer. Generally, we call ordinary integers rational integers. In the complex plane, the norm of a point $x = a + bi$ is defined as

$$\|x\| = a^2 + b^2 = (a + bi)(a - bi).$$

Obviously, the Gaussian integers with norm 1 are 1, -1, i, and $-i$, which are called units of $Z[i]$. The definition of the norm associates with the problem of representing integers as the sum of two squares.

According to the Fibonacci identity, it is easy to get

$$\|xy\| = \|x\|\|y\|.$$

The so-called Fibonacci identity is

$$(a^2 + b^2)(c^2 + d^2) = (ac \pm bd)^2 + (ad \mp bc)^2.$$

It was firstly recorded in Diophantus' *Arithmetica* as early as the 3rd century. In the 7th century, the Indian mathematician Brahmagupta (c. 598–670) rediscovered it and generalized it to

$$(a^2 + nb^2)(c^2 + nd^2) = (ac \pm nbd)^2 + n(ad \mp bc)^2.$$

Both of the formulas were recorded in Fibonacci's *Liber quadratorum* (1225). Moreover, we can generalize it to

$$(ma^2 + nb^2)(mc^2 + nd^2) = (mac \pm nbd)^2 + mn(ad \mp bc)^2$$

Theorem (Gauss). *We can use the Euclidean division of the integers on Gaussian integers, namely, for any $x, y \in A$ with $y \neq 0$, there exist $q, r \in A$, such that*

$$X = qy + r, \quad 0 \leq \|r\| < \|y\|.$$

Proof. Since $y \neq 0$,

$$\frac{x}{y} = \alpha + i\beta, \quad \alpha, \beta \in Q,$$

where Q and Z denote the set of rational numbers and integers, respectively. Thus, there exist $a, b \in Z$ such that $|a-\alpha|, |b-\beta| \leq 1/2$. Let $q = a + bi$. Then $\|x/y - q\| \leq 2/4$. On the other hand, note that $x = qy + (x - qy)$, so we have

$$x - qy = 0 \quad \text{or} \quad \|x - qy\| \leq \tfrac{1}{2}\|y\| < \|y\|.$$

A ring that can be endowed with the Euclidean division of the integers is called a Euclidean domain or a Euclidean ring. Thus the Gaussian ring is a Euclidean domain. In Euclidean domain, the fundamental theorem of arithmetic also holds, although the definition of prime varies in different rings. Primes in Gaussian ring are called Gaussian primes. According to a proposition in the next section, namely, a rational prime p can be represented as the sum of two squares if and only if it is of the form $4m + 1$, we can get the following theorem. \square

Theorem (Fermat–Euler). *Rational primes can be factorized as products of Gaussian primes:*

$$2 = (-i)(1+i)^2, \quad here \ 1+i \ is \ a \ Gaussian \ prime;$$

$$p = p, \quad here \ p \ is \ a \ prime, \ if \ p \equiv 3 \ (\mathrm{mod}\, 4);$$

$$p = q_1 q_2, \quad here \ q_1, q_2 \ are \ unassociated \ primes, \ if \ p \equiv 1 \ (\mathrm{mod}\, 4).$$

In the Gaussian ring, there are also some conjectures and unsolved problems about the Gaussian primes, for instance,

(1) From the theorem, there are infinitely many Gaussian primes $3, 7, 11, 19, \ldots$ on the real or imaginary axes. Are there any other lines that have infinitely many Gaussian primes on them? In particular, are there infinitely many Gaussian primes in the form $1 + ki$?

(2) Is it possible to walk to the infinity by walking on the Gaussian primes and taking steps of bounded length?

(3) There are corresponding Goldbach's conjecture and twin prime conjecture in Gaussian ring.

Here is a German stamp with Gaussian integers. The set of Gaussian primes constitutes a beautiful figure, which can be used to design floor or tablecloth. The second figure describes the distribution of primes in the Eisenstein–Jacobi integer domain. This kind of integers can be represented as $a + b\omega$, where a and b are rational integers, and ω satisfies $\omega^2 + \omega + 1 = 0$. In this ring, integers can be factorized in only one way as well. However, since there are 6 units (2 more than Gaussian ring), the graph shows up as a symmetrical hexagon.

It is worth mentioning that, besides the Gaussian ring, there are some other interesting number theory problems in other quadratic fields. For instance, Fermat's Last Theorem has some special solutions:

$$(18 + 17\sqrt{2})^3 + (18 - 17\sqrt{2})^3 = 42^3$$

$$\left(\tfrac{1+\sqrt{-7}}{2}\right)^4 + \left(\tfrac{1-\sqrt{-7}}{2}\right)^4 = 1^4.$$

Gaussian integers in a German stamp

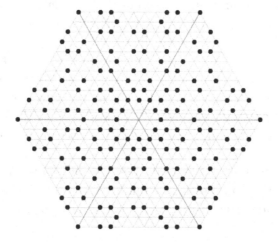

The Eisenstein–Jacobi prime table

The first equation ($n = 3$) has no solution in $Q(\sqrt{3})$, and no solution in $Q(\sqrt{2})$ when $n \geq 4$; the second equation ($n = 4$) has only one solution in quadratic fields.

17. The Legendre Symbol

Adrien-Marie Legendre is the youngest one of the famous "three L's" during the golden age of French mathematics. The other two are Lagrange and Laplace (1749–1827). He was born in Paris, which is really uncommon, although France is reputed for having many masters in mathematics. He was known by his contemporaries for a book entitled *Eléments de géométrie*, which made improvements on Euclid's book with the same name. It is also popular in the English-speaking world, and one of the version was translated by Scottish philosopher, writer, and historian Thomas Carlyle (1795–1881). Nowadays, Legendre is known for Legendre's function and Legendre's polynomial in differential equations as well as Legendre's symbol in number theory, which is defined as follows.

1820 watercolor caricature of Legendre found in 2004

Definition 4.2. The Legendre symbol $\left(\frac{a}{p}\right)$ is an arithmetic function to a given odd prime p with the following rules:

$$\left(\frac{a}{p}\right) = \begin{cases} 1 & \text{if } a \text{ is a quadratic residue modulo } p, \\ -1 & \text{if } a \text{ is a quadratic nonresidue modulo } p, \\ 0 & \text{if } p \,|\, a. \end{cases}$$

From the definition, if we can compute the value of the Legendre symbol, then we know whether the congruence

$$x^2 \equiv a \pmod{p}$$

has solutions. The computation of Legendre symbol is more efficient than Euler's criterion, however, the process is based on the latter.

Portrait of politician Legendre, who was known as mathematician Legrendre for almost two centuries

According to Theorem 4.1, we have

$$\left(\frac{a}{p}\right) \equiv a^{\frac{p-1}{2}} \pmod{p}. \tag{4.6}$$

Thus,

$$\left(\frac{1}{p}\right) = 1, \quad \left(\frac{-1}{p}\right) = (-1)^{\frac{p-1}{2}}.$$

From the definition,

$$\left(\frac{a+kp}{p}\right) = \left(\frac{a}{p}\right).$$

Similarly, by using (4.6) repeatedly, we have

$$\left(\frac{a_1 a_2 \ldots a_n}{p}\right) \equiv \left(\frac{a_1}{p}\right)\left(\frac{a_2}{p}\right)\cdots\left(\frac{a_n}{p}\right) \pmod{p}.$$

Then from the definition,

$$\left|\left(\frac{a_1 a_2 \ldots a_n}{p}\right) - \left(\frac{a_1}{p}\right)\left(\frac{a_2}{p}\right)\cdots\left(\frac{a_n}{p}\right)\right| \leq 2, \quad p > 2.$$

Thus,

$$\left(\frac{a_1 a_2 \ldots a_n}{p}\right) = \left(\frac{a_1}{p}\right)\left(\frac{a_2}{p}\right)\cdots\left(\frac{a_n}{p}\right),$$

namely, $\left(\frac{a}{p}\right)$ is a completely additive function. In particular,

$$\left(\frac{ab^2}{p}\right) = \left(\frac{a}{p}\right) \quad \text{if } p \nmid b.$$

Next, we consider the case where $a = 2$. We need the following lemma.

Lemma 4.1. (Gauss). *Let $(a,p) = 1$, and let the least nonnegative residue of $ak(1 \leq k \leq \frac{p-1}{2})$ modulo p be r_k. If the number of r_k which is greater than $\frac{p}{2}$ is t, then*

$$\left(\frac{a}{p}\right) = (-1)^t.$$

Proof. Let a_1, \ldots, a_s be all the r_k less than $\frac{p}{2}$, and let b_1, \ldots, b_s be all the r_k greater than $\frac{p}{2}$. From the assumption,

$$a^{\frac{p-1}{2}}\left(\frac{p-1}{2}\right)! = \prod_{k\leq\frac{p-1}{2}} ak \equiv \prod_{i=1}^{s} a_i \prod_{j=1}^{t} b_j \pmod{p}. \tag{4.7}$$

Since $P/2 < b_j < p$, we have $1 \leq p - b_j < p/2$ and $p - b_j \neq a_i$. Otherwise, there exists a pair of i and j such that $a_i + b_j = p$,

namely, there exist k_1 and k_2 such that $k_1 + k_2 \equiv 0 \pmod{p}$, which is impossible. Thus, according to (4.7) and $s + t = (p-1)/2$, we get

$$a^{\frac{p-1}{2}} \left(\frac{p-1}{2}\right)! \equiv (-1)^t \prod_{i=1}^{s} a_i \prod_{j=1}^{t} (p - b_j) = (-1)^t \left(\frac{p-1}{2}\right)! \pmod{p}.$$

This completes the proof of the lemma. □

Theorem 4.3. *For any odd prime p,*

$$\left(\frac{2}{p}\right) = (-1)^{\frac{p^2-1}{8}}.$$

If $(a, p) = 1$, and $2 \nmid a$, then

$$\left(\frac{a}{p}\right) = (-1)^{\sum_{1 \leq k \leq \frac{p-1}{2}} \left[\frac{ak}{p}\right]}.$$

Proof. We use the definition of r_k, a_j, b_j, s and t as in the above lemma. According to Proposition 3 in appendix of Section 7,

$$ak = p \left[\frac{ak}{p}\right] + r_k.$$

Thus, from the proof of Lemma 4.1,

$$a \frac{p^2 - 1}{8} = p \sum_{k=1}^{\frac{p-1}{2}} \left[\frac{ak}{p}\right] + \sum_{i=1}^{s} a_i + \sum_{j=1}^{t} b_j$$

$$= p \sum_{k=1}^{\frac{p-1}{2}} \left[\frac{ak}{p}\right] + \sum_{i=1}^{s} a_i + \sum_{j=1}^{t} (p - b_j) + 2 \sum_{j=1}^{t} b_j - tp$$

$$= p \sum_{k=1}^{\frac{p-1}{2}} \left[\frac{ak}{p}\right] + \sum_{k=1}^{\frac{p-1}{2}} k - tp + 2 \sum_{j=1}^{t} b_j$$

$$= p \sum_{k=1}^{\frac{p-1}{2}} \left[\frac{ak}{p}\right] + \frac{p^2 - 1}{2} - tp + 2 \sum_{j=1}^{t} b_j,$$

namely,

$$(a-1)\frac{p^2-1}{8} \equiv \sum_{k=1}^{\frac{p-1}{2}} \left[\frac{ak}{p}\right] + t \pmod 2. \tag{4.8}$$

If $a = 2$, then $0 \le \left[\frac{ak}{p}\right] \le \left[\frac{p-1}{p}\right] = 0$, it follows from (4.8) that

$$t \equiv \frac{p^2-1}{8} \pmod 2.$$

If $2 \nmid a$, then we get from (4.8) that

$$t \equiv \sum_{k=1}^{\frac{p-1}{2}} \left[\frac{ak}{p}\right] \pmod 2.$$

Thus, Theorem 4.3 holds according to Gauss' lemma. □

Corollary. *If $p = 8n \pm 1$, then 2 is a quadratic residue; if $p = 8n \pm 3$, then 2 is a quadratic non-residue.*

Now we can prove the conclusion in Section 3, namely, if there exist infinitely many Sophie Germain primes in the form $4m + 3$, then there are infinitely many Mersenne composite numbers. In fact, if $p \equiv 3 \pmod 4$ and if p and $q = 2p + 1$ are both primes, then by Fermat's little theorem,

$$2^{2p} \equiv 1 \pmod q,$$

hence $q|2^p - 1$ or $q|2^p + 1$. If the latter is true, then $2^{p+1} = \left(2^{\frac{p+1}{2}}\right)^2 \equiv -2 \pmod q$, which means that $\left(\frac{-2}{q}\right) = 1$. On the other hand, since $q \equiv 7 \pmod 8$, we have $\left(\frac{-1}{q}\right) = -1$, $\left(\frac{2}{q}\right) = 1$. A contradiction! Therefore $q|M_p = 2^p - 1$, i.e., M_p is composite.

Finally, we introduce the class number in imaginary quadratic field, which is used to describe the ideal factorization. If the class number is 1, it can be uniquely decomposed. The class number is also related with binomial $ax^2 + bx + c$. When its discriminant $\Delta = b^2 - 4ac \equiv 0$ or $1 \pmod 4$, let $h(\Delta)$ denote the number of simplified binomials ($|b| \le a \le c$). Gauss conjectured that there are 9 numbers with class number 1, namely $= -3, -4, -7, -8, -11, -19, -43, -67, -163,$

among them 7 are prime numbers, this has been confirmed after a century and half. For example, for binomial $x^2 + xy + y^2$, its discriminant $\Delta = -3, h(-3) = 1$, while for binomial both $x^2 + 3y^2$ and $2(x^2 + xy + y^2)$ their common discriminant $\Delta = -12, h(-12) = 2$. It is worth mentioning that for any odd prime number $p \equiv 3 \pmod 4$, $h(-p)$ is odd.

The famous Dirichlet class number formula refers to any prime p,

$$
h(-p) = \begin{cases} 1 & \text{if } p = 3, \\ -\dfrac{1}{p} \displaystyle\sum_{i=1}^{p-1} i \left(\dfrac{i}{p}\right) & \text{if } p > 3, \ p \equiv 3 \pmod 4, \\ \dfrac{1}{2} \displaystyle\sum_{i=1}^{p-1} (-1)^{\frac{i(i-2)}{2}} \left(\dfrac{i}{p}\right) & \text{if } p \equiv 1 \pmod 4. \end{cases}
$$

And for any prime $p > 3$, $p \equiv 3 \pmod 4$, we also have

$$
h(-p) = \frac{1}{2 - \left(\frac{2}{p}\right)} \sum_{i=1}^{(p-1)/2} \left(\frac{i}{p}\right).
$$

For any odd prime p, let $Q = \sum_{\left(\frac{i}{p}\right)=1} i$ be the sum of its quadratic residue. From Theorem 4.2, it is easy to see that $p|Q$. Further, by using the Dirichlet class number formula and the properties of quadratic residue, we have

$$
\frac{Q}{p} = \begin{cases} \dfrac{p-1}{4} & \text{if } p \equiv 1 \pmod 4, \\ \dfrac{p-1-2h(-p)}{4} & \text{if } p \equiv 3 \pmod 4. \end{cases}
$$

Remark. As for cubic residue and quartic residue, assuming their corresponding sums are Q_3 and Q_4, respectively, then we have

$$
\frac{Q_3}{p} = \begin{cases} \dfrac{p-1}{6} & \text{if } p \equiv 1 \pmod 6, \\ \dfrac{p-1}{2} & \text{if } p \equiv 5 \pmod 6, \end{cases}
$$

where \sim means approximately equal to, there might be positive and negative errors.

Representing integers as the sum of squares

It's easy to see that $2 = 1^2 + 1^2, 5 = 1^2 + 2^2, 13 = 2^2 + 3^2$, etc. In generally, 2 and any old prime in the form $4n+1$ can be expressed as a sum of two squares, and the expression is unique after the symmetry is excluded. While any odd prime in form of $4n+3$ can't be expressed as a sum of two squares. The latter is obvious since $x^2 + y^2 \equiv 0, 1$ or 2 (mod 4), there are several proofs for the former, one of them is the method of descent as Section 13. Here we present a brief proof by using the drawer principle in Section 3.

Since $p \equiv 1 \pmod 4$, we have $\left(\frac{-1}{p}\right) = 1$, there exists an integer t such that

$$t^2 + 1 \equiv 0 \pmod p, \quad (t, p) = 1. \tag{R}$$

Consider $tx - y$, when x and y pass through $0, 1, \ldots, [\sqrt{p}]$, there are totally $([\sqrt{p}] + 1)^2 > p$ numbers. By draw principle, we have two pairs $\{x_1, y_1\}$ and $\{x_2, y_2\}$ satisfying

$$tx_1 - y_1 \equiv tx_2 - y_2 \pmod p.$$

Noting $(t, p) = 1$, we have $x_1 \neq x_2, y_1 \neq y_2$. Suppose $y_1 > y_2$, $y = y_1 - y_2, x = \pm(x_1 - x_2)$, then.

$$ty \equiv x \pmod p,$$

here $0 < x, y < \sqrt{p}$.

Because of $(y, p) = 1$, there exists an integer y^{-1} such that $yy^{-1} \equiv 1 \pmod p$. Hence $t \equiv \pm xy^{-1} \pmod p$, put it into (R), then $x^2 + y^2 \equiv 0 \pmod p$. Since $0 < x^2 + y^2 < 2p$, we have $x^2 + y^2 = p$.

Now we show the uniqueness of representation. Let

$$p = x^2 + y^2 = a^2 + b^2, \quad x > 0, \ y > 0, \ a > 0, \ b > 0.$$

Then

$$(ax - by)(ax + by) = a^2 x^2 - b^2 y^2 = a^2(x^2 + y^2) - y^2(a^2 + b^2)$$
$$\equiv 0 \pmod p.$$

Hence $p|(ax - by)$, or $p|(ax + by)$. Using the Fibonacci identity in Section 16, we have

$$p^2 = (ax \mp by)^2 + (ay \pm bx)^2.$$

If $p|ax - by$, since $ay + bx > 0$, we have $ax - by = 0$. Noting $(a, b) = (x, y) = 1$, then $a = y$, $b = x$. If $p|ax + by$, since $ax + by > 0$, we have $ay - bx = 0$, then $a = x$, $b = y$. This completes the proof of uniqueness of representation.

There is also another proof which is constructive. Let a and b be a quadratic residue and a quadratic non-residue modulo p, respectively. Then

$$p = s_a^2 + s_b^2,$$

where

$$s_k = \frac{1}{2} \sum_{i=0}^{p-1} \left(\frac{i(i^2 + k)}{p} \right), \quad k = a, b.$$

Representing an integer as the sum of two squares is a special case of quadratic form $ax^2 + bxy + cy^2$, where a, b, and c are integers, and $d = b^2 - 4ac$ is called its discriminant. Whether and how a quadratic form can represent an integer is one of the main topics, here we are going to discuss the representation of positive integers as the sum of squares, and we have the following result.

Theorem. *Let $n = n_1^2 n_2$, $n > 0$, and let n_2 be square-free. Then n can be represented as the sum of two squares if and only if n_2 has no prime factors of the form $4m + 3$.*

Proof. If n_2 has no prime factors of the form $4m + 3$, then according to $2 = 1^2 + 1^2$ and other conclusions we got before, all prime factors of n_2 can be represented as the sum of two squares. By using the Brahmagupta–Fibonacci identity of the previous section, we prove the sufficiency.

Second, we do the necessity part of the proof. Let $p = 4m + 3$ divide n_2. Then we can find an r such that $p^r || n$, where r is the greatest integer such that p^r divides n. According to the assumption, r must be odd. Let $n = x^2 + y^2$, and set $(x, y) = d$. Then

$x = dx_1, y = dy_1, (x_1, y_1) = 1$, and $n = d^2(x_1^2 + y_1^2)$. Since r is odd, we have $p|x_1^2 + y_1^2$ and $(p, x_1) = 1$. Otherwise, $p|x_1$ and $p|y_1$, which contradict $(x_1, y_1) = 1$. Thus,

$$x_1^2 + y_1^2 \equiv 0 \pmod{p}, (p, x_1) = 1.$$

According to Example 2.5 in Section 6, there exists an integer x_1' such that $x_1 x_1' \equiv 1 \pmod{p}$. Hence,

$$(y_1 x_1')^2 \equiv -1 \pmod{p},$$

namely, $\left(\frac{-1}{p}\right) = 1$. On the other hand, according to the properties of the Legendre symbol, $\left(\frac{-1}{p}\right) = (-1)^{2m+1} = -1$. A contradiction! □

The Lagrangian points in the Sun–Earth system

In particular, a prime p can be represented as the sum of two squares if and only if it is of the form $4m + 1$. Moreover, let $n = 2^\alpha \prod p^r \prod q^s$, where p and q are prime factors of n of the form $4m+1$ and $4m + 3$, respectively. Let $\delta(n)$ be the number of representations of n as the sum of two squares. Then by using the properties of $Z[i]$, we can prove that

$$\delta(n) = 4 \prod (1 + r) \prod \frac{1 + (-1)^s}{2},$$

or

$$\delta(n) = 4 \sum_{d|n} k(d),$$

where

$$k(n) = \begin{cases} 0 & \text{if } n \text{ is even,} \\ (-1)^{\frac{n-1}{2}} & \text{if } n \text{ is odd} \end{cases}$$

is a complete multiplicative function.

As an application of the theorem, we have the following result. The equation

$$\begin{cases} n = a + b \\ ab = x^2 \end{cases} \tag{μ}$$

has positive integer solutions if and only if n is even, or n has prime factors of the form $4m + 1$.

Now, we are going to show two famous results.

Theorem (Lagrange). *Any positive integer can be represented as the sum of four squares.*

Theorem (Gauss). *A positive integer can be represented as the sum of three squares if and only if it is not of the form $4^k(8n + 7)$ for integers k and n.*

In particular, integers of the form $8n + 3$ can be represented as the sum of three squares, and from this point we can prove Lagrange theorem. First, we consider prime p. If $p = 2$, then it is obviously true. If $p \equiv 1 \pmod{4}$, then according to the above theorem, it can be represented as the sum of two squares. If $p \equiv 3 \pmod{4}$, then p or p^{-4} equals 3 modulo 8. Thus, it can be represented as the sum of three squares, while 4 is also a square. For general integers, we just note that if $m = x_1^2 + x_2^2 + x_3^2 + x_4^2$ and $n = y_1^2 + y_2^2 + y_3^2 + y_4^2$, then

$$mn = (x_1y_1 + x_2y_2 + x_3y_3 + x_4y_4)^2$$
$$+ (x_1y_2 - x_2y_1 + x_3y_4 - x_4y_3)^2$$
$$+ (x_1y_3 - x_3y_1 + x_4y_2 - x_2y_4)^2$$
$$+ (x_1y_4 - x_4y_1 + x_2y_3 - x_3y_2)^2.$$

Lagrange, Euler and Gauss are good examples of perfect mathematicians. On the one hand, they can discern amazing properties of

nature numbers; on the other hand, they were familiar with remote stars in the infinite universe.

If we generalize $\delta(n)$ to $\Delta(n) = \#\{n = a + b, ab = c^2\}$, then $\Delta(n) \geq \delta(n)$. In 1966, Beiler proved that, if n is an odd integer with $n = n_1 n_2$, where n_1 and n_2 are products of those prime factors which are congruent to 1 and 3 modulo 4 respectively, then

$$\Delta(n) = \frac{1}{2}\left\{\sum_{d|n_1} 2^{\omega(d)} - 1\right\},$$

where $\omega(d)$ denotes the number of distinct prime factors of d, which is also a multiplicative function. We (Tianxin Cai, Zhongyan Shen, Deyi Chen, The number of representations for $n = a+b$ with $ab = tc^2$ or $\binom{c}{2}$) investigated a more general situation. Meanwhile, we showed that, for even integer $n = 2^{\alpha} n_1 n_2$ ($\alpha \geq 1$, n_1 and n_2 are defined as above),

$$\Delta(n) = \sum_{d|n_1} 2^{\omega(d)}.$$

Here we need to point out that, as to why primes of the form $4k + 1$ can be represented as the sum of two squares, for instance, $13 = 2^2 + 3^3 = (2 + 3\sqrt{-1})(2 - 3\sqrt{-1})$, is that in quadratic filed $Q(\sqrt{-1})$, this type of primes can be represented as the product of two primitive elements, while primes of the form $4k + 3$ are primitive elements themselves in $Q(\sqrt{-1})$. In generally, problems such as prime factorization in a quadratic field and the unique prime ideal decomposition in an integral domain of a quadratic field belong to the theory of the class field. It is a branch of algebraic number theory introduced from Fermat's claim and Gauss's law of quadratic reciprocity which we will introduce in the next section.

Apart from (μ), we also considered the solvability of the equation

$$\begin{cases} n = a + b, \\ ab = \binom{c}{2}. \end{cases} \tag{ν}$$

Our result is: (ν) has solutions if and only if $2n^2 + 1$ is a composite.

In fact, (ν) has solutions if and only if $2n^2 + 1$ can be represented as $2x^2 + y^2$, $y \geq 3$. (\Rightarrow Set $x = a - b$ and $y = 2c - 1$. \Leftarrow Since y is odd, then n and x have the same parity, so set $a = (n + x)/2$, $b = (n - x)/2$ and $c = (y + 1)/2$.) Thus, taking $T = 2n^2 + 1$, the number of solutions of (ν) is the number of representations of T as $2x^2 + y^2$ minus 1. According to Mass Formula (J. P. Cook, *The Mass Formula for Binary Quadratic Forms*, Theorem 5.2), note that we require x and y to be positive integers, thus $\frac{1}{2} \sum_{d|2n^2+1} \left(\frac{-2}{d} \right) - 1$. Here $(\)$ is the Kronecker symbol (cf. Section 19), which is the same as the Jacobi symbol when d is an odd integer greater than 1. Therefore (ν) has no solutions if and only if $\sum_{d|2n^2+1} \left(\frac{-2}{d} \right) = 2$.

If $d = p_1^{\alpha_1} \ldots p_k^{\alpha_k} \mid 2n^2 + 1$, then $p_i \mid 2n^2 + 1$, where p_i ($1 \leq i \leq k$) are odd primes. Thus $p_i \nmid n$, and there exists n' such that $nn' = 1$ (mod p_i). Since $2n^2 + 1 \equiv o$ (mod p_i), we have $2 + (n')^2 \equiv o$ (mod p_i) and $\left(\frac{-2}{p_i} \right) = 1$. According to the definition of Jacobi (Kronecker) symbol, we have $\left(\frac{-2}{d} \right) = 1$, thus ($\nu$) has no solution if and only if $\sum_{d|2n^2+1} 1 = 2$, namely, $2n^2 + 1$ is prime.

Similarly, we can deduce that the equation (Landau's problem: are there infinitely many primes of the form $n^2 + 1$?)

$$\begin{cases} n = a + 2b \\ ab = \binom{c}{2} \end{cases}$$

has positive integer solutions if and only if $n^2 + 1$ is a composite.

If we replace 2 by an odd prime p, the result does not always hold. However, it holds when $p = 3, 5, 11$, and 29, namely, equation

$$\begin{cases} n = a + pb \\ ab = \binom{c}{2} \end{cases}$$

has positive integer solutions (a can be 0) if and only if $2n^2 + p$ is a composite.

Moreover, the author has the following two conjectures.

Conjecture 1. $\{prime\ p | 2n^2 + p\ are\ all\ primes,\ 0 \leq n \leq p - 1\} = \{3, 5, 11, 29\}$.

Conjecture 2. {*prime $p|$ there is no prime $q < p$, such that $q|2n^2 + p$, $n \in N$*} $= \{3, 5, 11, 29\}$.

In addition, we conjecture that the set of primes of the form $2n^2 + 29$ equals the set of the minimum factor of numbers of the form $2(29 + \Delta)^2 + 29$, and the set of prime factors of the numbers of the form $2n^2 + 29$ equals the set of factors of numbers of the form $2(29i + \Delta/i) + 29$. Here Δ runs over all non-negative triangular numbers, i is any positive integer, and Δ is a multiple of i.

Recently, we (Hao Zhong, Tianxin Cai, On the number of representations of $n = a + b$ with ab a multiple of a polygonal number generalize (ν) to the following

$$\begin{cases} n = a + b, \\ ab = tP(m, c). \end{cases}$$

Here t is a positive integer and $P(m, c) = \frac{c}{2}\{(m - 2)c - (m - 4)\}$ is the cth m-polygonal number.

Let $r_{m,t}(n)$ be the number of solutions of the above equation, what we prove includes the following results:

$$r_{3,1}(n) = [\{d(2n^2 + 1) - 1\}/2],$$

$$r_{5,1}(n) = [\{d(6n^2 + 1) - 1\}/2],$$

$$r_{7,1}(n) = [\{d_{(3)}(10n^2 + 9) - 1\}/2],$$

$$r_{3,2}(n) = d(n^2 + 1)/2 - 1,$$

$$r_{8,2}(n) = [\{d_{(2)}(3n^2 + 8) - 1\}/2].$$

It follows then,

$$r_{3,1}(n) = 0 \Leftrightarrow 2n^2 + 1 \in \mathbb{P},$$

$$r_{5,1}(n) = 0 \Leftrightarrow 6n^2 + 1 \in \mathbb{P},$$

$$r_{7,1}(n) = 0 \Leftrightarrow 10n^2 + 9 \in \mathbb{P} \cup 9\mathbb{P},$$

$$r_{3,2}(n) = 0 \Leftrightarrow n^2 + 1 \in \mathbb{P},$$

$$r_{8,2}(n) = 0 \Leftrightarrow 3n^2 + 8 \in \mathbb{P} \cup 4\mathbb{P} \cup 8\mathbb{P}.$$

In 1978, the Polish-born American mathematician Henryk Iwaniec (1947–) proved that (*Invent. Math.* **47**(2), 171–188) there are infinitely many integers n, such that the irreducible quadratic $an^2 + bn + c$ has at most two prime factors, here $a > 0$, c is odd. Hence, when $(m, t) = (3, 1)$, $(5, 1)$ or $(7, 1)$, then for infinitely n, $r_{m,t}(n) \leq 1$.

Finally, we ask you to consider the similar equation as (ν) with more variables, e.g.,

$$\begin{cases} n = a + b + c, \\ abc = \begin{pmatrix} d \\ 3 \end{pmatrix}. \end{cases}$$

We guess that there are positive integer solutions for $n \neq 1, 2, 4, 7, 11$.

Remark 1. In 2017, Hourong Qin (Representation integers by positive ternary quadratic forms, *Math. Res. Lett.* **24**(2), 535–548) proved the following conjecture raised by Ye Tian:

Let m be a positive integer, $m \equiv 1 \pmod 4$, $\left(\frac{m}{7}\right) = 1$, which means $m \equiv 1, 2$, or $4 \pmod 7$. Then there exist x, y, z such that

$$m = x^2 + 7y^2 + 49z^2.$$

Before that, Hourong Qin and Bing Du proved (*J. Pure Appl. Algebra* **216** (2012) 1637–1645): every prime in the form of $8n + 1$ could be expressed by some products and quotients of the sum of the 4th power of 2 integers, e.g., $17 = 2^4 + 1$, $41 = \frac{3^4 + 1}{1 + 1}$, $73 = \frac{7^4 + 3^4}{(2^4 + 1)(1 + 1)^7}$, $22513 = \frac{(55^4 + 56^4)(3^4 + 14^4)(3^4 + 7^4)}{(9^4 + 22^4)(1 + 10^4)(1 + 2^4)(1 + 1)}$.

Remark 2. In 2019, Zhiwei Sun (oeis.org/A308734) raised the following conjecture (checked up to 10^9): for any positive integer n, there are non-negative integers w, x, y, z, such that

$$n = w^2 + x^2 + y^2 + z^2,$$

where $y = 2^a 3^b$, $z = 2^a 5^b$. This conjecture is an enhanced version of Lagrange's theorem.

18. The Law of Quadratic Reciprocity

To compute the Legendre symbol $\left(\frac{a}{p}\right)$ for any a, we need to introduce the law of quadratic reciprocity, which as Gauss referred, is the "golden theorem". Now we will give two proofs.

Theorem 4.4 (Gauss, The Law of Quadratic Reciprocity). *Let p and q be odd primes such that $(p, q) = 1$. Then*

$$\left(\frac{q}{p}\right) = (-1)^{\frac{p-1}{2}\frac{q-1}{2}} \left(\frac{p}{q}\right).$$

Proof. Obviously, we only need to prove the following equality

$$\left(\frac{p}{q}\right)\left(\frac{q}{p}\right) = (-1)^{\frac{p-1}{2}\frac{q-1}{2}}. \tag{4.9}$$

Since p, q are odd, according to Theorem 4.3,

$$\left(\frac{q}{p}\right) = (-1)^{\sum\limits_{1\leq k\leq \frac{p-1}{2}}\left[\frac{qk}{p}\right]}, \quad \left(\frac{p}{q}\right) = (-1)^{\sum\limits_{1\leq k\leq \frac{q-1}{2}}\left[\frac{pk}{q}\right]}.$$

Thus, to prove (4.9), we only need to prove

$$\sum_{1\leq k\leq \frac{p-1}{2}}\left[\frac{qk}{p}\right] + \sum_{1\leq k\leq \frac{q-1}{2}}\left[\frac{pk}{q}\right] = \frac{p-1}{2}\frac{q-1}{2}. \tag{4.10}$$

Without loss of generality, let $p > q$, we will use two methods to prove (4.10).

A geometric method: In the following figure, we can see that the number of integer lattices in the rectangle $OABC$ (without sides OA and OC) is $\frac{p-1}{2}\frac{q-1}{2}$. On the other hand, since $(p, q) = 1$, there is no integer lattice in the line OK (except for the point O). Thus, it divides integer lattices in the rectangle $OABC$ into two parts: one is above OK, and the other is below OK. According to the definition

of $[x]$, the numbers of integer lattices in each part is

$$\sum_{k=1}^{\frac{q-1}{2}} \left[\frac{pk}{q}\right], \quad \sum_{k=1}^{\frac{p-1}{2}} \left[\frac{qk}{p}\right],$$

respectively. Thus (4.10) holds. □

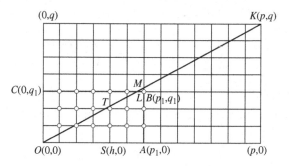

The geographic proof of the quadratic reciprocal law

An algebraic method: Denote the binary function $f(x,y) = qx - py$. Substitute x with $1, 2, \ldots, \frac{p-1}{2}$ for x, and $1, 2, \ldots, \frac{q-1}{2}$ for y. Then $f(x,y)$ has $\frac{p-1}{2}\frac{q-1}{2}$ values, which are all distinct each other. Otherwise, if

$$f(x,y) = f(x',y')$$

namely,

$$(x - x')q = (y - y')p.$$

We have $p|x - x'$ and $q|y - y'$, thus $x = x'$ and $y = y'$. Then we will use another method to get the number of values of $f(x,y)$, namely, computing the number of positive and negative values. For a given $1 \le x \le \frac{p-1}{2}$, the inequality $f(x,y) > 0$ is true if and only if $y < \frac{qx}{p}$, or more precisely, $y \le [\frac{qx}{p}]$. Thus, the number of positive values is $\sum_{k=1}^{\frac{p-1}{2}} [\frac{qk}{p}]$. Similarly, the number of negative values is $\sum_{k=1}^{\frac{q-1}{2}} [\frac{pk}{q}]$. This proves (4.10).

From Theorem 4.3, Theorem 4.4 and other properties of the Legendre symbol, we can find an efficient way to compute the value of $\left(\frac{a}{p}\right)$, thus telling whether congruence (4.2) has solutions or not.

Example 4.3. Let $p = 421$ and $q = 383$. We compute $\left(\frac{383}{421}\right)$.

Using the law of the quadratic reciprocity and other properties of the Legendre symbol, we have

$$\left(\frac{383}{421}\right) = \left(\frac{421}{383}\right)(-1)^{\frac{382\cdot420}{4}} = \left(\frac{38}{383}\right) = \left(\frac{2}{383}\right)\left(\frac{19}{383}\right)$$

$$= \left(\frac{19}{383}\right) = \left(\frac{383}{19}\right)(-1)^{\frac{18\cdot382}{4}} = -\left(\frac{383}{19}\right) = -\left(\frac{3}{19}\right)$$

$$= -\left(\frac{19}{3}\right)(-1)^{\frac{2\cdot18}{4}} = \left(\frac{19}{3}\right) = \left(\frac{1}{3}\right) = 1.$$

Thus 383 is a quadratic residue modulo 421.

Example 4.4. If $p \equiv \pm 1 \pmod{12}$, then 3 is a quadratic residue modulo p; if $p \equiv \pm 5 \pmod{12}$, then 3 is a quadratic nonresidue modulo p.

N-gonal number and Fermat

In Section 17, the Lagrange theorem states that any positive integer can be represented as the sum of four squares. This problem was first introduced by Fermat. The 18th note he described at the margin of a copy of Diophantus' *Arithmetica* is as follows.

Theorem (Fermat Polygonal Number Theorem). *Every positive integer is a sum of at most n n-gonal numbers when $n \geq 3$.*

Here the n-gonal number is the number of black points in the n-gon described by the following figure.

The term figurate numbers is used for members of different sets of numbers, generalizing from triangular numbers to different shapes. When Pythagoras started to study the concept of numbers, he was fond of describing numbers as little stones in a sandbeach. Stones can constitute different shapes, thus they generate a series of figurate numbers. For instance, 1, 3, 6, 10, ... are triangular numbers. In fact, they are the following binomial coefficients:

$$\binom{m+1}{2} = \frac{m(m+1)}{2}, \quad m \geq 1.$$

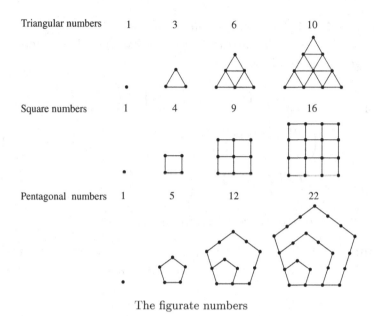

The figurate numbers

Numbers such as $1, 4, 9, 16, \ldots$ are squares. In generally, n-gonal numbers can be written as

$$\frac{m}{2}\{(n-2)m - (n-4)\}, \quad m \geq 1.$$

As what Fermat mentioned in his notes, the above statement disclosed the mysterious and incisive aspect of number theory. He also mentioned that he would write a book on it, which was never written. After a century, when Euler read this note, he was really excited. Euler also felt regretful for Fermat did not leave any proof. We might say that it is Fermat who elevated this old mathematical game to the height of theory. Since then, Euler became a successor of Fermat in the field of number theory. He proved many statements of Fermat, including the famous Fermat–Euler theorem. However, he did not find a proof to the n-gonal number problem. Later in 1770, Lagrange proved the case of $n = 4$ based on Euler's work. In 1796, Gauss gave a proof of the case $n = 3$. And in 1813, the French mathematician Augustin Louis Cauchy (1789–1857) completely solved this problem.

To the side of the proof of the case $n = 3$, a masterpiece of a young student, Gauss wrote the word "Eureka" (meaning "I have found it") in his diary, which was proclaimed by Archimedes when he found the Archimedes principle. This shows that Gauss was very proud of his proof. He used a proposition in Section 17, namely, a positive integer of the form $8n + 3$ can be represented as the sum of three squares, to give the proof. Suppose $8n + 3 = a^2 + b^2 + c^2$. Then it is easy to see that a, b, and c are all odd. Let $a = 2x - 1$, $b = 2y - 1$, and $c = 2z - 1$. According to the following identity

$$\frac{(2x - 1)^2 + (2y - 1)^2 + (2z - 1)^2 - 3}{8}$$
$$= \frac{x(x - 1)}{2} + \frac{y(y - 1)}{2} + \frac{z(z - 1)}{2},$$

we have that

$$n = \frac{x(x - 1)}{2} + \frac{y(y - 1)}{2} + \frac{z(z - 1)}{2} = \binom{x}{2} + \binom{y}{2} + \binom{z}{2}.$$

To the case of $n = 4$, in 1828, the German mathematician Jacobi (1804–1851) gave a new and powerful proof. For any non-negative integer n, let the number of representations as the sum of four squares be $a(n)$, Jacobi used the fact that

$$\sum_{n=0}^{\infty} a(n)e^{2\pi i n z}$$

is an automorphic form. For more details on automorphic forms, see Section 29.

It is worth mentioning that Zhiwei Sun (*J. Combin. Number Theory* 1 (2009), 65–78) conjectured that, except for 216, every nature number is either a prime, or a triangular number, or the sum of a prime and a triangular number. The number of triangular numbers less than x is $O(\sqrt{x})$. Comparing with Conjecture F in Section 5, the number of true figurate primes less than x is $O(\sqrt{x}/\ln x)$.

Remark. In addition to n-gonal number, there are also numbers of polyhedrons. For example, the octahedral numbers $\sum_{k=1}^{n} \frac{k(k+1)}{2}$, $1, 4, 10, 20, 35, 56, \ldots$ the octahedral numbers $\frac{n(2n^2+1)}{3}$, $1, 6, 19, 44, 85, 146, \ldots$.

In 1850, the British mathematician Frederick Pollock (1783–1870) raised the following two conjectures, which have not yet been proved. Like his earlier compatriot Wilson, Pollock was a lawyer and politician, he is also knighted.

Conjecture 1 (Pollock). *Every positive integer is the sum of at most five octahedral numbers.*

Conjecture 2 (Pollock). *Every positive integer is the sum of at most seven octahedral numbers.*

19. The Jacobi symbol

With the introduction of the law of quadratic reciprocity, the computation of the Legendre symbol can be realized. However, such computation requires to know the standard factorization of an integer, which does not have a common method. In this sense, the Legendre symbol is somewhat imperfect. Jacobi, who also named the terminology "determinant", noticed the defection. He changed the odd prime p in the Legendre symbol $\left(\frac{a}{p}\right)$ to an ordinary odd integer m, and made the following definition.

Portrait of German mathematician Jacobi, who died of smallpox at 46

Definition 4.3. Jacobi symbol $\left(\frac{a}{m}\right)$ is a function to any odd integer m greater than 1 for all integers a. Its value is

$$\left(\frac{a}{m}\right) = \left(\frac{a}{p_1}\right)\left(\frac{a}{p_2}\right)\cdots\left(\frac{a}{p_r}\right),$$

where $m = p_1 p_2 \ldots p_r$, each p_i is a prime, and $\left(\frac{a}{p_i}\right)$ is Legendre symbol modulo p_i for a.

Now we give some properties of the Jacobi symbol. First, for any $a \equiv a_1 \pmod{m}$,

$$\left(\frac{a}{m}\right) = \left(\frac{a_1}{m}\right).$$

This is because from $a \equiv a_1 \pmod{m}$, we have $a \equiv a_1 \pmod{p_i}$ for $1 \leq i \leq r$. Thus, according to the definition,

$$\left(\frac{a}{m}\right) = \left(\frac{a}{p_1}\right)\left(\frac{a}{p_2}\right)\cdots\left(\frac{a}{p_r}\right) = \left(\frac{a_1}{p_1}\right)\left(\frac{a_1}{p_2}\right)\cdots\left(\frac{a_1}{p_r}\right) = \left(\frac{a_1}{m}\right).$$

Then,

$$\left(\frac{1}{m}\right) = 1, \quad \left(\frac{-1}{m}\right) = (-1)^{\frac{m-1}{2}}.$$

The second formula can be deduced by Definition 4.3 and Lemma 4.2 below, namely,

$$\left(\frac{-1}{m}\right) = \prod_{i=1}^{r}\left(\frac{-1}{p_i}\right) = (-1)^{\sum\limits_{i=1}^{r}\frac{p_i-1}{2}} = (-1)^{\frac{m-1}{2}}.$$

Similarly, it is easy to prove that

$$\left(\frac{a_1 a_2 \ldots a_n}{m}\right) = \left(\frac{a_1}{m}\right)\left(\frac{a_2}{m}\right)\cdots\left(\frac{a_n}{m}\right). \qquad (4.11)$$

In particular, if $(b, m) = 1$, then

$$\left(\frac{ab^2}{m}\right) = \left(\frac{a}{m}\right) \quad \text{if } (b, m) = 1.$$

This is because from the definition of the Jacobi symbol and the properties of the Legendre symbol,

$$\left(\frac{a_1 a_2 \ldots a_n}{m}\right) = \prod_{i=1}^{n} \left(\frac{a_1 a_2 \ldots a_n}{p_i}\right) = \prod_{i=1}^{n} \prod_{j=1}^{r} \left(\frac{a_j}{p_i}\right)$$

$$= \prod_{j=1}^{n} \prod_{i=1}^{r} \left(\frac{a_j}{p_i}\right) = \prod_{j=1}^{n} \left(\frac{a_j}{m}\right).$$

In order to get the formula of $\left(\frac{2}{m}\right)$ and the law of quadratic reciprocity for the Jacobi symbol, we need the following lemma.

Lemma 4.2. *Let* a_1, a_2, \ldots, a_n *be odd integers. Then we have*

$$\frac{a_1 a_2 \ldots a_n - 1}{2} \equiv \sum_{i=1}^{n} \frac{a_i - 1}{2} \pmod{2},$$

$$\frac{(a_1 a_2 \ldots a_n)^2 - 1}{8} \equiv \sum_{i=1}^{n} \frac{a_i^2 - 1}{8} \pmod{8}.$$

To prove it, we use induction on n as well as the following method. Note that

$$\frac{a_1 a_2 \ldots a_n - 1}{2} = \frac{\left(1 + 2\frac{a_1 - 1}{2}\right)\left(1 + 2\frac{a_2 - 1}{2}\right) \cdots \left(1 + 2\frac{a_n - 1}{2}\right) - 1}{2}$$

$$= \sum_{i=1}^{n} \frac{a_i - 1}{2} + 2N_1,$$

$$\frac{a_1^2 a_2^2 \ldots a_n^2 - 1}{8} = \frac{\left(1 + 8\frac{a_1^2 - 1}{8}\right)\left(1 + 8\frac{a_2^2 - 1}{8}\right) \cdots \left(1 + 8\frac{a_n^2 - 1}{8}\right) - 1}{8}$$

$$= \sum_{i=1}^{n} \frac{a_i^2 - 1}{8} + 2N_2.$$

According to the above lemma, we obtain that

$$\left(\frac{2}{m}\right) = \prod_{i=1}^{r} \left(\frac{2}{p_i}\right) = (-1)^{\sum_{i=1}^{r} \frac{p_i^2 - 1}{8}} = (-1)^{\frac{m^2 - 1}{8}}.$$

Finally, we are going to prove that the Jacobi symbol also satisfies the law of quadratic reciprocity, namely, for any distinct odd integers m and n greater than 1,

$$\left(\frac{n}{m}\right) = (-1)^{\frac{m-1}{2}\frac{n-1}{2}}\left(\frac{m}{n}\right). \tag{4.12}$$

Actually, if $(m,n) \neq 1$, then $\left(\frac{n}{m}\right) = \left(\frac{m}{n}\right) = 0$. If $(m,n) = 1$, then by letting $n = q_1 q_2 \ldots q_s$ and using Definition 4.3, (4.11), and the law of quadratic reciprocity for the Legendre symbol,

$$\left(\frac{n}{m}\right) = \prod_{i=1}^{r}\left(\frac{n}{p_i}\right) = \prod_{i=1}^{r}\prod_{j=1}^{s}\left(\frac{q_j}{p_i}\right)$$

$$= \prod_{i=1}^{r}\prod_{j=1}^{s}(-1)^{\frac{p_i-1}{2}\frac{q_j-1}{2}}\left(\frac{p_i}{q_j}\right)$$

$$= (-1)^{\sum_{i=1}^{r}\sum_{j=1}^{s}\frac{p_i-1}{2}\frac{q_j-1}{2}}\prod_{i=1}^{r}\prod_{j=1}^{s}\left(\frac{p_i}{q_j}\right).$$

By using Lemma 4.2, we get

$$\sum_{i=1}^{r}\sum_{j=1}^{s}\frac{p_i-1}{2}\frac{q_j-1}{2} = \left(\sum_{i=1}^{r}\frac{p_i-1}{2}\right)\left(\sum_{j=1}^{s}\frac{q_j-1}{2}\right)$$

$$= \left(\frac{m-1}{2}+2N_1\right)\left(\frac{n-1}{2}+2N_2\right)$$

$$= \frac{m-1}{2}\frac{n-1}{2}+2N.$$

Thus,

$$\left(\frac{n}{m}\right) = (-1)^{\frac{m-1}{2}\frac{n-1}{2}}\prod_{i=1}^{r}\prod_{j=1}^{s}\left(\frac{p_i}{q_j}\right) = (-1)^{\frac{m-1}{2}\frac{n-1}{2}}\left(\frac{m}{n}\right).$$

It is worth noting that although the Jacobi symbol is a generalization of the Legendre symbol, we cannot use it to determine whether the corresponding quadratic congruence has a solution or not.

For instance, on the one hand,

$$\left(\frac{2}{9}\right) = \left(\frac{2}{3}\right)^2 = 1.$$

On the other hand, $x^2 \equiv 2 \pmod 9$ has no solutions. In this sense, we can regard the Jacobi symbol as a computational tool when dealing with the Legendre symbol.

In the end, we are going to introduce the Kronecker symbol. It was defined by the German mathematician Leopold Kronecker (1823–1891), who was also a critic to Cantor's set theory. It is a further generalization of the Jacobi symbol, which is useful in the theory of quadratic forms.

Definition 4.4. Let $m = u p_1^{\alpha_1} \ldots p_k^{\alpha_k}$ be non-negative integers, $u = \pm 1$, and let a be an integer. Then the value of the Kronecker symbol $\left(\frac{a}{m}\right)$ is

$$\left(\frac{a}{m}\right) = \left(\frac{a}{u}\right) \prod_{i=1}^{k} \left(\frac{a}{p_i}\right)^{\alpha i},$$

Portrait of German mathematician Kronecker

where $\left(\frac{a}{p_i}\right)$ is the Legendre symbol for odd prime p_i, and

$$\left(\frac{a}{1}\right) = 1, \quad \left(\frac{a}{-1}\right) = \begin{cases} 1 & \text{if } a < 0, \\ -1 & \text{if } a \geq 0, \end{cases}$$

$$\left(\frac{a}{0}\right) = \begin{cases} 1 & \text{if } a = \pm 1, \\ 0 & \text{otherwise}, \end{cases} \quad \left(\frac{a}{2}\right) = \begin{cases} 0 & \text{if } a \text{ is even}, \\ 1 & \text{if } a \equiv \pm 1 \;(\text{mod } 8), \\ 1 & \text{if } a \equiv \pm 3 \;(\text{mod } 8). \end{cases}$$

The Hadamard matrix and Hadamard conjecture

In 1893, three years before proving the prime number theorem, Hadamard raised the following question: If A be a real square matrix of order n such that all its entries a_{ij} satisfy $-1 \leq a_{ij} < 1$, then what is the maximum value of its determinant? The answer given by himself is the following theorem.

Theorem (Hadamard). *Let A be a real square matrix of order n such that all its entries a_{ij} satisfy $-1 < a_{ij} \leq 1$. Then $|\det A| \leq (\sqrt{n})^n$. The equality holds if and only if $a_{ij} = \pm 1$, and the rows (columns) of A are all orthogonal to each other.*

Definition. Let A be a real square matrix of order n such that all its entries a_{ij} satisfy $a_{ij} = \pm 1$ and the rows (columns) of A are all orthogonal to each other. Then we call A an Hadamard matrix.

Hadamard lived to 99, more than 50 days longer than the British philosopher Russell. His best student must be André Weil. Hadamard visited China twice in 1936. The place he traveled include Shanghai, Peking, Hangzhou, etc. He lectured in Tsinghua University for two months. He also visited Jiao Tong University and Zhejiang University. The Hadamard matrix problem is a famous problem in combinatorics. It is easy to see that $HH^T = nI_n$, and $\det H = \pm(\sqrt{n})^n$. Moreover, the order of an Hadamard matrix must be 1, 2, or a multiple of 4. Examples of Hadamard matrices of order 2 and 4 are showed below:

$$\begin{bmatrix} 1 & 1 \\ -1 & 1 \end{bmatrix}, \quad \begin{bmatrix} 1 & 1 & 1 & 1 \\ -1 & 1 & -1 & 1 \\ -1 & -1 & 1 & 1 \\ -1 & 1 & 1 & -1 \end{bmatrix}.$$

Hadamard Conjecture. *For any positive integer n being a multiple of 4, there exists a Hadamard matrix of order n.*

Actually, this problem has a long history. In 1867, Sylvester found a way to construct Hadamard matrices. Let H be a Hadamard matrix of order n. Then

$$\begin{bmatrix} H & H \\ H & -H \end{bmatrix}$$

is an Hadamard matrix of order $2n$. Thus, we can construct Hadamard matrices of order $2^k (k \geq 1)$. In 1933, the British mathematician Raymond Paley (1907–1933) found a more efficient way by using the theory of quadratic residue.

Theorem. *Let p be a prime of the form $4k + 3$, and let R and N denote the set of quadratic residues and nonresidues modulo p, respectively. Define a square matrix B of order p, with its entries*

$$b_{ij} = \begin{cases} 1, & j - i \in R, \\ -1, & j - i \in N. \end{cases}$$

Let A be a square matrix of order $p + 1$ such that all entries in its first row and column are 1, and let its lower right submatrix of order p be B. Then A is an Hadamard matrix of order $p + 1$.

It is easy to prove the theorem by the theory of quadratic residues and the Legendre symbol. Similarly, Paley constructed Hadamard matrices of order $2(q + 1)$, where q is a prime of the form $4k + 1$. We can say that Paley made the biggest contribution to the theory of the existence of Hadamard matrices. Regretfully, in the same year, he was killed at the age of 26 by an avalanche at Deception Pass, Fossil Mountain, in the Canadian Rockies. At that time, he was skiing alone at an altitude of 9600 ft in Banff, Alberta. His death was witnessed by companions lower down the mountainside.

So far, there are 13 multiples of 4 less than or equal to 2000 for which no Hadamard matrix of that order is known. They are: 668, 716, 892, 1004, 1132, 1244, 1388, 1436, 1676, 1772, 1916, 1948, and 1964.

20. Congruences Modulo a Composite

In the previous sections, we have discussed the solvability of congruences modulo odd primes. In this section, we are going to discuss the solvability of congruences modulo a composite, i.e.,

$$x^2 \equiv a \pmod{m}, \quad (a, m) = 1 \tag{4.13}$$

and the number of their solutions.

According to the fundamental theorem of arithmetic, m can be written as the standard form of a prime factorization $m = 2^\alpha p_1^{\alpha_1} \ldots p_k^{\alpha_k}$. Then by Theorem 3.4, we can see that (4.13) has solutions if and only if the following system of congruences

$$x^2 \equiv a \pmod{2^\alpha}, \quad x^2 \equiv a \pmod{p_i^{\alpha_i}}, \quad 1 \le i \le k \tag{4.14}$$

has solutions. And if this system has solutions, the number of solutions of (4.13) is the product of the numbers of solutions of each congruence in (4.14). Therefore, we first discuss the following congruence:

$$x^2 \equiv a \pmod{p^\alpha}, \quad (a, p) = 1, \alpha \ge 1. \tag{4.15}$$

The tomb of Paley in winter in Banff

Theorem 4.5. *Equation* (4.15) *has solutions if and only if* $\left(\frac{a}{p}\right) = 1$. *If it has solutions, then the number of solutions is 2.*

Proof. If $\left(\frac{a}{p}\right) = -1$, then $x^2 \equiv a \pmod{p}$ has no solution, and this is the necessity. Now we show the sufficiency. If $\left(\frac{a}{p}\right) = 1$, then according to Theorem 4.1, the congruence has exactly two solutions. Let $x \equiv x_1 \pmod{p}$ be one of the solutions. It is easy to see that $(x_1, p) = 1$. Thus, $(2x_1, p) = 1$. Let $f(x) = x^2 - a$. Then $p \nmid 2f'(x_1)$. According to Theorem 3.5, we get exactly one solution of (4.15). Thus, two solutions of $x^2 \equiv a \pmod{p}$ give exact two solutions of (4.15).

Now we are going to discuss the congruence

$$x^2 \equiv a \pmod{2^\alpha}, \quad (2, a) = 1, \ \alpha \geq 1. \tag{4.16}$$

It is easy to see that (4.16) has solutions when $\alpha = 1$, and it has only one solution. We will give the result for $\alpha > 1$ without proof. \square

Theorem 4.6. *Let* $\alpha > 1$, *then* (4.16) *has solutions if and only if* (i) *when* $\alpha = 2, a \equiv 1 \pmod{4}$; (ii) *when* $\alpha \geq 3, a \equiv 1 \pmod{8}$. *If the above conditions are satisfied, then* (4.16) *has two solutions when* $\alpha = 2$, *and four solutions when* $\alpha \geq 3$.

According to Theorems 4.5, 4.6, and 3.4, we have the following theorem.

Theorem 4.7. *The congruence*

$$x^2 \equiv a \pmod{m}, m = 2^\alpha p_1^{\alpha_1} \cdots p_k^{\alpha_k}, (a, m) = 1$$

has solutions if and only if (i) $\left(\frac{a}{p_i}\right) = 1, 1 \leq i \leq k$; (ii) *if* $\alpha = 2$, $a \equiv 1 \pmod{4}$; *if* $\alpha \geq 3, a \equiv 1 \pmod{8}$. *If the above conditions are satisfied, then the number of solutions is* $2^k (\alpha = 0, 1), 2^{k+1} (\alpha = 2)$, $2^{k+2} (\alpha \geq 3)$.

Finally, we present an interesting property of adjacent quadratic residues and non-residues. Let p be a prime, Z_p^* denote a reduced

residue system modulo p, and R and N the set of quadratic residues and non-residues modulo p, respectively. Define

$$\begin{aligned}
RR &= \{a \in Z_p^* : a \in R, a+1 \in R\}, \\
RN &= \{a \in Z_p^* : a \in R, a+1 \in N\}, \\
NR &= \{a \in Z_p^* : a \in N, a+1 \in R\}, \\
NN &= \{a \in Z_p^* : a \in N, a+1 \in N\}.
\end{aligned} \tag{4.17}$$

Example 4.5. Let $p = 13$. Then $R = \{1, 3, 4, 9, 10, 12\}$ and $N = \{2, 5, 6, 7, 8, 11\}$. Thus $RR = \{3, 9\}$, $RN = \{1, 4, 10\}$, $NR = \{2, 8, 11\}$, and $NN = \{5, 6, 7\}$.

Note that, if $a \neq p-1$, then

$$\left\{\left(\frac{a}{p}\right) + 1\right\}\left\{\left(\frac{a+1}{p}\right) + 1\right\} = \begin{cases} 4 & \text{if } a \in RR, \\ 0 & \text{if } a \notin RR. \end{cases}$$

Thus,

$$|RR| = \frac{1}{4} \sum_{a=1}^{p-2} \left\{\left(\frac{a}{p}\right) + 1\right\}\left\{\left(\frac{a+1}{p}\right) + 1\right\}. \tag{4.18}$$

Through *Maple*, we can count that, when $p = 1999$, $|RR| = 499$. We can prove that the cardinalities of all four sets defined in (4.17) are closed to $|Z_p^*/4 = (p-1)/4$. Here, we take $|RR|$ as an example.

Theorem 4.8. *Let p be an odd prime. Then*

$$|RR| = \frac{p - 4 - \left(\frac{-1}{p}\right)}{4}.$$

Proof. According to (4.18) and the properties of the Legendre symbol,

$$\begin{aligned}
|RR| &= \frac{1}{4}\left\{\sum_{a=1}^{p-2}\left(\frac{a}{p}\right)\left(\frac{a+1}{p}\right) + \sum_{a=1}^{p-2}\left(\frac{a}{p}\right) + \sum_{a=1}^{p-2}\left(\frac{a+1}{p}\right) + \sum_{a=1}^{p-2} 1\right\} \\
&= \frac{1}{4}\left\{\sum_{a=1}^{p-2}\left(\frac{1 + a^{-1}}{p}\right) - \left(\frac{-1}{p}\right) + p - 3\right\}. \tag{4.19}
\end{aligned}$$

Here a^{-1} denotes the inverse of a, namely, $aa^{-1} \equiv 1 \pmod{p}$. When a runs over $Z_p^* \backslash \{-1\}$, so does a^{-1}. Hence, $1 + a^{-1}$ runs over $Z_p^* \backslash \{1\}$. Thus,

$$\sum_{a=1}^{p-2} \left(\frac{1 + a^{-1}}{p} \right) = - \left(\frac{1}{p} \right) = -1.$$

Substituting it into (4.19), we complete the proof of Theorem 4.8. \square

Corollary. *Let p be an odd prime. Then*

$$|RN| = \frac{p - \left(\frac{-1}{p} \right)}{4}.$$

Remark. It is well-known that

$$\sum_{a \in N} a - \sum_{b \in Q} b = \begin{cases} 0 & \text{if } p \equiv 1 \pmod{4}, \\ ph(-p) & \text{if } p \equiv 3 \pmod{4}. \end{cases}$$

It follows from the Dirichlet class number formula in Section 17

$$\sum_{a \in N} a^2 - \sum_{b \in Q} b^2 = p^2 h(-p) \quad \text{if } p \equiv 3 \pmod{4}$$

and

$$\sum_{a \in N} a^2 \equiv \sum_{b \in Q} b^2 \pmod{4p} \quad \text{if } p \equiv 5 \pmod{8},$$

$$\sum_{a \in N} a^2 \equiv \sum_{b \in Q} b^2 \pmod{8p} \quad \text{if } p \equiv 1 \pmod{8}.$$

Constructibility of the regular 17-gon

In 1769, the 19-year-old Gauss showed that a regular 17-gon can be constructed with straightedge and compass. It is the first break-through since ancient Greek. Due to this, Gauss made his decision of his major from a tangled choice between mathematics and philosophy (philology). For simplicity, we apply Gauss's method on regular pentagon.

First, let $z = e^{2\pi i/5}$ be one of a 5th roots of unity which is not 1. Hence,

$$z^4 + z^3 + z^2 + z + 1 = 0.$$

Since 2 is a primitive root modulo 5,

$$z^{2^0} + z^{2^1} + z^{2^2} + z^{2^3} + 1 = 0.$$

Let $\alpha = z + z^4$ and $\beta = z^2 + z^3$. Then $\alpha + \beta = \alpha\beta = -1$. According to Vieta's formulas, α and β satisfy

$$x^2 + x - 1 = 0.$$

Thus,

$$\alpha, \beta = \frac{-1 \pm \sqrt{5}}{2}.$$

Substitute it into $z + z^4 = \frac{-1+\sqrt{5}}{2}$, since $z \cdot z^4 = z^5 = 1$, then z satisfies

$$z^2 - \left(\frac{-1 + \sqrt{5}}{2} \right) z + 1 = 0.$$

Therefore,

$$z = \frac{1}{2} \left\{ \frac{-1 + \sqrt{5}}{2} \pm \sqrt{\left(\frac{-1+\sqrt{5}}{2} \right)^2 - 4} \right\}$$

$$= \frac{-1 + \sqrt{5}}{4} \pm \frac{\sqrt{10 + 2\sqrt{5}}}{4} i.$$

By checking the complex plane, we know that the sign is positive, namely,

$$z = \frac{-1 + \sqrt{5}}{4} + \frac{\sqrt{10 + 2\sqrt{5}}}{4} i,$$

where

$$\cos \frac{2\pi}{5} = \frac{-1 + \sqrt{5}}{4}.$$

The 17-angle star in the base of Gauss' monument in Braunschweig. Photo taken by the author

Therefore, the side length of pentagon is

$$s_5 = \sqrt{\left(1 - \frac{-1 + \sqrt{5}}{4}\right)^2 + \left(\frac{\sqrt{10 + 2\sqrt{5}}}{4}\right)^2} = \frac{\sqrt{10 - 2\sqrt{5}}}{2}.$$

Since the operations like addition, subtraction, multiplication, division, and square roots can be constructed with straightedge and compass, thus point z, and moreover, all vertices of the pentagon can be constructed with straightedge and compass (see the following figure).

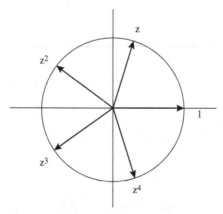

Graphing method for regular pentagon

Similarly, we can construct a regular 17-gon with straightedge and compass, since 3 is a primitive root modulo 17. Gauss obtained

$$\cos\frac{2\pi}{17} = -\frac{1}{16} + \frac{1}{16}\sqrt{17} + \frac{1}{16}\sqrt{34 - 2\sqrt{17}}$$
$$+ \frac{1}{8}\sqrt{17 + 3\sqrt{17} - \sqrt{34 - 2\sqrt{17}} - 2\sqrt{34 + 2\sqrt{17}}}.$$

More generally, as Gauss pointed out, a regular n-gon is constructible with straightedge and compass if and only if

$$n = 2^a p_1 \ldots p_k,$$

where $a \geq 0$, $k \geq 0$, and p_i are distinct Fermat primes. It is worth mentioning that Gauss only proved the sufficiency.

Exercises 4

1. If p is odd prime, show that the number of solutions of the equation $x^2 \equiv a \pmod{p}$ is $1 + \left(\frac{a}{p}\right)$; is moreover if $p \nmid a$, then the number of of solutions of the equation $ax^2 + bx + c \equiv 0 \pmod{p}$ is $1 + \left(\frac{b^2 - 4ac}{p}\right)$.
2. If $p = 4n + 3$ ($n > 1$) and $q = 2p + 1$ are both prime, show that the Mersenne number M_p is not prime.
3. Show that there are infinitely many prime in the form of $4n + 1$.
4. Show that there are infinitely many prime in the form of $8n + 3$.
5. The number 563 is prime, determine whether the congruence equation $x^2 \equiv 429 \pmod{563}$ has solutions or not?
6. If $(x, y) = 1$, find the form of odd prime factor of $x^2 - 3y^2$.
7. The number 1013 is prime. Determine that whether the congruent equation $x^2 \equiv 1503 \pmod{1013}$ has solutions?
8. If p, q are two distinct odd primes and $p = q + 4a$, show that $\left(\frac{a}{p}\right) = \left(\frac{a}{q}\right)$.
9. Find the solutions of $x^2 \equiv 59 \pmod{125}$.
10. If p is odd prime, show that

$$|NR| = |NN| = \frac{p - 2 + \left(\frac{-1}{p}\right)}{4}.$$

11. If p is odd prime, show that the number of quadratic residues modulo p is $\frac{p-1}{2}$; if $l \geq 3$, how many quadratic residues modulo 2^l are there?

12. If p is odd prime, show that the congruence

$$1 + x^2 + y^2 \equiv 0 \pmod{p}$$

has a solution which satisfies $0 \leq x, y < \frac{p}{2}$.

Chapter 5

Nth Power Residues

21. Definition of Order

We have discussed quadratic congruences in the previous chapter, and obtained the method to determine whether the congruence

$$x^2 \equiv a \pmod{m}, \quad (a, m) = 1$$

has solutions. Now, we will discuss further when the congruence

$$x^n \equiv a \pmod{m}, \quad (a, m) = 1 \tag{5.1}$$

is solvable. In order to do that, we need to introduce the concepts of order, primitive root and index, which are all established on the basis of linear congruence theory, and have no connection with the quadratic residue.

From Euler's theorem, if $(a, m) = 1$, $m > 1$, then $a^{\phi(m)} \equiv 1 \pmod{m}$. That is to say, there exists a positive integer γ satisfying $a^\gamma \equiv 1 \pmod{m}$. For example, the congruence

$$a^4 \equiv 1 \pmod 8$$

is true for $a = 1, 3, 5$ and 7. However, when the exponent of a changes from 4 to smaller 2 in the above congruence, the congruence

$$a^2 \equiv 1 \pmod 8,$$

is also true for $a = 1$, 3, 5 and 7. Hence, we give the following definition.

Definition 5.1. Let $m > 1$, $(a, m) = 1$, be the smallest positive integer γ such that

$$a^\gamma \equiv 1 \pmod{m}$$

is called the order of a modulo m. Further, if the order of a modulo m is $\phi(m)$, then a is a primitive root of m.

Cover of the Chinese edition of Gauss' *Disquisitiones Arithmeticae* (2011)

Example 5.1. 2 has order 4 modulo 5, order 3 modulo 7, order 6 modulo 9. Thus, 2 is primitive root modulo 5 and modulo 9, but not modulo 7.

For any positive integer $m > 1$, does there exists a primitive root modulo m? In the case that primitive roots exist, how many are the primitive roots exactly? These are the two questions we want to ask. In order to find the answers, we need to establish some properties of order.

Theorem 5.1. *If a has order δ modulo m, then the integers $a^0(=1)$, $a^1, \ldots, a^{\delta-1}$ are incongruent modulo m.*

Proof. We prove it by negation. Suppose there are two integers k and l such that

$$a^l \equiv a^k \pmod{m}, \quad 0 \le l, k < \delta.$$

Let's say $l > k$, since $(a, m) = 1$, it is true that $a^{l-k} \equiv 1 \pmod{m}$, $0 < l - k < \delta$, which makes a contradiction to the order δ of m. This completes the proof of Theorem 5.1. □

Theorem 5.2. *If a has order δ modulo m, then $a^\gamma \equiv a^{\gamma_1} \pmod{m}$ if and only if $\gamma \equiv \gamma_1 \pmod{\delta}$. In particular, $a^\gamma \equiv 1 \pmod{m}$ if and only if $\delta | \gamma$.*

Corollary 5.1. *If a has order δ modulo m, then $\delta | \phi(m)$.*

Theorem 5.3. *If x has order $ab(a > 0, b > 0)$ modulo m, then x^a has order b modulo m.*

Proof. First, since $(x, m) = 1$, we have $(x^a, m) = 1$. The order of x^a modulo n exists, say δ. Then $(x^a)^\delta \equiv 1 \pmod{m}$. Theorem 5.2 asserts that $ab | a\delta$, that is $b | \delta$.

On the other hand, since x has order ab modulo m, then $(x^a)^b \equiv 1 \pmod{m}$ and $\delta | b$. Thus, $b = \delta$. This completes the proof of Theorem 5.3. □

Theorem 5.4. *If x and y have order a and b modulo m, respectively, and $(a, b) = 1$, then xy has order ab modulo m.*

Proof. Since $(x, m) = 1$ and $(y, m) = 1$, then $(xy, m) = 1$. Suppose xy has order δ. Then $(xy)^\delta \equiv 1 \pmod{m}$. We have

$$1 \equiv (xy)^{b\delta} \equiv x^{b\delta} y^{b\delta} \equiv x^{b\delta} \pmod{m}.$$

Thus $a | b\delta$ by Theorem 5.2, as $(a, b) = 1$. It follows that $a | \delta$. Similarly, $b | \delta$, hence $ab | \delta$. On the other hand, $(xy)^{ab} = (x^a)^b (y^b)^a \equiv 1 \pmod{m}$. Thus $\delta | ab$ by Theorem 5.2. Therefore $\delta = ab$. This completes the proof of Theorem 5.4. □

Obviously, the results of Theorem 5.4 can be extended to the case of arbitrarily many integers.

Example 5.2. Let $a > 1$ and p be an odd prime. Then the odd prime factor of $a^p \pm 1$ is either the factor of $a \pm 1$ or of the form $2px + 1$.

Example 5.3. Let $F_n = 2^{2^n} + 1$ be the Fermat number. Then 2 has order 2^{n+1} modulo F_n. If p is a factor of F_n, then 2 has order 2^{n+1} modulo p too.

Example 5.4. If a has order δ modulo m, then a^x has order $\frac{\delta}{(x,\delta)}$ modulo m. In particular, if $m = p$ is an odd prime and $\delta < p - 1$, then none of a^γ $(1 \leq \gamma \leq \delta)$ is the primitive root modulo p.

From the corollary of Theorem 5.2, we naturally ask whether there exists an integer a with order δ modulo m for each divisor δ of $\phi(m)$? The answer is no. For example, let $m = 15$. Then $\phi(15) = 8$. It is easy to verify that $a^4 \equiv 1 \pmod{15}$ for each integer a coprime to m, which means that there is no integer a having order 8. On the other hand, we will show that in Section 23, if there exist primitive roots modulo m, then for any divisor δ of $\phi(m)$, there exists an integer a having order δ modulo m.

Egyptian fraction

There is an old story. A wealthy chieftain announced before his death that he will give his 11 luxury cars to his 3 daughters. He agreed to have $\frac{1}{2}$ to the eldest, $\frac{1}{3}$ to the second, and $\frac{1}{6}$ to the youngest. A problem arose that how to distribute the cars to his 3 daughters in strict accordance with the Emirates, without smashing the car? Finally, a car dealer helped to solve the problem. He provided a car to have a total of 12 cars, so the 3 daughters were left by will of 6 vehicles, 3 vehicles and 2 vehicles, respectively. When they took the cars away, there was still one left, which was the dealer's. That is to say, he did a favor, without paying anything.

This story actually proposed such a mathematical problem, how to decompose the rational number $\frac{n}{n+1}$ into the sum of 3 unit fractions or the sum of 3 Egyptian fractions,

$$\frac{n}{n+1} = \frac{1}{x} + \frac{1}{y} + \frac{1}{z}.$$

The above relevant vehicle allocation problem is equivalent to the solution of $n = 11$. The answer is $(x, y, z) = (2, 3, 6)$.

Egyptian fraction problems concern the representation of a rational number as the sum of some unit fractions. That is to say, given positive integers m, n, solve the equation

$$\frac{m}{n} = \frac{1}{x_1} + \frac{1}{x_2} + \cdots + \frac{1}{x_k}. \tag{5.2}$$

Egyptian fraction: Eye of Horus

The conclusion is affirmative, but problems arise from some special specified value.

Take $m = 4$, $k = 1$, 2 or 3, it's easy to see that (5.2) has solutions for any positive integer n. Take $m = 4$, $k = 3$. In 1948, Erdős and Germany-born American mathematician Ernst Gabor Straus (1922–1983), who was Einstein's assistant, conjectured that for any $n > 1$, the equation

$$\frac{4}{n} = \frac{1}{x} + \frac{1}{y} + \frac{1}{z}$$

always has solutions.

When $n \equiv 2 \pmod 3$, the conjecture is true, since

$$\frac{4}{n} = \frac{1}{n} + \frac{1}{\frac{n-2}{3}+1} + \frac{1}{n\left(\frac{n-2}{3}+1\right)}.$$

The American-born English mathematician Louis Mordell (1888–1972) proved that the conjecture is true when $n \not\equiv 1 \pmod{24}$. In 1982, Xunqian Yang (Proc. *Amer. Math. Soc.* **85**(4), 496–498) discovered that the conjecture is true when one of n, $n+1$, $n+4$ or $4n+1$ has a factor $d \equiv 3 \pmod 4$. This is because

$$\frac{4}{n} = \begin{cases} \dfrac{1}{nk(k+1)} + \dfrac{1}{n(k+1)} + \dfrac{1}{kv}, & \text{if } n = (4k-1)v, \\[2mm] \dfrac{1}{nk} + \dfrac{1}{nkv} + \dfrac{1}{kv}, & \text{if } n+1 = (4k-1)v, \\[2mm] \dfrac{1}{nk} + \dfrac{1}{nk(kv-1)} + \dfrac{1}{kv-1}, & \text{if } n+4 = (4k-1)v, \\[2mm] \dfrac{1}{nk} + \dfrac{1}{k(kv-n)} + \dfrac{1}{n(kv-n)}, & \text{if } 4n+1 = (4k-1)v. \end{cases}$$

This shows that the conjecture is true for almost all n. So far it is known to be true for all $n \leq 10^{14}$.

Take $m = 5$, $k = 3$. In 1956, the Polish mathematician Waclaw Sierpiński (1882–1969) conjectured that for any $n > 1$, the equation

$$\frac{5}{n} = \frac{1}{x} + \frac{1}{y} + \frac{1}{z}$$

always has solutions.

In 1966, Bonnie M. Stewart (1915–1994) showed that the conjecture is true when $n \not\equiv 1 \pmod{278468}$. Meanwhile, he verified that the conjecture is true for all $n \leq 10^9$.

However, these two conjectures have not been proved or disproved. It remains an open problem.

If we take in (5.2) that $m = n = 1$ and $x_1 < x_2 < \cdots < x_k$, then for any positive integer k, try to determine the minimum value

of x_k. Denote it by $m(k)$, it is known that $m(3) = 6$, $m(4) = 12, \ldots,$ $m(12) = 30$. In fact,

$$1 = \frac{1}{2} + \frac{1}{3} + \frac{1}{6}, 1 = \frac{1}{2} + \frac{1}{4} + \frac{1}{6} + \frac{1}{12},$$

$$1 = \frac{1}{6} + \frac{1}{7} + \frac{1}{8} + \frac{1}{9} + \frac{1}{10} + \frac{1}{14} + \frac{1}{15} + \frac{1}{18} + \frac{1}{20} + \frac{1}{24} + \frac{1}{28} + \frac{1}{30},$$

but the general result is unknown. The problem was raised by Erdős and Graham, Erdős also asked if there exists a constant c such that $m(k) < ck$.

22. The Existence of Primitive Roots

For a positive integer n, the primitive root might not exist modulo n. Gauss pointed out that both Euler and Swiss-German mathematician Johann Heinrich Lambert (1728–1777) knew primitive roots. Lambert's most famous work is the proof of the irrationality of π, he is also a physicist, astronomer, and philosopher, and his famous book *New Tool Law* influenced the great philosopher Immanuel Kant (1724–1804), the latter planned to dedicate his most important academic work *Critique of Pure Reason* (1781) to Lambert, but it did not complete because the delay of publication and the early death of Lambert. Gauss is the first to prove the existence of primitive roots, now we give necessary and sufficient conditions for the existence of primitive roots. To save space, we only give the proof of some results.

Theorem 5.5. *Let p be an odd prime. There exist primitive roots modulo p.*

Proof. Let's consider the reduced residue system $1, 2, \ldots, p-1$ modulo p, each of them has its own order. Select all the different orders and denoted them by $\delta_1, \delta_2, \ldots, \delta_r$. Let $\lambda = [\delta_1, \delta_2, \ldots, \delta_r]$. We only need to prove (i) there exists g whose order is λ modulo p; (ii) $\lambda = p - 1$.

(i) Let $\lambda = q_1^{\alpha_1} q_2^{\alpha_2} \ldots q_s^{\alpha_s}$ be the canonical factorization of λ. Then for each $1 \leq t \leq s$, there exists δ such that $\delta = a q_t^{\alpha_t}$. Therefore, there exists an integer $1 \leq x \leq p - 1$ with order δ modulo p.

Portrait of German mathematician Lambert, his idea influenced his fellow philosopher Immanuel Kant

By Theorem 5.3, $x_t = x^a$ has order $q_t^{\alpha_t}$. Let $g = x_1 x_2 \dots x_s$. Then g has order λ modulo p by Theorem 5.4.

(ii) It is obvious that the congruence $x^\lambda \equiv 1 \pmod{p}$ has $p - 1$ solutions $1, 2, \dots, p - 1$. By Lagrange theorem, $p - 1 \leq \lambda$. On the other hand, by the corollary of Theorem 5.2, $\delta_s | p - 1$ for $1 \leq s \leq p-1$, thus $\lambda \leq p - 1$. Therefore, $\lambda = p - 1$. This completes the proof of Theorem 5.5. $\qquad\square$

The following theorems can be derived from the definition of primitive root, Theorem 5.5 and Euler theorem, the proof of which is constructive.

Theorem 5.6. *Let g be a primitive root modulo p. Then there exists an integer t such that $g + pt$ is a primitive root modulo $p^\alpha (\alpha \geq 1)$, where u is determined by the equation $(g + pt)^{p-1} = 1 + pu$.*

Theorem 5.7. *Let g be a primitive root modulo $p^\alpha (\alpha \geq 1)$. Then the odd number of g and $g + p^\alpha$ is a primitive root modulo $2p^\alpha$.*

Theorem 5.8. *There exists a primitive root modulo m if and only if $m = 2, 4, p^\alpha$ or $2p^\alpha$, where p is an odd prime.*

In particularly, 1 is a primitive root modulo 2 and 3 is a primitive root modulo 4. We have the criterion for the determination of the primitive root modulo m when m has primitive roots.

Theorem 5.9. *Let $m > 1$, $(g, m) = 1$, and q_1, q_2, \ldots, q_s, be all the different factors of $\phi(m)$. Then g is a primitive root modulo m if and only if*

$$g^{\frac{\phi(m)}{q_i}} \not\equiv 1 \pmod{m}, \quad 1 \le i \le s. \tag{5.3}$$

Proof. First we prove the necessity. If g is a primitive root modulo m, then g has order $\phi(m)$. However, $0 < \frac{\phi(m)}{q_i} < \phi(m)$, therefore (5.3) is true.

Next we prove the sufficiency. If (5.3) is true, then g has order δ modulo m. We use reduction to absurdity to prove $\delta = \phi(m)$. If $\delta < \phi(m)$, then $\delta | \phi(m)$ by the corollary of Theorem 5.2. Accordingly $\frac{\phi(m)}{\delta} > 1$, so there exists $q_i | \frac{\phi(m)}{\delta}$. Let $\frac{\phi(m)}{\delta} = q_i u$, i.e., $\frac{\phi(m)}{q_i} = \delta u$. Then $g^{\frac{\phi(m)}{q_i}} = (g^\delta)^u \equiv 1 \pmod{m}$. This is a contradiction to (5.3). Therefore $\delta = \phi(m)$, i.e., g is a primitive root modulo m. This completes the proof of Theorem 5.9. □

Example 5.5. Let $m = 41$. Then $\phi(m) = \phi(41) = 2^3 5$, $q_1 = 2$, $q_2 = 5$. Thus $\frac{\phi(m)}{q_1} = 20$, $\frac{\phi(m)}{q_2} = 8$. Hence g is a primitive root modulo 41 if and only if

$$g^8 \not\equiv 1 \pmod{41}, \quad g^{20} \not\equiv 1 \pmod{41}, \quad (g, 41) = 1.$$

After verifying one by one, 6 is the least primitive root modulo 41.

Now we consider $m = 41^2 = 1681$. When $t = 0$, $41 \nmid u$, thus $g = 6$ is also a primitive root modulo 1681 by using the method of Theorem 5.6. Then by Theorem 5.7, $6 + 41^2 = 1687$ is a primitive root modulo $3362 (= 2 \times 41^2)$.

Finally, we have to find all the primitive roots modulo 41. Since 2 has order 20 modulo 41, we can rule out 20 integers: $2, 4, 8, 16, 32, 23, 5, 10, 20, 40, 39, 33, 25, 9, 18, 36, 31, 21, 1$ from the reduced system modulo 41. Since 3 has order 8 modulo 41, we can again rule out $3, 9, 27, 40, 38, 32, 14, 1$, where $1, 9, 32, 40$ had been ruled out. Therefore, by Theorem 5.12 the remaining 16 integers: $6, 7, 11, 12,$

$13, 15, 17, 19, 22, 24, 26, 28, 29, 30, 34, 35$ are exactly all the primitive roots modulo 41.

In Section 10, we have introduced the RSA system of the public key cryptosystem, it mainly makes use of the Da Yan Shu, Euler's theorem and the difficulty of integer factorization. In addition, another safe and reliable method is the so-called discrete logarithm, the public key system and the so-called digital signature scheme that makes use of the property and theory of primitive root. Specifically, for a bigger primitive root g, if $h = g^x$, the calculation of x by h is still a difficult problem so far. Therefore, it has strong security.

Artin's conjecture

In 1927, the Austria–Armenia mathematician Emil Artin (1898–1962) put forward such a problem to the German mathematician Helmut Hasse (1898–1979): for a given integer $a \neq -1$ which is not a perfect square then there exist infinitely many primes p such that a is a primitive root modulo p. Further, let $a = a_0 b^2, a_0 > 1$ without square factors, $S(a)$ the set of prime numbers p such that a is a primitive root modulo p. Artin conjectured:

(1) $S(a)$ has a positive density inside the set of primes. In particular, $S(a)$ is an infinite set.
(2) If a is not a perfect power and $a_0 \not\equiv 1 \pmod 4$, then the density constant (Artin's constant)of $S(a)$ is

$$C_{\text{Artin}} = \prod_p \left(1 - \frac{1}{p(p-1)}\right) = 0.3739558136\ldots.$$

Here p is taken over all primes. In other cases, the density constant is a rational multiple of Artin's constant.

For example, take $a = 2$. Then

$$S(2) = \{3, 5, 11, 13, 19, 29, 37, 53, 59, 61, 67, 83, 101, 107, 131, 139,$$
$$149, 163, 173, 179, 181, 197, 211, 227, 269, 293, 317, 347, 349,$$
$$373, 379, 389, 419, 421, 443, 461, 467, 491, \ldots\}.$$

Artin (left) and Hasse

There are 95 primes less than 500, 38 of them have 2 as primitive root with the ratio is $\frac{38}{95} = \frac{2}{5} = 0.4$ which is close to the Artin constant.

Many number theorists made efforts for Artin conjecture, in 1967, the English mathematician Christopher Hooley (1928–) published a conditional proof for the conjecture, under certain kind of the Generalized Riemann Hypothesis. The British mathematician Roger Heath-Brown (1952–) improved on their result and showed unconditionally that there are at most two exceptional prime numbers among 3, 5, and 7 for which Artin's conjecture fails, which means that one of them is a primitive root modulo p for infinitely many p. Unfortunately, Heath-Brown's proof is not constructive. Up to now, we are not sure whether an integer a is a primitive root modulo infinitely many primes. Two years ago before that, Indian mathematicians Gupta and Murty (1953–) proved that there are infinitely many a as primitive root modulus infinitely many primes. It is interesting that Michael Artin, the son of Emil Artin, is also mathematician, whose major is algebraic geometry and who won a Wolf prize in 2013.

On the other hand, Erdős proposed an open problem: when the prime number p is large enough, whether there exists a prime $q < p$ such that q is a primitive root of p.

Finally, there is the so-called smallest primitive root problem. Let g_p be the smallest primitive root modulo p. Fridlander (1949) and Salie (1950) proved that there exists a constant $C_1 > 0$ such that $g_p > C_1 \log p$. Burgess (1962) proved that, for any $\varepsilon > 0$, there exists a constant C_2 such that $g_p < C_2 p^{\frac{1}{4}+\varepsilon}$; Shoup (1992) proved that there exists a constant C_3 such that $g_p < C_3 \log^6 p$ under the Generalized Riemann hypothesis.

23. The Nth Power Residues

In order to study the nth power congruent equation (5.1), we need to introduce the concept of indicies. In this section, we restrict the modulo to $m = p^\alpha$ or $2p^\alpha$, namely the primitive root exists modulo m, let g be a primitive root modulo m. We want to find the condition for solvability and the number of solutions, as well as the number of the primitive roots modulo m. Before giving the definition of an index, we need the following property, which can be derived from the definition of primitive root and reduced residue system directly:

If γ goes through the least non-negative complete residue system modulo $\phi(m)$, then g^γ goes through a reduced residue system modulo m.

Definition 5.2. Let a be an integer and g be a primitive root modulo m. Then there exists a non-negative integer γ such that

$$a \equiv g^\gamma \ (\text{mod } m),$$

γ is called the index relative to g modulo m.

Obviously, the index of a is related not only to modulo m, but also the selected primitive root g. For example, 2 and 3 are the primitive roots modulo 5, 3 is the index of 3 relative to 2 modulo 5, and 1 is the index of 3 relative to 3 modulo 5. For any integer, a coprime to m and any primitive root g modulo m, the index of a always exists. For any integer a not coprime to m and any primitive root g modulo m,

the index of a does not exist. It is easy to prove the following properties by Theorem 5.2 and the related properties.

Property 5.1. There exists the index $0 \leq \gamma < \phi(m)$ of a related to primitive root g modulo m, denoted by $\text{ind}_g a$ or $\text{ind}\, a$. Let γ' be any index of a modulo m. Then $\gamma' \equiv \gamma \pmod{\phi(m)}$. On the other hand, if g is a primitive root modulo m, then the set $\{g^0 = 1, g^1, \ldots, g^{\phi(m)-1}\}$ forms a reduced residue system modulo m.

Property 5.2. If a_1, a_2, \ldots, a_n integers coprime to m, then

$$\text{ind}(a_1 a_2 \ldots a_n) \equiv \text{ind}\, a_1 + \text{ind}\, a_2 + \cdots + \text{ind}\, a_n \pmod{\phi(m)}.$$

In particular,

$$\text{ind}\, a^n \equiv n \, \text{ind}\, a \pmod{\phi(m)}.$$

Property 5.3. If g, h are two primitive roots modulo m, then

(i) $\text{ind}_h g \cdot \text{ind}_g h \equiv 1 \pmod{\phi(m)}$,
(ii) $\text{ind}_g h \cdot \text{ind}_h a \equiv \text{ind}_g a \pmod{\phi(m)}$.

Example 5.6. Taking 6 as the primitive root, make the two index tables (Tables 5.1 and 5.2) modulo 41 as follows:

Table 5.1 Index table modulo 41

	0	1	2	3	4	5	6	7	8	9
0		0	26	15	12	22	1	39	38	30
1	8	3	27	31	25	37	24	33	16	9
2	34	14	29	36	13	4	17	5	11	7
3								32	35	6

Table 5.2 Anti-index table modulo 41

	0	1	2	3	4	5	6	7	8	9
0	1	6	36	11	25	27	39	29	10	19
1	32	28	4	24	21	3	18	26	33	34
2	40	35	5	30	16	14	2	12	31	22
3	9	13	37	17	20	38	23	15	8	7

Here Table 5.1 is used to check the index and Table 5.2 is used to check the corresponding number by index. For example, the index

of 33 is 18 from Table 5.1, and the corresponding number of index 18 is 33 from Table 5.2. Both leads to $\text{ind}_6\ 33 = 18$. Now, we use the index to solve the nth power congruent equation (5.1). For this, we first introduce the concepts of nth power residue and nth power nonresidue.

Definition 5.3. Let m be a positive integer, if (5.1) has a solution, then a is called an nth power residue modulo. Otherwise, a is called an nth power nonresidue modulo m.

Theorem 5.10. *If $(n, \phi(m)) = d, (a, m) = 1$, then (5.1) is solvable, that is, a is an nth power residue modulo m if and only if $d|\text{ind}\ a$, and d is the number of solutions if there exist. In the reduced residues system modulo m, the number of nth power residue is $\frac{\phi(m)}{d}$.*

Proof. It is easy to verify that (5.1) is equivalent to the following congruence:

$$n \operatorname{ind} x \equiv \operatorname{ind} a \pmod{\phi(m)}. \tag{5.4}$$

By Theorem 3.1 of Section 11, equation (5.4) has a solution $X = \operatorname{ind} x$ if and only if $d|\operatorname{ind} a$, and in the case of solvable, the number of solutions is d. Furthermore, the number of nth power residue modulo m is the number of multiples of d in the sequence $0, 1, 2, \ldots, \phi(m)-1$, i.e., $\frac{\phi(m)}{d}$. This completes the proof of Theorem 5.10. □

Corollary 5.2. *The number a is an nth power residue modulo m if and only if*

$$a^{\frac{\phi(m)}{d}} \equiv 1 \pmod{m}, \quad d = (n, \phi(m)).$$

Example 5.7. Find all the solutions of the following congruences

$$x^8 \equiv \pm 16 \pmod{41}. \tag{5.5}$$

Since $n = 8$, $\phi(41) = 40$, $d = (8, 40) = 8$, $\operatorname{ind} 16 = 24$, $8|24$, by Theorem 5.10 and its proof, the congruence $x^8 \equiv 16 \pmod{41}$ has 8 solutions, they satisfy

$$8 \operatorname{ind} x \equiv 24 \pmod{40},$$

i.e.,

$$\text{ind}\, x \equiv 3 \pmod 5$$

that is $\text{ind}\, x \equiv 3, 8, 13, 18, 23, 28, 33, 38 \pmod{40}$. Hence, $x \equiv 8, 10, 11, 17, 24, 30, 31, 33 \pmod{41}$ by Table 5.2. The congruence $x^8 \equiv -16 \pmod{41}$ has no solution due to $\text{ind}(-16) = \text{ind}\, 25 = 4$, $8 \nmid 4$.

Theorem 5.11. *If* $(a, m) = 1$, *then the order of a modulo m is* $\delta = \frac{\phi(m)}{(\text{ind}\, a, \phi(m))}$. *In particular, a is a primitive root of modulo m if and only if* $(\text{ind}\, a, \phi(m)) = 1$.

Proof. By $a^\delta \equiv 1 \pmod m$ and Property 5.12, we have $\delta\, \text{ind}\, a \equiv 0 \pmod{\phi(m)}$. Note that $\delta | \phi(m)$. Then $\text{ind}\, a \equiv 0 \pmod{\frac{\phi(m)}{\delta}}$. Since $\frac{\phi(m)}{\delta}$ is a common divisor of $\text{ind}\, a$ and $\phi(m)$, then $\frac{\phi(m)}{\delta} | (\text{ind}\, a\phi(m))$, i.e.,

$$\frac{\phi(m)}{(\text{ind}\, a, \phi(m))} \bigg| \delta.$$

Let $d = (\text{ind}\, a, \phi(m))$. Then

$$\frac{\phi(m)}{d}\, \text{ind}\, a \equiv 0 \pmod{\phi(m)},$$

So $a^{\frac{\phi(m)}{d}} \equiv 1 \pmod{\phi(m)}$, by the definition of δ,

$$\delta \bigg| \frac{\phi(m)}{d} = \frac{\phi(m)}{(\text{ind}\, a, \phi(m))}.$$

Therefore

$$\delta = \frac{\phi(m)}{(\text{ind}\, a, \phi(m))}.$$

This completes the proof of Theorem 5.11. $\qquad\qquad\square$

Theorem 5.12. *In the reduced residues system modulo m, the number of integers with index δ is $\phi(\delta)$. In particular, the number of primitive root is $\phi(\phi(m))$.*

Proof. By Theorem 5.11, in the reduced residues system modulo m, the number of integers with index δ equals the number those a satisfying the condition

$$(\text{ind } a, \phi(m)) = \frac{\phi(m)}{\delta}.$$

This is equivalent to the number of integers u satisfying the condition

$$(u, \phi(m)) = \frac{\phi(m)}{\delta}, \quad 0 \leq u < \phi(m).$$

Let $u = \frac{\phi(m)}{\delta} v$. This is equivalent again to the number of integer v satisfying the condition

$$(v, \delta) = 1, \quad 0 < v < \delta,$$

i.e., $\phi(\delta)$. This completes the proof of Theorem 5.12. □

Remark. By the proof of Theorem 5.12 and $a \equiv g^{\text{ind}_g a} \pmod{m}$, the set of all primitive roots modulo m is

$$\{g^j, \ 1 \leq j \leq \phi(m), \ (j, \phi(m)) = 1\}.$$

Example 5.8. In the reduced residue system modulo 41, the a's with index 8 are $3, 14, 27, 38$, which satisfy that $(\text{ind } a, 40) = \frac{40}{8} = 5$, while primitive root a satisfy $(\text{ind } a, 40) = 1$ and the number of such a's is $\phi(40) = 16$, they are $6, 7, 11, 12, 13, 15, 17, 19, 22,$ $24, 26, 28, 29, 30, 34, 35$.

Obviously, the method to find all the primitive roots given in Example 5.8 is better than that in Example 5.5.

Finally, we consider two questions that Gauss studied in *Disquisitiones Arithmeticae*. That is, for a fixed positive modulus or prime power modulus, find the relationship between the sum or product of all the primitive roots and the modulus.

Theorem 5.13. *Let the set of primitive roots modulo m is* $\{g_1, g_2, \ldots, g_r | r = \phi(\phi(m))\}$.
Then

$$\prod_{i=1}^{r} g_i \equiv (-1)^r \pmod{m}.$$

Proof. The above congruence is equivalent to

$$\prod_{i=1}^{r} g_i \equiv \begin{cases} -1 \ (\text{mod } m), & m = 2, 3, 4, 6, \\ 1 \ (\text{mod } m), & \text{otherwise.} \end{cases}$$

When $m \leq 6$, it could be verified directly. When $m > 6$, the number r is even. If x is a primitive root modulo m, then it is easy to see that its inverse y ($xy \equiv 1 \ (\text{mod } m)$) is also a primitive root and $y \not\equiv x \ (\text{mod } m)$. So the elements of $\{g_1, g_2, \ldots, g_r\}$ can be arranged in pairs according to the reciprocal relationship, thus the product is congruent to 1 modulo m. This completes the proof of Theorem 5.13. □

Corollary (Gauss). *For any prime $p \neq 3$, the product of its all primitive roots is congruent to 1 modulo p.*

Example 5.9. If p is a prime, then for any positive integer k,

$$1^k + 2^k + \cdots + (p-1)^k \equiv \begin{cases} 0 \ (\text{mod } p) & \text{if } p-1 \nmid k, \\ -1 \ (\text{mod } p) & \text{if } p-1 \mid k. \end{cases}$$

Example 5.10. If p is an odd prime, then

(1) $1^p + 2^p + \cdots + (p-1)^p \equiv 0 \ (\text{mod } p)$.
(2) If m is a positive integer, and $2^m \not\equiv 1 \ (\text{mod } p)$, then

$$1^m + 2^m + \cdots + (p-1)^m \equiv 0 \ (\text{mod } p).$$

Theorem 5.14. *If p is an odd prime and $\{g_1, g_2, \ldots, g_r \mid r = \phi(\phi(p^\alpha))\}$ is the set of primitive roots modulo p^α, then*

$$\sum_{i=1}^{r} g_i \equiv \mu(p-1)\phi(p^{\alpha-1}) \ (\text{mod } p^\alpha).$$

Proof. Let $n = \phi(p^\alpha)$ and g be a primitive root modulo $p^\alpha (\alpha \geq 1)$. By the remark to Theorem 5.12 and using

$$\sum_{d \mid m} \mu(d) = \begin{cases} 1 & \text{if } m = 1, \\ 0 & \text{if } m > 1. \end{cases}$$

Twice, we have

$$\sum_{i=1}^{r} g_i = \sum_{\substack{j=1 \\ (j,n)=1}}^{n} g^j = \sum_{j=1}^{n} g^j \sum_{d|(j,n)} \mu(d) = \sum_{d|n} \mu(d) \sum_{k=1}^{n/d} g^{kd}$$

$$= \sum_{d|n} \mu(d) g^d \frac{g^n - 1}{g^d - 1} = \sum_{d|n} \mu(d) \frac{g^n - 1}{g^d - 1}$$

$$\equiv \sum_{0 \le i \le 1} \mu(p^i(p-1)) \frac{g^n - 1}{g^{p^i(p-1)} - 1}$$

$$\equiv \mu(p-1) \left(\frac{g^n - 1}{g^{p-1} - 1} - \frac{g^n - 1}{g^{p(p-1)} - 1} \right) \pmod{p^\alpha}. \quad \text{(A)}$$

This is because g is a primitive root, so for any $i \ge 1, e|p-1, e < p-1$, and

$$g^{p^i e} - 1 \equiv g^e - 1 \not\equiv 0 \pmod{p}.$$

Now we need to prove, for any $a > 1$ and coprime to p and $1 \le i \le \alpha$,

$$\frac{a^{\phi(p^\alpha)} - 1}{a^{\phi(p^i)} - 1} \equiv p^{\alpha-i} \pmod{p^\alpha}. \quad \text{(B)}$$

By induction on α, it is obvious when $\alpha = 1$. Assume that the congruence is established for $\alpha \ge 1$, i.e.,

$$a^{\phi(p^\alpha)} - 1 = (a^{\phi(p^i)} - 1)(p^{\alpha-i} + K_\alpha p^\alpha),$$

where K_α and the following $K'_\alpha, K_{\alpha+1}$ are all integers. Consider the case $\alpha + 1$. By Euler theorem, $p^i|(a^{\phi(p^i)} - 1)$, using the expression of binomial coefficients, we have

$$a^{\phi(p^{\alpha+1})} = (a^{\phi(p^\alpha)})^p = (1 + (a^{\phi(p^i)} - 1)(p^{\alpha-i} + K_\alpha p^\alpha))^p$$

$$= 1 + (a^{\phi(p^i)} - 1)(p^{\alpha+1-i} + K_\alpha p^{\alpha+1})$$

$$+ (a^{\phi(p^i)} - 1)K'_\alpha p^{2\alpha+1-i}$$

$$= 1 + (a^{\phi(p^i)} - 1)(p^{\alpha+1-i} + K_{\alpha+1} p^{\alpha+1}).$$

Therefore (B) holds. Take $a = g$ and $i = 1, 2$ in (B) to (A) successively, we obtain

$$\sum_{i=1}^{r} g_i \equiv \mu(p-1)(p^{\alpha-1} - p^{\alpha-2}) = \mu(p-1)\phi(p^{\alpha-1}) \pmod{p^\alpha}.$$

This completes the proof of Theorem 5.14. \square

Corollary (Gauss.) *For any prime p, the sum of its all primitive roots is congruent to $\mu(p-1)$ modulo p.*

Finally, we use the Carmichael theorem in Section 12 to show that how Gauss theorem and the existing condition of primitive roots can be deduced from each other. Here we must point out that the proof of the Carmichael theorem relies on Qin Jiushao theorem, hence a in the definition of the Carmichael function does exist, which means that it can be taken. However, we only need the following two lemmas, they are consequences of the Carmichael theorem. Here we give their simple and direct proofs.

Lemma 5.1. *If $m > 1$ and there is no primitive root modulo m, then for any integer a with $(a, m) = 1$,*

$$a^{\frac{\phi(m)}{2}} \equiv 1 \pmod{m}.$$

Proof. Let $p^\alpha \| m$ and $m = p^\alpha m_1$, $m_1 > 2$. Then $\phi(m_1)$ must be even. By Euler's theorem,

$$a^{\frac{\phi(m)}{2}} = a^{\frac{\phi(p^\alpha m_1)}{2}} = (a^{\phi(p^\alpha)})^{\frac{\phi(m_1)}{2}} \equiv 1 \pmod{p^\alpha},$$

so

$$a^{\frac{\phi(m)}{2}} \equiv 1 \pmod{m}. \qquad \square$$

Lemma 5.2. *For any positive integer m, there exists an integer a with $(a, m) = 1$ and δ the index of a modulo m, such that for any integer x with $(x, m) = 1$, there holds that $x^\delta \equiv 1 \pmod{m}$. In particular, if there exists a primitive root modulo m, then a is one of them.*

Proof. We use the method of the proof for Theorem 5.5. In a reduced residue system modulo m, each number has an index. From the $\phi(m)$ indices we take all the distinct ones, denoted by $\delta_1, \delta_2, \ldots, \delta_r$. Take $\delta = [\delta_1, \delta_2, \ldots, \delta_r]$ and let $\delta = q_1^{\alpha_1} q_2^{\alpha_2} \ldots q_s^{\alpha_s}$ be the standard prime factorization. Then for each $1 \le t \le s$, there exists some δ_i such that $\delta_i = a q_t^{\alpha_t}$. Therefore, there is an integer x with $(x, m) = 1$ and index δ_i modulo m. By Theorem 5.3, the index of $x_t = x^a$ modulo m is $q_t^{\alpha_t}$. Let $X = x_1 x_2 \ldots x_3$. Then by Theorem 5.4, the index of X modulo m is δ. Since δ is a multiple of each δ_j, this completes the proof of Lemma 5.2. \square

Now we give a new proof of Gauss theorem.

First of all, assume that there exists a primitive root g modulo m. Then, for $m = 2$ or 4, Gauss' theorem is obviously true. For $m = p^\alpha$ or $2p^\alpha$, by Theorem 4.7, $x^2 \equiv 1 \pmod{m}$ has exactly two solutions, i.e., $x \equiv \pm 1 \pmod{m}$. Hence $g^{\frac{\phi(m)}{2}} \equiv -1 \pmod{m}$, and

$$\prod_{\substack{1 \le i \le m \\ (i,m)=1}} i = \prod_{j=1}^{\phi(m)} g^j = g^{\sum_{j=1}^{\phi(m)} j} = g^{\frac{\phi(m)(\phi(m)+1)}{2}} = \left(g^{\frac{\phi(m)}{2}}\right)^{\phi(m)+1}$$

$$\equiv (-1)^{\phi(m)+1} \equiv -1 \pmod{m}$$

Next, assume that there is no primitive root modulo m. Take a with $(a, m) = 1$ in Lemma 5.2 and $n = \frac{\phi(m)}{\delta}$. Then n is a positive integer. Set $A_0 = \{a^0 = 1, a, \ldots, a^{\delta-1}\}$. Its elements are distinct in the reduced residue system modulo m. In the rest of the system, take arbitrary one as b_1, set $A_1 = \{b_1, b_1 a, \ldots, b_1 a^{\delta-1}\}$. Obviously, these δ elements are also distinct in the reduced residue system modulo m. Note that $a^\delta \equiv 1 \pmod{m}$. It is easy to deduce that they are different from those in A_0. Continuing this way, we finally take one in the rest as b_{n-1}. Set $A_{n-1} = \{b_{n-1}, b_{n-1} a, \ldots, b_{n-1} a^{\delta-1}\}$. Then these δ elements are still distinct in the system modulo m. By $a^\delta \equiv 1 \pmod{m}$ again we can deduce that, they are different from the elements in $A_0, A_1, \ldots, A_{n-2}$.

Thus, all the elements of the reduced residue system modulo m are represented by the new form. Using Lemmas 5.2 and 5.1 gives

(assuming $b_0 = 1$) that

$$\prod_{\substack{1 \le i \le m \\ (i,m)=1}} i \equiv \prod_{i=0}^{n-1} \prod_{j=0}^{\delta-1} b_i a^j = \prod_{i=0}^{n-1} b_i^\delta \left(\prod_{j=0}^{\delta-1} a^j \right)^{\frac{\phi(m)}{\delta}}$$

$$\equiv \left(a^{\sum_{0 \le j \le \delta-1} j} \right)^{\frac{\phi(m)}{\delta}} = (a^{\frac{\phi(m)}{2}})^{\delta-1} \equiv 1 \pmod{m}.$$

Problem. Let $m > 1$ and $\lambda(m)$ be the Carmichael function. For each positive factor d of $\lambda(m)$, if there are $\phi'(d)$ incongruent numbers in the reduced residue system modulo m, then $\sum_{d|\lambda(m)} \phi'(d) = \phi(m)$. How about the properties of $\phi'(d)$? It should be a multiplicative function, but the value might be greater than d. Can we establish a kind of function like Dirichlet character?

Pell's equation

Pell's equation is very important in the study of number theory. Many Diophantine equations can be transformed into Pell's equation or quasi Pell equation.

Let d be a positive integer that is not a perfect square. The Diophantine equation of the form

$$x^2 - dy^2 = 1 \qquad\qquad \text{(P1)}$$

is called Pell's equation.

John Pell (1611–1685) is the British mathematician in 17th century. He had no special contributions to mathematics, and the equation named after him is entirely due to a memory errors of Euler (and due to it there came Pell's number and Pell's primes later on). In fact, another British mathematician at the same time with Pell (he had no special contribution to mathematics, too) did the research earlier in Europe for this kind of equation. Euler who was always humble, named this equation after Pell in writing a monograph discussing such equations.

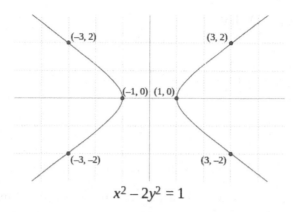

$$x^2 - 2y^2 = 1$$

Pell's equation for $d = 2$ and six of its integral solutions

In India, the first man who studied the Pell equation is Brahmagupta, who used a method of wisdom to give a kind of special solution referred to as "Nava Blatty". Five centuries later, his countryman Bhaskara II (c.1114–1185) found the general solutions of Pell equation for some d.

Obviously, $(x, y) = (\pm 1, 0)$ are the trivial solutions of all Pell equations. Fermat conjectured that for each d, there exists a non-trivial solution. In 1768, Lagrange proved that the Pell equation has a non-trivial solution. The following theorem is a basic conclusion for the solutions of Pell's equation.

Theorem A. *Pell's equation has infinitely many integer solutions* (x, y). *Let* (x_0, y_0) *be the positive integer solution such that* $x + y\sqrt{d}$ *takes the minimum value (basic solution), then all the solutions* (x, y) *can be expressed by*

$$x + y\sqrt{d} = \pm(x_0 + y_0\sqrt{d})^n,$$

where n is an arbitrary integer.

Let d be a positive integer that is not a perfect square. The following equation is also known as the Pell equation

$$x^2 - dy^2 = -1. \tag{P2}$$

Theorem B. *If Pell's equation (P2) has solutions with* (x'_0, y'_0) *be the positive integer solution such that* $x + y\sqrt{d}$ *takes the minimum*

Brahmagupta indulging in computing

value (*basic solution*), *then all the solutions* (x, y) *can be expressed by*

$$x + y\sqrt{d} = \pm(x_0' + y_0'\sqrt{d})^{2n+1},$$

where n is an arbitrary integer.

Let k, l be integers greater than 1, $(k, l) = 1$, and kl not perfect square, the following equation is also known as the Pell equation

$$kx^2 - ly^2 = 1. \tag{P3}$$

Theorem C. *If Pell's equation (P3) has solutions with let (x_0', y_0') the positive integer solution such that of $x + y\sqrt{d}$ takes the minimum value (basic solution), then all the solutions can be expressed by*

$$x\sqrt{k} + y\sqrt{l} = (x_0\sqrt{k} + y_0\sqrt{l})^{2n+1},$$

where n is an arbitrary integer.

It needs to be pointed out that the basic solution of Theorems A and B satisfies the relation:

$$x_0 + y_0 d = (x_0' + y_0' \sqrt{d})^2.$$

There are usually infinitely many solutions of Pell's equation. When $d = 2$, the ratio $\frac{x}{y}$ trends to $\sqrt{2}$, and this kind of approximation is very useful in the calculation of irrational numbers and continued fractions, which is probably the reason why people are interesting in Pell's equation. By the way, the continued fraction is also an important branch of number theory.

If we express the fraction $\frac{1}{a+\frac{1}{b}}$ by $\frac{1}{a+}\frac{1}{b}$, then the fraction $a_0 + \frac{1}{a_1} + \frac{1}{a_2} + \cdots + \frac{1}{a_n}$ is known as a continued fraction, where a_0 is an integer and the other a_i are positive integers. We denote a continued fraction as $[a_0, a_1, a_2, \ldots, a_n]$ (the same symbol with the least common multiple). The continued fraction can be infinite, as long as the numbers of positive integers in the brackets is infinity. For example,

$$1 + \sqrt{5} = [3, 4, 4, 4, \ldots].$$

Each continued fraction expresses a unique real number. On the other hand, every real number can be uniquely expressed as a continued fraction. Accurately speaking, each irrational number can be uniquely expressed as infinite continued fractions, and each rational number can be expressed exactly in two ways as a finite continued fraction. This is because, if $a_n > 1$, then

$$[a_0, a_1, a_2, \ldots, a_n] = [a_0, a_1, a_2, \ldots, a_n - 1, 1].$$

When an infinite continued fraction $[a_0, a_1, \ldots a_n, \ldots]$ satisfies $a_{s+i} = a_{s+kt+i}, 1 \leq i \leq t, k \geq 0$, where s, t are fixed non negative integer and positive integer, respectively, it is a called cycle continued fraction, denoted as $[a_0, a_1, \ldots a_s, \dot{a}_{s+1}, \ldots, \dot{a}_{s+t}]$.

Each cycle continued fraction is a real root of a quadratic irreducible equation with integer coefficients, and the converse is true. That is to say, the coefficient d in Pell's equation can be expressed into a cycle continued fraction.

Theorem (Fermat–Lagrange). *If d is a positive integer that is not a perfect square, then Pell's equation has a positive integer solution.*

Finally, we give an example of application for Pell's equation, which we finished in the *Number Theory Class and Seminar*.

Example. The equation $x^3 + y^3 + z^3 = x + y + z = 3$ has exactly four solutions:

$$(x, y, z) = (1, 1, 1), (4, 4, -5), (4, -5, 4), (-5, 4, 4).$$

Moreover, what about the solutions of $x^3 + y^3 + z^3 + w^3 = x + y + z + w = 4$?

We need the following identity,

$$(x + y + z)^3 = x^3 + y^3 + z^3 + 3(x + y)(y + z)(z + x) \qquad (S)$$

substituting the known conditions, we have

$$(x + y)(y + z)(z + x) = 8$$

or

$$(3 - x)(3 - y)(3 - z) = 8.$$

Let $x' = 3 - x$, $y' = 3 - y$, $z' = 3 - z$, then

$$x'y'z' = 8, \quad x' + y' + z' = 6.$$

We have the solutions $(2,2,2, (-1,-1,8), (-1,8,-1), (8,-1,-1),$ and the solutions of original equation are $(1,1,1)$, $(4,4,-5)$, $(4,-5,4)$, $(-5,4,4)$.

Unfortunately, we cannot generalize (S) to the case of four variables. However, (S) has another variant which is more complicate, namely

$$(x+y+z)^3 = (x^2+y^2+z^2((x+y+z)+2\{(x+y)(y+z)(z+x)+xyz\}.$$

We can generalize this identity to the case of four variabes as follows:

$$(x + y + z + w)^4$$
$$= (x^3 + y^3 + z^3 + w^3)(x + y + z + w)$$
$$+ 3\{(x + y + z)(x + y + w)(x + z + w)(y + z + w) - xyzw\}.$$
$$(T)$$

Substituting the known conditions, we have

$$(x + y + z)(x + y + w)(x + z + w)(y + z + w) - xyzw = 80.$$

Let $x' = 4 - x$, $y' = 4 - y$, $z' = 4 - z$, $w' = 4 - w$. Then

$$\begin{cases} (4 - x')(4 - y')(4 - z')(4 - w') = c, \\ x'y'z'w' = 80 + c, \\ x' + y' + z' + w' = 12. \end{cases}$$

Take $(x', y', z', w') = (0, 0, 3, 9)$, we get the solution $(4, 4, 1, -5)$; take $(x', y', z', w') = (6, -6, 15, -3)$, we get the solution $(-2, 10, -11, 7)$. In generally, taking $y' = -x'$, and let $z' = 6 + \sqrt{u}$, $w' = 6 - w$, we have

$$\begin{cases} (16 - x'^2)(4 - u) = c, \\ -x'^2(36 - u) = 80 + c. \end{cases}$$

Subtracting these two equations, we get $u = 2x'^2 + 9$, here u must be a square. Let $u = v^2$, v satisfying

$$v^2 - 2x' = 9$$

or

$$\left(\frac{v}{3}\right)^2 - 2\left(\frac{x'}{3}\right) = 1$$

This is a Pell equation and there are infinitely many solutions. But they might be only partial solutions of the given equations.

24. The Case of Composite Modulus

In the previous section, we have discussed the basic properties of index and their applications to the equation of nth power residue. But the index is defined only with modulo m which has primitive roots, and most positive integers do not have a primitive root. Therefore, it is necessary for us to generalize the concept of index to index group. It is meaningful for all positive integer moduli greater than 1.

Firstly, we note that when $\alpha \geq 3$ modulo 2^α does not have primitive root. Moreover, any integer a coprime to 2^α satisfies the following congruence:

$$a^{2^{\alpha-2}} \equiv 1 \pmod{2^\alpha}. \tag{5.6}$$

Next we want to ask whether there exists an integer a, whose order is $2^{\alpha-2}$ modulo $2^{\alpha}(\alpha \geq 3)$. The answer is yes, we have the following theorem.

Theorem 5.15. *Let $\alpha \geq 3$. Then 5 has order $2^{\alpha-2}$ modulo 2^{α}, and*

$$\pm 5^0, \pm 5^1, \ldots, \pm 5^{2^{\alpha-2}-1} \tag{5.7}$$

form the reduced residue system modulo 2^{α}.

Proof. By (5.6) and Theorem 5.2, 5 has order $\delta | 2^{\alpha-2}$ modulo 2^{α}. On the other hand, it is not difficult to verify that by using the mathematical induction, when $\alpha \geq 3$, $5^{2^{\alpha-3}} \equiv 1 + 2^{\alpha-1} \pmod{2^{\alpha}}$. Hence $\delta = 2^{\alpha-2}$, both $5^0, 5^1, \ldots, 5^{2^{\alpha-2}-1}$ and $-5^0, -5^1, \ldots, -5^{2^{\alpha-2}-1}$ are incongruent each other modulo 2^{α}, and these $2^{\alpha-1}$ numbers are coprime to modulus 2^{α}. Moreover, they are all incongruent modulo 2^{α} too. In fact, each one in the first group is congruent to 1 modulo 4, and each one in the second is congruent to 3 modulo 4. Therefore, the array (5.7) constitutes a reduced residue system modulo 2^{α}. This completes the proof of Theorem 5.15. \square

We need to point out that if we use 3 instead of 5, the conclusion of Theorem 5.15 is still valid, but the proof is slightly troublesome.

Corollary. Let

$$c_{-1} = \begin{cases} 1, & \alpha = 1, \\ 2, & \alpha \geq 2, \end{cases} \qquad c_0 = \begin{cases} 1, & \alpha = 1, \\ 2^{\alpha-2}, & \alpha \geq 2. \end{cases}$$

If γ_{-1}, γ_0 go through the least non-negative complete residue system modulo c_{-1} and modulo c_0 respectively, then $(-1)^{\gamma_{-1}} 5^{\gamma_0}$ goes through a reduced residue system modulo 2^{α}.

The congruence

$$(-1)^{\gamma_{-1}} 5^{\gamma_0} \equiv (-1)^{\gamma'_{-1}} 5^{\gamma'_0} \pmod{2^{\alpha}}$$

is true if and only if

$$\gamma_{-1} \equiv \gamma'_{-1} \pmod{c_{-1}}, \gamma_0 \equiv \gamma'_0 \pmod{c_0}.$$

Now, we introduce the concept of index group modulo 2^{α}.

Definition 5.4. If

$$a \equiv (-1)^{\gamma_{-1}} 5^{\gamma_0} \pmod{2^\alpha},$$

then γ_{-1}, γ_0 is called a index group of a modulo 2^α.

It is easy to see that any integer coprime to 2^α has a unique index group γ_{-1}, γ_0 satisfying $0 \le \gamma_{-1} < c_{-1}, 0 \le \gamma_0 < c_0$; in addition, the set of integers with the same index group forms a residue class coprime to 2^α. Using Definition 5.4, we can also deduce the following theorem.

Theorem 5.16. *Let a_1, a_2, \ldots, a_n be n integers coprime to 2^α, and $\gamma_{-1}(a_i), \gamma_0(a_i)$ $(1 \le i \le n)$ be an index group of a_i modulo 2^α. Then*

$$\sum_{i=1}^{n} \gamma_{-1}(a_i), \sum_{i=1}^{n} \gamma_0(a_i)$$

are an index group of $a_1 a_2 \ldots a_n$ modulo 2^α.

Now we extend the definition of index groups to general positive integers.

Definition 5.5. Let $m = 2^\alpha p_1^{\alpha_1} \ldots p_k^{\alpha_k}$, g_i be a primitive root modulo $p_i^{\alpha_i} (1 \le i \le k)$, and $(a, m) = 1$. If

$$a \equiv (-1)^{\gamma_{-1}} 5^{\gamma_0} \pmod{2^\alpha}, \quad a \equiv g_i^{\gamma_i} \pmod{p_i^{\alpha_i}}, \ 1 \le i \le k,$$

then $\gamma_{-1}, \gamma_0, \gamma_1, \ldots, \gamma_k$ is called an index group of a modulo m.

By this definition, it is not difficult for us to derive the following properties.

Property 5.4. If a is an integer coprime to m, then a has only one index group $\gamma_{-1}, \gamma_0, \gamma_1, \ldots, \gamma_k$ satisfying $0 \le \gamma_i < c_i, -1 \le i \le k$, where $c_i = \phi(p_i^{\alpha_i})$ $(1 \le i \le k)$.

Property 5.5. The set of integers with the same index group $\gamma_{-1}, \gamma_0, \gamma_1, \ldots, \gamma_k$ modulo m constitutes a residue class coprime to modulo m. (The derivation of this property requires Qin Jiushao's theorem).

Property 5.6. Let a_1, a_2, \ldots, a_n be integers coprime to m, $\gamma_{-1}(a_i), \gamma_0(a_i), \ldots, \gamma_k(a_i)$ $(1 \leq i \leq n)$ be an index group of a_i modulo m. Then $\sum_{i=1}^n \gamma_{-1}(a_i), \sum_{i=1}^n \gamma_0(a_i), \ldots, \sum_{i=1}^n \gamma_k(a_i)$ is an index group of $a_1 a_2 \ldots a_n$ modulo m.

Diophantine arrays

The ancient Greek mathematician Diophantus not only left a classic book *arithmetic*, but he was also the first man to propose the so-called Diophantine array problem. Diophantus found, the four rational numbers $\{\frac{1}{16}, \frac{33}{16}, \frac{17}{4}, \frac{105}{16}\}$ satisfies such property that the product of any two distinct elements with a sum 1 is a rational number square. Long time after, Fermat found a positive integer array, i.e., $\{1, 3, 8, 120\}$, which satisfies the property. As a matter of fact,

$$1 \cdot 3 = 1 = 2^2, \; 1 \cdot 8 + 1 = 3^2, \; 1 \cdot 120 + 1 = 11^2,$$
$$3 \cdot 8 + 1 = 5^2, \; 3 \cdot 120 + 1 = 19^2, \; 8 \cdot 120 + 1 = 31^2.$$

Definition 5.6. The set $\{a_1, a_2, \ldots, a_m\}$ of positive integers (positive rational numbers) is called an order m Diophantine array (rational array), if every $a_i a_j + 1$ $(1 \leq i < j \leq m)$ is a perfect square.

Euler found infinite many order 4 Diophantine integer array

$$\{a, b, a + b + 2r, 4r(a + r)(b + r)\},$$

as long as they satisfy $ab + 1 = r^2$. For example, $r = 3$, then both $\{1, 8, 15, 528\}$ and $\{2, 4, 12, 300\}$ are order 4 Diophantine arrays. At the same time, Euler also found the 5th Fermat array, but it is a rational number, i.e.,

$$\frac{777480}{8288641}.$$

In 1999, Gibbs found the first groups of order 6 Diophantine rational array,

$$\left\{ \frac{11}{192}, \frac{35}{192}, \frac{155}{27}, \frac{512}{27}, \frac{1235}{48}, \frac{180873}{16} \right\}.$$

About the Diophantine integer array, there is a famous conjecture.

Conjecture 1. *There does not exist a order 5 Diophantine integer array.*

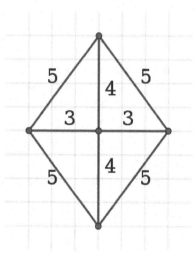

Erdōs-Diophantine graph in graph theory

In 1969, by using the algorithm theory of logarithmic algebraic linear form he himself created, the English mathematician Alan Baker (1939–2018) and his teacher Harold Davenport (1907–1969) proved that, if $\{1, 3, 8, d\}$ is Diophantine integer array, then d must be 120. In other words, the Fermat array cannot be extended. In 1998, the Croatia mathematician Andrej Dujella (1966–2012) improved this result by saying that $\{1, 3\}$ has an infinite number of ways to be extended to order 4, but cannot be extended to order 5. The best results so far are also obtained by Dujella. He proved in 2004 that there is no Diophantine integer array of order 6 or more. There are at most finitely many sets of 5 order Diophantine integer array.

On the other hand, Artin *et al.*, proved in 1979 that each of order 3 Diophantine integer array can be extended to order 4. Their method is as follows. Let $ab + 1 = r^2, ac + 1 = s^2, bc + 1 = t^2$, and

$$d_+ = a + b + c + 2abc + 2rst.$$

Then $\{a, b, c, d_+\}$ is a Diophantine integer array.

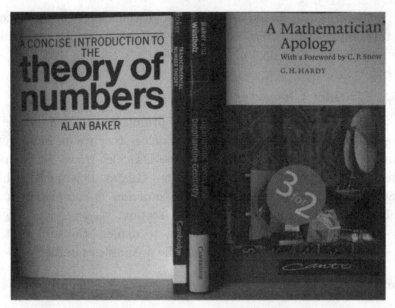

A Concise Introduction to the Theory of Numbers by Baker and *A Mathematician's Apology* by Hardy. Photo taken by the author in Cambridge

As a matter of fact, Dujella proposed the following stronger conjecture, from which Conjecture 1 can be derived directly.

Conjecture 2. *If $\{a, b, c, d\}$ is a Diophantine integer array and $d > \max\{a, b, c\}$, then $d = d_+$.*

If in the definition $ab + 1$ is replaced by $ab + s$, where s is any integer denoted by $P_s\{a_1, a_2, \ldots, a_m\}$, then the situation will be different. For example, it has been found that $P_9\{1, 7, 40, 216\}$, $P_{-7}\{1, 8, 11, 16\}$, $P_{256}\{1, 33, 105, 320, 18240\}$, and $P_{2985984}\{99, 315, 9920, 32768, 44460, 19534284\}$. Is s here bounded? This is the question we want to know the answer.

Suppose we define the Diophantine prime array $\{a_1, a_2, \ldots, a_m\}$, i.e., let each one of $a_i a_j + 1$ $(1 \leq i < j \leq m)$ be prime. For example, $\{1, 2, 6, 18, 30, 270, 606\}$ is the order 7 Diophantine prime array. In general, we have the following conjecture.

Conjecture 3. *There exists infinitely many Diophantine prime array with arbitrary length.*

This conjecture should be more difficult than the twin prime conjecture, but it can be derived by the k twin prime conjecture which we introduce in Section 15.

25. Dirichlet Character

In this section, we use the properties of primitive roots and indices to establish the characteristic function which is very useful in analytic number theory. It is also known as the Dirichlet character, named after the German mathematician Peter Gustav Lejeune Dirichlet (1805–1859). The Dirichlet character function is completely multiplicative. With the notation of last section, $\gamma_{-1}, \gamma_0, \gamma_1, \ldots, \gamma_k$ is an index group modulo m. We use ρ_i to express the unit root of $c_i(-1 \le i \le k)$th power, which might be a complex number.

Definition 5.7. For any group $\rho_{-1}, \rho_0, \rho_1, \ldots, \rho_k$, the function

$$\chi(a) = \begin{cases} \rho_{-1}^{\gamma_{-1}} \rho_0^{\gamma_0} \rho_1^{\gamma_1}, \ldots, \rho_k^{\gamma_k}, & (a, m) = 1, \\ 0, & (a, m) > 1 \end{cases}$$

is called a character function modulo m. In particular, if $\rho_{-1} = \rho_0 = \rho_1 = \cdots = \rho_k = 1$, then the corresponding character function is called principal character function.

It is obvious that for any positive integer m greater than 1, there are $c_{-1}c_0c_1 \ldots a_k = \phi(m)$ groups of unit roots, i.e., there are at most $\phi(m)$ different characters modulo m. Further, we have the following theorem.

Theorem 5.17. *There are exactly $\phi(m)$ different characters modulo m.*

Proof. Let $\rho_{-1}, \rho_0, \rho_1, \ldots, \rho_k$ and $\rho'_{-1}, \rho'_0, \rho'_1, \ldots, \rho'_k$ be two different groups of unit roots. The corresponding character functions are $\chi(a), \chi'(a)$. Assume that $-1 \le j \le k$, and $\rho_j \ne \rho'_j$. By Property 5.5 and the definition of character function, there exists an integer a such that it has index group modulo m with $\gamma_j = 1$, and the rest $\gamma_i = 0$. Therefore, $\chi(a) = \rho_j, \chi'(a) = \rho'_j, \chi(a) \ne \chi'(a)$. □

The tomb of Dirichlet reminds us of his pigeon cage principle. Photo taken by the author in Göttingen

Theorem 5.18. *Let* $\chi(a)$ *be a character function modulo* m. *Then*

$$\sum_{a=0}^{m-1} \chi(a) = \begin{cases} \phi(m) & \text{if } \chi(a) \text{ is a principal character,} \\ 0 & \text{if } \chi(a) \text{ is not a principal character.} \end{cases}$$

Proof. When $\chi(a)$ is a principal character, the conclusion is obvious. Let $\chi(a)$ be a non-principal character. Then there exists j with $-1 \le j \le k$, $\rho_j \ne 1$, but $\rho_j^{c_j} = 1$, i.e., $(\rho_j - 1)(\rho_j^{c_j-1} + \cdots + \rho + 1) = 0$. Thus $\sum_{\gamma_j=0}^{c_j-1} \rho_j^{\gamma_j} = 0$. By Definition 5.6,

$$\sum_{a=0}^{m-1} \chi(a) = \sum_{\gamma_{-1}=0}^{c_{-1}-1} \sum_{\gamma_0=0}^{c_0-1} \cdots \sum_{\gamma_k=0}^{c_k-1} \rho_{-1}^{\gamma_{-1}} \rho_0^{\gamma_0} \cdots \rho_k^{\gamma_k} = \prod_{j=-1}^{k} \sum_{\gamma_j=0}^{c_j-1} \rho_j^{\gamma_j} = 0.$$

This completes the proof of Theorem 5.18. $\qquad\square$

Theorem 5.19. *Let a be any positive integer. Then*

$$\sum_{\chi \ (\text{mod } m)} \chi(a) = \begin{cases} \phi(m) & \text{if } a \equiv 1 \ (\text{mod } m), \\ 0 & \text{if } a \not\equiv 1 \ (\text{mod } m). \end{cases}$$

Proof. If $a \equiv 1 (\text{mod } m)$, then $\chi(a) = 1$ by Definition 5.6, therefore the conclusion is true. If $(a, m) > 1$, then $\chi(a) = 0$, so the conclusion is also true. While $(a, m) = 1$, a has an index group $\gamma_{-1}, \gamma_0, \gamma_1, \ldots, \gamma_k$. By the definition of character,

$$\sum_{\chi (\text{mod } m)} \chi(a) = \prod_{i=-1}^{k} \sum_{\rho_i} \rho_i^{\gamma_i}.$$

Here ρ_i runs through the unit roots of degree $c_i(-1 \leq i \leq k)$.

When $a \not\equiv 1 (\text{mod } m)$, a has an index group $\gamma_j > 0$ modulo m, thus $c_j > 1$. For such c_j, there exists a primitive unit root $\varepsilon(\varepsilon^{\gamma_j} \neq 1)$ of degree c_j (that can generate all unit roots of degree c_j). Therefore,

$$\sum_{\rho_j} \rho_j^{\gamma_j} = \sum_{i=0}^{c_j-1} (\varepsilon^i)^{\gamma_j} = \frac{1 - (\varepsilon^{c_j})^{\gamma_j}}{1 - \varepsilon^{\gamma}} = 0.$$

So $\chi(a) = 0$. This completes the proof of Theorem 5.19. □

Finally, we want to point out that if $\psi(a)$ is a complex function defined on integers, then $\psi(a)$ is a character function modulo n if and only if it satisfies the following properties:

(i) $\psi(a) = 0$ if $(a, m) > 1$;
(ii) $\psi(a)$ is not all 0; if $(a, m) = 1$;
(iii) $\psi(a_1)\psi(a_2) = \psi(a_1 a_2)$;
(iv) if $a_1 \equiv a_2 \ (\text{mod } m)$, then $\psi(a_1) = \psi(a_2)$.

Example 5.11. Let p be an odd prime. Then the Legendre symbol $\left(\frac{a}{p}\right)$ is a character function modulo p; if $a > 1$ is an odd, then the Jacobi symbol $\left(\frac{a}{m}\right)$ is a character function modulo m.

The Dirichlet *L*-function, which corresponds to the Dirichlet character, is defined by

$$L(s, \chi) = \sum_{n=1}^{\infty} \frac{\chi(n)}{n^s}.$$

It is the generalization of the so-called Riemann zeta function

$$\zeta(s) = \sum_{n=1}^{\infty} \frac{1}{n^s}, \quad s = \sigma + it, \sigma > 1$$

in the whole complex plane. They both play a very important role in analytic number theory. In 1837, Dirichlet proved the following conjecture of Legendre, using *L*-function theory creating by himself.

Theorem 5.20 (Dirichlet). *Let a, b be positive integers that are relatively prime. Then in arithmetic progression*

$$an + b, \quad n = 0, 1, 2, \ldots,$$

there exist infinitely many primes. Moreover, let $\pi_a(x) = \sum_{\substack{p \leq x \\ p \equiv b \, (\mathrm{mod} \, a)}} 1$.
Then

$$\pi_a(x) \sim \frac{\pi(x)}{\phi(a)}.$$

This theorem is said to be the beginning of analytic number theory. Later the Dirichlet function played a more and more important role that even generalized the Riemann conjecture (also known as the Riemann hypothesis).

The Riemann conjecture (generalized Riemann conjecture). *All non-trivial zeros of function $\zeta(s)$ (function $L(s, \chi)$) lie on the critical line $\frac{1}{2} + it$.*

Maybe, this is the greatest conjecture in mathematical history. The proof of this conjecture would lead to solve a series of problems in number theory and function theory. Here is an example. For any positive integer m, consider the Diophantine equation

$$ab + bc + ca = m. \tag{5.8}$$

Let $E(x)$ be the set of positive integers not greater than x that make (5.8) no solution. In 1994, the author (*Publ. Math. Debrecen*, **44**, 131–132) proved by using sieve methods: for any $\varepsilon > 0$, there always holds

$$E(x) = O(2^{-(1-\varepsilon)\log x/\log\log x}).$$

A few years later, Tianxin Cai, Rongzheng Jiao (*Chinese Sci. Bull.* **44** (1999) 1093–1095), Xigeng Chen, Maohua Le (*Acta Math. Sinica* **41** (1998) 577–582) proved independently that under the generalized Riemann hypothesis there always exists a solution, unless $m \in \{1, 2, 4, 6, 10, 18, 22, 30, 42, 58, 70, 78, 102, 130, 190, 210, 330, 462\}$.

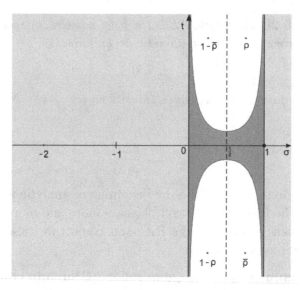

Symmetry of zero point distribution of Riemann's zeta function

Whether the conclusion is true or not without the generalized Riemann hypothesis is the problem we hope to be solve.

Saying about the Riemann hypothesis, Hilbert once put it as the eighth problem in the 23 math problems, together with the Goldbach conjecture and the twin prime conjecture, and later the hypothesis becomes one of the "7 Big Difficult Millennium Problems". However, we guess that there were contradictions in Hilbert's heart at that time. On the one hand, he said when being asked, if he relived after

500 years, what he most wants to know is whether the Riemann conjecture has been proved. On the other hand, he believed that certain numbers to be transcendental (the seventh problem) will be more difficult. The fact turned out, his latter idea was wrong.

Three special exponent sums

In 1808, Gauss defined

$$S(m,n) = \sum_{k=0}^{n-1} e^{2\pi i k^2 m/n} = \sum_{k=0}^{n-1} e\left(\frac{k^2 m}{n}\right).$$

It is called the Gaussian sum. Obviously, when $k_1 \equiv k_2 \pmod{n}$, then $e\left(\frac{k_1^2 m}{n}\right) = e\left(\frac{k_2^2 m}{n}\right)$, which is denoted as

$$S(m,n) = \sum_{k(n)} e\left(\frac{k^2 m}{n}\right).$$

Next we prove that $S(m,n)$ is a multiplicative function of n, i.e., when $(n_1, n_2) = 1$,

$$S(m, n_1 n_2) = S(mn_2, n_1) S(mn_1, n_2).$$

Let $k = n_2 k_1 + n_1 k_2$, by Theorem 2.6 of Section 7, when k_1, k_2 runs through complete residue system of modulo n_1, n_2, respectively, k runs through the complete residue system modulo $n_1 n_2$, and

$$mk^2 = m(n_2 k_1 + n_1 k_2)^2 \equiv mn_2^2 k_1^2 + mn_1^2 k_2^2 \pmod{n_1 n_2}.$$

Thus,

$$S(mn_2, n_1) S(mn_1, n_2) = \sum_{k_1(n_1)} e\left(\frac{k_1^2 mn_2}{n_1}\right) \sum_{k_2(n_2)} e\left(\frac{k_2^2 mn_1}{n_2}\right)$$

$$= \sum_{\substack{k_1(n_1) \\ k_2(n_2)}} e\left(\frac{k_1^2 mn_2}{n_1} + \frac{k_2^2 mn_1}{n_2}\right)$$

$$= \sum_{\substack{k_1(n_1) \\ k_2(n_2)}} e\left(\frac{m(k_1^2 n_2^2 + k_2^2 n_1^2)}{n_1 n_2}\right)$$

$$= \sum_{k(n_1 n_2)} e\left(\frac{mk^2}{n_1 n_2}\right) = S(m, n_1 n_2).$$

Portrait of Indian mathematician Ramanujan

In 1918, the Indian mathematician Ramanujan (1887–1920) defined the following sum (Ramanujan sum):

$$c_q(m) = \sum_{k'(q)} e\left(\frac{km}{q}\right),$$

where k goes through the residue system modulo q. The Ramanujan sum also can be expressed as

$$c_q(m) = \sum_{\rho} \rho^m,$$

where ρ goes through all qth power original root of the unit element 1.

If $(q_1, q_2) = 1$, then

$$c_{q_1 q_2}(m) = c_{q_1}(m) c_{q_2}(m).$$

Using Theorem 5.8 of Section 8, we obtain

$$c_{q_1}(m)c_{q_2}(m) = \sum_{\substack{k_1(n_1) \\ k_2(n_2)}} e\left(m\left(\frac{k_1}{q_1} + \frac{k_2}{q_2}\right)\right)$$

$$= \sum_{\substack{k_1(n_1) \\ k_2(n_2)}} e\left(\frac{m(k_1 q_2 + k_2 q_1)}{q_1 q_2}\right) = c_{q_1 q_2}(m).$$

In 1926, the Holland mathematician Kloosterman (1900–1968) defined the following sum (Kloosterman sum):

$$K(a, b, n) = \sum_{k'(n)} e\left(\frac{ak + b\bar{k}}{n}\right),$$

where k goes through the reduced residue system modulo n and $\bar{k}k \equiv 1 \pmod{n}$. It is not difficult to prove that, when $(n_1, n_2) = 1$,

$$K(a, b_1, n_1)K(a, b_2, n_2) = K(a, b, n_1 n_2),$$

where $b = b_1 n_2^2 + b_2 n_1^2$.

The above three sums play an important role in the study of number theory.

Finally, we introduce three congruences on partitions discovered by Ramanujan. Denoted $p(n)$ as the way of writing n as a sum of positive integers, e.g., $p(1) = 1$, $p(2) = 2$, $p(3) = 3$, $p(4) = 5 \ldots$. The five partitions of 4 are

$$4 = 1 + 3 = 2 + 2 = 1 + 1 + 2 = 1 + 1 + 1 + 1.$$

We call $p(n)$ the number of partitions or the partition function. The partition number problem constitutes a branch of number theory. It is easy to prove that $p(n)$ satisfies the following identity:

$$\sum_{n=0}^{\infty} p(n)x^n = \prod_{k=1}^{\infty} (1 - x^k)^{-1}.$$

In 1918, Ramanujan and the British mathematician G. H. Hardy (1877–1947) proved

$$p(n) \approx \frac{1}{4\sqrt{3}n} e^{\pi\sqrt{2n/3}}.$$

Before and after 1919, Ramanujan discovered and proved

$$p(5n + 4) \equiv 0 \pmod{5},$$
$$p(7n + 5) \equiv 0 \pmod{7},$$
$$p(11n + 6) \equiv 0 \pmod{11}.$$

He also claimed that there is no more similar simple congruence as $p(qn + b) \equiv 0 \pmod{q}$. In 2003, this judge were proved by Ahlgren and Boylan (*Invent. Math.* **153**(3), 487–502). They used the method of modular forms in their method.

On the other hand, as early as the 1960's, the English American computational number theorist A. O. L. Atkin (1925–2008) proved that

$$p(17303n + 237) \equiv 0 \pmod{13}.$$

In 2000, the Japanese-American mathematician Ken Ono (1968–) proved that there exists a similar congruence for any prime modulus. For example,

$$p(4063467631n + 30064597) \equiv 0 \pmod{31}.$$

The following year, he and his colleagues proved that there exists similar congruence as long as the modulus is positive integer coprime to 6.

Finally, let's introduce what we call the square partition. If we denote $p_2(n)$ the number of the ways to express n as the sum of positive integers whose product is a square, e.g., $p_2(1) = p_2(2) = p_2(3) = 1, p_2(4) = p_2(5) = 3, p_2(6) = p_2(7) = 4$, here the four partition of 6 are

$$3 + 3 = 4 + 1 + 1 = 2 + 2 + 1 + 1 = 1 + 1 + 1 + 1.$$

It's easy to show that when $n > 7$, $p_2(n)$ is increasing. Let $q_2(n) = p_2(n) - p_2(n-1)$. We can show that $q_2(n)$ goes to the infinity. Here we raise two questions:

(1) When $n > 50$, is $q_2(n)$ is increasing?

(2) Does $p_2(n)$ have the congruence modulo prime like $p(n)$?

On the other hand, we can consider the partition such that the partial sums are distinct. This is more strict than the partition into the sum of different positive integers. For example, there are only three ways for the partition of 6: $6 = 5 + 1 = 4 + 2$, while $3 + 2 + 1$ is excluded, since $3 = 2 + 1$. What properties does it have?

We find that such a partition is related with the so-called Golomb ruler. Let's consider the set of n non-negative integers

$$a_1 < a_2 < \cdots < a_n,$$

with distinct differences of any two elements and the least length $a_n - a_1$. For instance, $\{0\}$, $\{0, 1\}$, $\{0, 1, 3\}$, $\{0, 1, 4, 6\}$, $\{0, 1, 4, 9, 11\}$, $\{0, 1, 4, 10, 15, 17\}$. For any positive integer n, try to find the Golomb ruler (maybe not unique). In February 2014, the 27th Golomb ruler was found with the length 553.

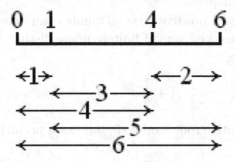

Golomb ruler of order 4 and length 6

Remark. In 1919, Ramanujan also discovered and proved some other congruences, including

$$p(25n + 24) \equiv 0 \pmod{5^2},$$
$$p(49n + 47) \equiv 0 \pmod{7^2},$$
$$p(121n + 116) \equiv 0 \pmod{11^2}.$$

Meanwhile he raised some conjectures, including

$$p(125n + 99) \equiv 0 \pmod{5^3}$$
$$p(343n + 243) \equiv 0 \pmod{7^3}$$

the former one was proved by Krecmarin (1933), while the latter one was negated because of $7^2\|p(243)$.

Recently, we discovered the following congruences

$$p(49n - 7i - 2) \equiv 0 \ (\text{mod } 7^2), \quad i = 0, 1, 2, 4,$$

$$p(125m - 25i - 1) \equiv 0 \ (\text{mod } 5^2), \quad i = 0, 1, 2.$$

Exercises 5

1. If p and $q = 2p+1$ are primes, and $p \equiv 1 \ (\text{mod } 4)$, show that 2 is a primitive root modulo q; if $3 < p \equiv 3 \ (\text{mod } 4)$, show that there are three consecutive integers that are primitive roots modulo q. Give an example.

2. Show that a primitive root modulo m must be non-quadratic residue, and the converse is not true. If p is prime, show that a primitive root modulo p is a non-quadratic residue if and only if p is a Fermat prime.

3. Show that each positive rational number can be a sum of distinct unit fractions, i.e., sum of finitely many distinct terms from the harmonic series

$$1 + \frac{1}{2} + \cdots + \frac{1}{n} + \cdots .$$

4. Find a primitive root modulo 5, but not a primitive root modulo 25.

5. If p is prime, show that for any positive integer k,

$$1^k + 2^k + \cdots + (p-1)^k \equiv \begin{cases} 0 \ (\text{mod } p) & \text{if } p-1 \nmid k, \\ -1 \ (\text{mod } p) & \text{if } p-1 \mid k. \end{cases}$$

6. If p is odd prime, show that
 (1) $1^p + 2^p + \cdots + (p-1)^p \equiv 0 \ (\text{mod } p)$;
 (2) if m is a positive integer and $2^m \not\equiv 1 \ (\text{mod } p)$, then $1^m + 2^m + \cdots + (p-1)^m \equiv 0 \ (\text{mod } p)$.

7. Construct a table of indices modulo 23 using 5 as primitive root.

8. Find the solution of $x^{15} \equiv 14 \pmod{41}$ by the table of indices.

9. Construct a table of indices modulo 2^5.

10. If p is prime, show that $x^8 \equiv 16 \pmod{p}$ has solutions.

11. Determine the number of solutions of $x^4 \equiv 61 \pmod{117}$.

12. Evaluate the character function $\chi(a)$ modulo 3 and modulo 5.

Chapter 6

Congruence Modulo Integer Powers

26. Bernoulli Number and Bernoulli Polynomial

We have presented several famous congruences modulo prime powers or integer powers in the previous chapters, such as Fermat Little Theorem, Wilson Theorem, Lagrange Theorem and Lucas Theorem belong to the former ones, while Euler Theorem, Gauss Theorem and Qin Jiushao Theorem, etc., belong to the latter ones. Now we consider some congruences modulo prime powers, which are mainly related to unit fractions (reciprocal of integers) or binomial (multinomial) coefficients). These congruences are beautiful and profound, which do not appear in the present textbooks of elementary number theory or even monographs of number theory. Some of them are got in the 21 century. Firstly, we introduce Bernoulli numbers and Bernoulli polynomials.

The Bernoulli numbers was created to calculate the sum

$$S_k(n) = 1^k + 2^k + \cdots + (n-1)^k.$$

As early as in the 5th century, the Indian mathematician Aryabhata got the formulas of the above sum for $k = 1, 2$ and 3, i.e.,

$$S_1(n) = \frac{(n-1)n}{2}, \quad S_2(n) = \frac{(n-1)n(2n-1)}{6},$$

$$S_3(n) = S_1^2(n) = \frac{(n-1)^2 n^2}{4}.$$

In order to get the sum $S_k(n)$ for higher power, we need to introduce the definition of Bernoulli numbers and Bernoulli polynomials.

Definition 6.1. Let x be a complex number. The polynomial $B_n(x)$ is called the nth Bernoulli polynomial if it satisfies

$$\frac{ze^{xz}}{e^z - 1} = \sum_{n=0}^{\infty} \frac{B_n(x)}{n!} z^n, \quad |z| < 2\pi. \tag{6.1}$$

In particular, $B_n(0) = B_n$ is called the nth Bernoulli number.

Portrait of Swiss mathematician Jacob Bernoulli

The Bernoulli numbers were discovered around the same time by the Swiss mathematician Jacob Bernoulli (1654–1705), after whom it is named, and independently by the Japanese mathematician Seki Takakazu (1642–1708). Seki's discovery was posthumously published in 1712; Bernoulli's, also posthumously, in 1713. The first Bernoulli numbers are

$$B_0 = 1, B_1 = -\frac{1}{2}, B_2 = \frac{1}{6}, B_{2n+1} = 0 \ (n \geq 1), B_4 = -\frac{1}{30}, B_6 = \frac{1}{42},$$

$$B_8 = -\frac{1}{30}, B_{10} = \frac{5}{66}, B_{12} = -\frac{691}{2730}, B_{14} = \frac{7}{6}, B_{16} = -\frac{3617}{510},$$

$$B_{18} = \frac{43867}{798}, B_{20} = -\frac{174611}{330}, B_{22} = \frac{854513}{138}, B_{24} = -\frac{236364091}{2730}, \dots.$$

Portrait of Japanese mathematician Seki Takakazu

In general, the B_n's satisfy the recursion formula

$$B_n = \sum_{k=0}^{n} \binom{n}{k} B_k,$$

where the denominator of B_n is $\Pi_{p-1|n}\, p$. The first five Bernoulli polynomials are as follows:

$$B_0(x) = 1, B_1(x) = x - \frac{1}{2}, B_2(x) = x^2 - x + \frac{1}{6},$$

$$B_3(x) = x^3 - \frac{3}{2}x^2 + \frac{1}{2}x, B_4(x) = x^4 - 2x^3 + x^2 - \frac{1}{30}.$$

By Definition 6.1, expanding the product into the power series and comparing the coefficients of both sides, we have the following theorem that relates Bernoulli numbers with Bernoulli polynomials.

Theorem 6.1. *For any non-negative integer* n,

$$B_n(x) = \sum_{k=0}^{n} \binom{n}{k} B_k x^{n-k}.$$

Meanwhile, there is some relation between Bernoulli polynomials.

Theorem 6.2. *For any non-negative integer n,*

$$B_{n+1}(x+1) - B_{n+1}(x) = (n+1)x^n.$$

Moreover, if $n \geq 1$, then

$$B_{n+1}(0) = B_{n+1}(1).$$

Proof. Expanding the following identity into the power series of z,

$$z\frac{e^{(x+1)z}}{e^z - 1} - z\frac{e^{xz}}{e^z - 1} = ze^{xz},$$

one has

$$\sum_{n=0}^{\infty} \frac{B_n(x+1) - B_n x}{n!} z^n = \sum_{n=0}^{\infty} \frac{x^n}{n!} z^{n+1}.$$

Comparing the coefficients of both sides gives the results. □

Corollary. *Let m and n be positive integers. Then*

$$1^n + 2^n + \cdots + (m-1)^n = \frac{B_{n+1}(m) - B_{n+1}}{n+1}$$

$$= \frac{1}{n+1} \sum_{k=0}^{n} \binom{n+1}{k} B_k m^{n-k+1}.$$

Let $\sigma > 1$ be the real part of complex numbers. The Riemann zeta function $\zeta(s)$ and the Hurwitz zeta function $\zeta(s+a)$ are defined by

$$\zeta(s) = \sum_{n=1}^{\infty} \frac{1}{n^s}, \quad \zeta(s,a) = \sum_{n=0}^{\infty} \frac{1}{(n+a)^s},$$

where a is an arbitrary complex number. These two functions are analytic in the half-plane with $\sigma > 1$ and can be continued analytically to all complex values. The values of them at the integral points

on real axis are related to Bernoulli numbers and Bernoulli polynomials. By the function equation for the $\zeta(s)$ function, we can get

$$\zeta(-n) = -\frac{B_{n+1}}{n+1},$$

$$\zeta(2k) = (-1)^{k+1}\frac{(2\pi)^{2k}B_{2k}}{2(2k)!}, \tag{6.2}$$

$$\zeta(-n, a) = -\frac{B_{n+1}(a)}{n+1}.$$

In particular,

$$\zeta(0, a) = -B_1(a) = \frac{1}{2} - a, \zeta(0) = \zeta(0, 1) = -\frac{1}{2}.$$

From 6.2 and the fact that $\zeta(s)$ takes zero at negative odd integers and positive values at positive even integers on the real axis,

$$B_{2n+1} = \zeta(-2n) = 0 \ (n \geq 1), (-1)^{k+1}B_{2k} > 0.$$

On the other hand, from $\zeta(2k) \to 1 \ (k \to \infty)$, we see that $|B_{2k}| \to \infty \ (k \to \infty)$. In fact, when $k \to \infty$,

$$(-1)^{k+1}B_{2k} \sim \frac{2(2k)!}{(2\pi)^{2k}}.$$

But for the values of $\zeta(s)$ function at positive odd integers, there is no formula like (6.2). In 1979, the Greek–French mathematician Roger Apéry (1916–1994) proved that $\zeta(3)$ is an irrational number. We do not know whether $\zeta(2k+1)(k \geq 2)$ is a rational or irrational. However, in 2003, the Russian number theorist Wadim Zudilin showed that at least one of the numbers $\zeta(5), \zeta(7), \zeta(9), \zeta(11)$ is irrational.

There are many generalizations of ζ-function, from one variable to several, including the Mordell–Tornheim ζ-function and the Witten ζ-function. Here we introduce a kind of multiple ζ-function

$$\zeta(s_1, s_2, \ldots, s_k) = \sum_{n_1 > n_2 > \cdots > n_k > 0} \frac{1}{n_1^{s_1} n_2^{s_2} \cdots n_k^{s_k}},$$

which was first defined by Euler in 1742 (for $k = 2$). Later Zagier generalized it to any k, so it is also called Euler–Zagier zeta function.

Euler also got a beautiful result:

$$\zeta(2,1) = \zeta(3).$$

In general, for any positive s,

$$\zeta(s,1) = \frac{1}{2}s\zeta(s+1) - \frac{1}{2}\sum_{k=2}^{s-1}\zeta(k)\zeta(s-k+1).$$

In 2006, Gangl, Kanelo, and Zagier proved that

$$\sum_{\substack{a+b=n \\ a,b,\geq 1}} \zeta(2a,2b) = \frac{3}{4}\zeta(2n), \quad n > 1,$$

$$\sum_{\substack{a+b=n \\ a,b\geq 1}} \zeta(2a-1,2b-1) = \frac{1}{4}\zeta(2n), \quad n > 1.$$

In 2012, we (Zhongyan Shen and Tianxin Cai, *J. Number Theory* **132**, 314–323) showed that

$$\sum_{\substack{a+b+c=n \\ a,b,c\geq 1}} \zeta(2a,2b,2c) = \frac{5}{8}\zeta(2n) - \frac{1}{4}\zeta(2)\zeta(2n-2), \quad n > 2,$$

$$\sum_{\substack{a+b+c+d=n \\ a,b,c,d\geq 1}} \zeta(2a,2b,2c,2d) = \frac{35}{64}\zeta(2n) - \frac{5}{16}\zeta(2)\zeta(2n-2), \quad n > 3.$$

Before Wiles proved Fermat's Last Theorem, Bernoulli numbers and related congruences played an important role in the research of this problem. For example, mathematicians found that if $x^p + y^p + z^p = 0$, $p \nmid xyz$ have a solution, then p must divide each of $B_{p-3}, B_{p-5}, B_{p-7}, B_{p-9}$. This is the effective method to verify the first case of Fermat's Last Theorem.

Kummer's congruence

In 1840, one of Gauss' students, German mathematician von Staudt (1798–1867) and self-learned Danish mathematician Clausen (1801–1885) proved the following theorem, independently.

Theorem (von Staudt–Clausen). *Let $n = 1$ or be any even integers. Then*

$$-B_n \equiv \sum_{(p-1)|n} \frac{1}{p} \;(\mathrm{mod}\;1).$$

That is to say, $B_n + \sum_{(p-1)|n} \frac{1}{p}$ is an integer, where B_n is the nth Bernoulli number.

When $n = 1$, only $p = 2$ satisfies $(p-1)\,|\,1$, and $B_1 = -\frac{1}{2}$. When $n = 2$, only $p = 2$ and 3 satisfy $(p-1)\,|\,2$, $\frac{1}{2} + \frac{1}{3} = \frac{5}{6}$, and $B_2 = \frac{1}{6}$. When $n = 4$, $p = 2$, 3 and 5 satisfy $(p-1)\,|\,4$, $\frac{1}{2} + \frac{1}{3} + \frac{1}{5} = \frac{31}{30}$, and $B_4 = -\frac{1}{30}$.

Corollary. *If $n = 1$ or even, p is a prime. Then*

$$pB_k \equiv \begin{cases} -1 \\ 0 \end{cases} (\mathrm{mod}\;p), \qquad \begin{cases} if \;\; p-1 | k, \\ if \; p-1 \nmid k. \end{cases}$$

The proof of the von Staudt–Clausen theorem depends on the following lemma.

Lemma. *For any non-negative integer k,*

$$\sum_{m=1}^{p-1} m^k \equiv \begin{cases} 1 & if \;\; p-1|k, \\ 0 & if \; p-1 \nmid k. \end{cases}$$

Proof. If $p - 1\,|\,k$, then the lemma follows from Fermat's Little Theorem. If $p - 1 \nmid k$, let g be a primitive root modulo p, then by Theorem 6.2, we have $g^k \not\equiv 1(\mathrm{mod}\;p)$. From $(g, p) = 1$ and Theorem 2.7, we know that $g, 2g, \ldots, (p-1)g$ go through the reduced system of residues modulo p. Then

$$\sum_{m=1}^{p-1} (gm)^k \equiv \sum_{m=1}^{p-1} m^k \;(\mathrm{mod}\;p),$$

i.e.,

$$(g^k - 1) \sum_{m=1}^{p-1} m^k \equiv 0 \;(\mathrm{mod}\;p),$$

hence

$$\sum_{m=1}^{p-1} m^k \equiv 0 \pmod{p}.$$

This completes the proof of the lemma. □

The following two theorems were given by German-born British mathematician Rado (1906–1989) and German mathematician Kummer (1810–1893). Rado was known for Erdõs–Ko–Rado theorem in combinatorics. Kummer taught for 10 years in a high school (Kronecker was his student) and research field was ballistics when he first worked in university, but he was the main founder of algebraic number theory. One of his Ph.D. students was Cantor who was known as the inventor of set theory. It is worth mention that Cantor's wife is the cousin of both Dirichlet's wife and musician Mendelssohn's wife.

Portrait of German mathematician Kummer

Theorem (Rado). *Let q be a prime in the form $3n + 1$. Then*

$$B_{2q} \equiv \frac{1}{6} \pmod{1}.$$

Since the only possible prime p satisfying $(p-1)|2q$ are 2, 3, $q+1$ and $2q + 1$, but $q + 1$ is even, $2q + 1$ is divisible by 3, then $p = 2$ or 3. By using von Staudt–Clausen theorem, we complete the proof of Rado theorem.

Theorem (Kummer's congruence). *Let p be an odd prime, a and b positive integers satisfying $a \equiv b \not\equiv 0 \pmod{p-1}$. Then*

$$\frac{B_a}{a} \equiv \frac{B_b}{b} \pmod{p},$$

where $\frac{1}{a}$ denotes the integer a' such that $a'a \equiv 1 \pmod{p}$, and B_n is the nth Bernoulli number.

Remark. Rado's theorem can be generalized to: Let q_i $(1 \leq i \leq r)$ be primes in the form of $3n + 1$. Then

$$B_{2q_1 \ldots q_r} \equiv \frac{1}{6} \pmod{1}.$$

And also, for any positive integer k, if $q = (2k + 1)n + k$ is a prime, then (R) still holds.

Remark. Under the condition of Kummer's congruence, let $a \equiv b \equiv 0 \pmod{p-1}$. Then

$$\frac{B_a}{a} - \frac{B_b}{b} \equiv \left(\frac{1}{a} - \frac{1}{b} \right) \left(1 - \frac{1}{p} \right) \pmod{p},$$

which means

$$p\{\beta_{a(p-1)} - \beta_{b(p-1)}\} \equiv \frac{1}{a} - \frac{1}{b} \pmod{p^2}.$$

Remark. The fraction $\beta_n = \frac{B_n}{n}$ is called the divided Bernoulli number. We discovered (Peng Yang, Tianxin Cai, Hao Zhong, on several

sums of divided Bernoulli numbers) the following: for any prime p,

$$p \sum_{k=1}^{p-1} \frac{B_k}{k} \equiv (-1)^{p-1} (p^2).$$

27. Wolstenholme Theorem

From this section on many congruences we discussed are related to binomial coefficients $\binom{m}{n}$. We find that it has the function to promote the congruences from modulo primes (integers) to modulo prime (integer) powers.

By Wilson's theorem (1770), it's easy to get a congruence about binomial coefficients, i.e., for any positive integer n and prime p,

$$\binom{np-1}{p-1} \equiv 1 \ (\mathrm{mod}\ p).$$

In 1819, the English mathematician Babbage (1791–1871) who has the name "father of the computer" proved that for any prime $p \geq 3$,

$$\binom{2p-1}{p-1} \equiv 1 \ (\mathrm{mod}\ p^2).$$

Its generalized form is, for any positive integers $a \geq b$ and primes $p \geq 3$,

$$\binom{ap}{bp} \equiv \binom{a}{b} \ (\mathrm{mod}\ p^2). \tag{6.3}$$

When $a = 2, b = 1$, this is Babbage's congruence.

We use the method of expanding the binomial coefficients to give a simple proof of this congruence.

It is obviously that $\binom{ap}{bp}$ is the coefficient of x^{bp} in the expanding of $(1+x)^{ap}$. Noting that

$$(1+x)^p = 1 + x^p + \sum_{k=1}^{p-1} \binom{p}{k} x^k = 1 + x^p + pf(x),$$

where $f(x)$ is polynomial with integer coefficients, we have

$$(1+x)^{ap} = (1 + x^p + pf(x))^a$$
$$\equiv (1+x^p)^a + a(1+x^p)^{a-1}pf(x) \,(\text{mod } p^2).$$

Since the right-hand side equals to

$$(1+x^p)^a + a(1+x^p)^{a-1}((1+x)^p - 1 - x^p)$$
$$= (1-a)(1+x^p)^a + a(1+x^p)^{a-1}(1+x)^p,$$

the coefficient of x^{bp} in the above formula is

$$(1-a)\binom{a}{b} + a\left(\binom{a-1}{b} + \binom{a-1}{b-1}\right) = \binom{a}{b}.$$

Here we use the property of the sum of binomial coefficients, then (6.3) holds.

In 1862, the English mathematician Wolstenholme (1829–1891) improved Babbage's congruence, and he showed the following theorem.

Portrait of English mathematician Wolstenholme

Theorem 6.3 (Wolstenholme). *For any primes $p \geq 5$,*

$$\binom{2p-1}{p-1} \equiv 1 (\text{mod } p^3). \tag{6.4}$$

There is an unproved conjecture that Wolstenholme's congruence holds if and only if p is a prime, i.e., (6.4) is a criterion to test primes. But for modulo integer or integer square, there are counterexamples satisfying (6.4), they are $n = 29.937$ and 16843^2, respectively. If (6.4) holds mod p^4 instead of mod p^3, then such p is called Wolstenholme prime. So far people just found two Wolstenholme primes: 16843 and 2124679 in the range $p \leq 10^9$.

Now we use the drawer principle to prove the theorem. Consider the counting problem to put the $2p$ elements in two groups, each group with p elements and select p ones from them (each group at least take one), then there are two ways to count the number of methods, comparing them we get

$$\binom{2p}{p} - 2 = \sum_{\substack{i+j=p \\ i,j>0}} \binom{p}{i}\binom{p}{j}.$$

It's easy to prove that

$$\binom{2p}{p} - 2 \equiv 0 \,(\text{mod } p^3) \quad p > 5.$$

This is the equivalent form to (6.4).

In our seminar, Deyi Chen generalized this method, putted the $4p$ or $6p$ elements in 4 or 6 equal groups and selected p ones in them, then we have

$$2\binom{3p}{p} - 9\binom{2p}{p} + 12 \equiv 0 \,(\text{mod } p^5), \quad p \geq 7,$$

$$3\binom{4p}{p} - 20\binom{3p}{p} + 54\binom{2p}{p} - 60 \equiv 0 \,(\text{mod } p^7), \quad p \geq 11.$$

It is worth to note that Wolstenholme was a student in St John's College, Cambridge University. He was a close friend of Leslie Stephen, whose daughter Virginia Woolf is a famous authoress. At the time of 30 years after Wolstenholme's death, Woolf published her novel *To the Lighthouse*, his parents appear as the main characters and Wolstenholme is a prototype of a character in the book.

In 1898, the English mathematician J. W. L. Glaisher (1848–1928) proved the following: for any positive integer n and prime $p \geq 5$,

$$\binom{np-1}{p-1} \equiv 1 \,(\mathrm{mod}\ p^3).$$

In 1952, the Norwegian mathematician Ljunggren (1905–1973) found the equivalent form of (6.4), i.e., for any integers $u > v > 0$ and prime p,

$$\binom{up}{vp} \equiv \binom{u}{v} \,(\mathrm{mod}\ p^3). \tag{6.5}$$

On the other hand, Wolstenholme himself discovered that Theorem 6.3 could be expressed as the form of harmonic series, i.e., for any prime $p \geq 5$,

$$\sum_{i=1}^{p-1} \frac{1}{i} \equiv 0 \,(\mathrm{mod}\ p^2), \quad \sum_{i=1}^{p-1} \frac{1}{i^2} \equiv 0 \,(\mathrm{mod}\ p). \tag{6.6}$$

Here we want to point out the meaning of congruence for rational numbers,

$$\frac{a}{b} \equiv \frac{c}{d} \,(\mathrm{mod}\ m)$$

if and only if $(bd, m) = 1$, $ad - bc = 0 \,(\mathrm{mod}\ m)$.

The first congruence of (6.6) is also called Wolstenholme's theorem (the two theorems could be transformed each other). Here we give an elementary proof, let

$$(x-1)(x-2) \cdots (x-(p-1))$$
$$= x^{p-1} - s_1 x^{p-2} + \cdots - s_{p-2} x + (p-1)!, \tag{6.7}$$

where $s_j (1 \leq j \leq p-2)$ are integers, $s_{p-2} = \sum_{i=1}^{p-1} \frac{(p-1)!}{i}$. According to Theorem 3.10 and Wilson's theorem, we have $p | s_j$. Let $x = p$ in (6.7), then

$$p^{p-1} - s_1 p^{p-2} + \cdots - p s_{p-2} = 0.$$

As for $p > 3$, take modulo p^3 in the above formula, then $s_{p-2} \equiv 0 \,(\mathrm{mod}\ p^2)$. This is the first congruence of (6.6).

For the second congruence of (6.6), we can use the following method. For $1 < i < p - 1$, there exists an unique j, $j \neq i$, $1 < j < p - 1$, such that $ij \equiv 1 \pmod{p}$. Then

$$\sum_{i=1}^{p-1} \frac{1}{i^2} \equiv 1 + \sum_{\substack{1<i\neq j<p-1 \\ ij\equiv 1(\text{mod } p)}} \frac{i^2 + j^2}{i^2 j^2} + (p-1)^2$$

$$\equiv \sum_{i=1}^{p-1} i^2 \equiv 0(\text{mod } p).$$

The usual generalization of (6.6) is

$$\sum_{i=1}^{p-1} \frac{1}{i} \equiv B_{p-3}p^2(\text{mod } p^3), \quad \sum_{i=1}^{p-1} \frac{1}{i^2} \equiv \frac{2p}{3}B_{p-3}(\text{mod } p^2).$$

On the other hand, as early as in 1888, the English mathematician Leudesdorf proved

$$\sum_{\substack{i=1 \\ (i,n)=1}}^{n-1} \frac{1}{i} \equiv 0 \pmod{n^2}, \quad (6, n) = 1.$$

It is worth to point out that we have considered the product of binomial coefficients, for example

$$\binom{36}{2} = \binom{3}{2}\binom{21}{2} = \binom{4}{2}\binom{15}{2},$$

$$\binom{1225}{2} = \binom{15}{2}\binom{120}{2} = \binom{12}{2}\binom{85}{2},$$

where $36 = 6^2 = \binom{9}{2}$. In general, we (Deyi Chen and Tianxin Cai, *Rocky Mountain J. Math.* **43**(1), (2013), 75–81) proved the following.

Let $r > 1$ be a triangular square number (both triangular and square number), $\binom{r}{2}$ has two different ways to be decomposed as the product of two non-1 triangular numbers.

There are infinitely many positive integers which have three different ways to be decomposed as the product of two positive integers of the form $n^2 - 1$.

For example,

$$40320 = (3^2 - 1)(71^2 - 1) = (5^2 - 1)(41^2 - 1)$$

$$= (7^2 - 1)(29^2 - 1),$$

$$47980800 = (3^2 - 1)(2449^2 - 1) = (29^2 - 1)(239^2 - 1)$$

$$= (41^2 - 1)(169^2 - 1).$$

Meanwhile, we have (verified to $\binom{r}{2}, r \leq 3 \times 10^5$)

Conjecture. There is no triangular number which has more than two different ways to be decomposed as the product of two non-1 triangular numbers.

Elliptic curve

Besides the conjectures about square numbers, there are some conjectures about cubic numbers in Fermat's comments. For example, he asserted that except 1 all triangular numbers are not cubic numbers. This is equivalent to that the following Diophantine equation

$$\frac{1}{2}y(y - 1) = x^3$$

has no positive integer solution. Rewrite it as

$$(2y - 1)^2 = (2x)^3 + 1,$$

change $2y - 1$ to y, and $2x$ to x, the above equation turns to the special form of

$$y^2 = x^3 + ax + b,$$

where $a, b \in \mathbb{Q}$, the cubic polynomial on the right-hand side does not have multiple roots. (To remove the multiple roots is to guarantee that there is no singular point), i.e., its discriminant $\Delta = 4a^3 + 27b^2 \neq 0$. The above equation is called elliptic curve defined over \mathbb{Q},

which is not for the curve denoting ellipse, but for it is related with the perimeter of an ellipse.

Although there is no idea of elliptic curve in Fermat's times, but he was the first one to seriously consider elliptic curves. In fact, to understand the integral and rational points on elliptic curves is one of the most interesting problems to Fermat. By Siegel's theorem, there are only finite many integral points on elliptic curve. In general, the number of rational points on elliptic curve is finite or infinite, but Mordell–Weil theorem tells us: the group of all rational points on elliptic curve is finite generated. In the margin of Diophantus' *Arithmetica*, Fermat asserted and proved that the elliptic curve E

$$y^2 = x^3 - x \tag{6.8}$$

has only three rational points, i.e., (0,0), (1,0) and $(-1,0)$.

The proof of Fermat used the method of descent: For rational number $a = \frac{m}{n}$, $(m, n) = 1$, define $H(a) = \max(|m|, |n|)$ be the height of a. Assume that (6.8) has rational solutions except (0,0), (1,0) and $(-1,0)$, select the one with the least height, denoted by (x_0, y_0). The idea of the proof is to construct a new rational solution such that $H(x_1, y_1) < H(x_0, y_0)$. Then there is a contradiction.

$$y^2 = x^3 - x$$

The elliptic curve

The result about the rational points on (6.8) shows that Fermat's Last Theorem holds for $n = 4$. In fact, if the equation

$$x^4 + y^4 = z^4$$

has integral solution (x, y, z), move y^4 to the right-hand side and multiple by $\frac{z^2}{y^6}$ on both sides, we get

$$\left(\frac{x^2 z}{y^3}\right)^2 = \left(\frac{z^2}{y^2}\right)^3 - \frac{z^2}{y^2}.$$

Then there exists a rational solution with $y \neq 0$ satisfying (6.8), a contradiction!

Elliptic curve is the algebraic curve with genus 1, this is the reason it has infinitely many rational points. The famous Mordell conjecture (1922) says that any algebraic curve with genus great than 1 over rational filed (including Fermat curve) has only finitely many rational points. In 1983, the German mathematician Gerd Faltings (1954–) proved this conjecture and generalized it to arbitrary field. He was awarded the Fields Medal in 1986.

Elliptic curve has rank which is an important invariant of it. Let $E(\mathbb{Q})$ denote all the rational points on elliptic curve E and the point at infinity. We can introduce an operator on $E(\mathbb{Q})$ such that it is an Abelian group (the point at infinity is zero in $E(\mathbb{Q})$). In 1922, Mordell proved that this group is finite generated, then we have

$$E(\mathbb{Q}) = E(\mathbb{Q})_f \bigoplus E(\mathbb{Q})_t$$

where $E(\mathbb{Q})_f$ is a free group and $E(\mathbb{Q})_t$ is a finite subgroup consti-tuted by points with finite order. The number of generators of $E(\mathbb{Q})_f$ is called the rank of the elliptic curve, denoted by $r(E)$. $r(E)$ is a non-negative integers and $r(E)=0$ if and only if E has only finitely many rational points.

In 1955, Japanese mathematicians Yutaka Taniyama (1927–1958) and Goro Shimura (1930–2019) raised the famous Taniyama–Shimura conjecture: Each rational elliptic curve is a modular curve. If Fermat's theorem does not hold, there are prime numbers p and positive integer arrays (a, b, c) such that $a^p + b^p = c^p$. In 1986, the American mathematician Kenneth Ribet proved the Serre's con-jecture which was raised 10 years earlier: the Frey's elliptic curve $y^2 = x(x - a^p)(x + b^p)$ is not a modular curve. This conclusion is inconsistent with Taniyama–Shimura conjecture. In other words, what Ribet proved is that Fermat's theorem can be deduced from

Taniyama–Shimura conjecture. In 1995, the British mathematician Andrew Wiles (1953–) proved the Taniyama–Shimura conjecture, therefore he proved the Fermat's Last Theorem.

At last, we introduce an application of the elliptic curve in the Diophantine equations. In 1996, Polish mathematician Andrzej Schinzel (1937–) showed that the Diophantine equation

$$x + y + z = xyz = 6$$

has infinitely many positive rational solutions. In 2011, we (Yong Zhang and Tianxin Cai, *Math. Comput.*, **81** (2013) 617–623) extended this result and proved the following theorem.

Theorem. *For any positive integer $n \geq 3$, the Diophantine equation*

$$x_1 + x_2 + \cdots + x_n = x_1 x_2 \ldots x_n = 2n \tag{A}$$

has infinitely many positive rational solutions.

The idea of the proof is like this: taking $x_1 = x_2 = \cdots = x_{n-3} = 1$, (A) reduces to

$$\begin{cases} x_{n-2} + x_{n-1} + x_n = n - 3, \\ x_{n-2} x_{n-1} x_n = 2n. \end{cases}$$

Eliminating x_n, putting $u = \frac{x_{n-2}}{x_{n-1}}, v = \frac{1}{x_{n-1}}$, we have

$$u^2 + u - (n+3)uv + 2nv^3 = 0.$$

Let $x = -72nv + 3(n+3)^2, y = 216n(2u + 1 - (n+3)v)$, we get a family of elliptic curves (the following figure for $n = 4$)

$$E_n : y^2 = x^3 + a_n x + b_n,$$

where

$$a_n = -27(n+3)(n^3 + 9n^2 - 21n + 27),$$
$$b_n = 54n^6 + 972n^5 + 3402n^4 - 5832n^3 + 7290n^2 - 26244n + 39366.$$

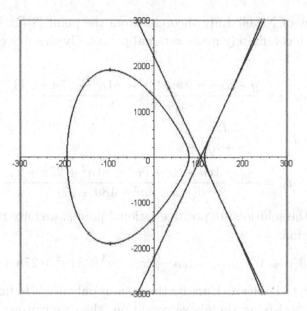

The elliptic curve E_4

It is easy to see that $R = (3(n+3)^2 - 72n, 216n(n-2))$ lies on E_n, by the group law, we have

$$[2]R = \left(\frac{3(n^4 + 2n^3 + 13n^2 - 36n + 36)}{(n-2)^2}, -\frac{216(2n^3 - 6n^2 + 7n - 2)}{(n-2)^3} \right),$$

$$[3]R = \left(\frac{3(n^6 - 36n^4 + 126n^3 - 180n^2 + 108n - 15)}{(n^2 - 3n + 3)^2}, \right.$$
$$\left. \frac{108(n-1)(n-2)(7n^4 - 33n^3 + 67n^2 - 66n + 28)}{(n^2 - 3n + 3)^3} \right).$$

The proof also needs the following two important theorems.

Theorem (Nagell–Lutz). *Let Δ be the discriminant of elliptic curve. Then any rational point of finite order on it satisfies that x and y are integers; and either $y = 0$ or else $y \,|\, \Delta$.*

Theorem (Poincaré–Hurwitz). *If the elliptic curve has infinitely many rational points, then it has infinitely many rational points in every neighborhood of any one of them.*

In view of Nagell–Lutz theorem, from the point $[3]R$, the elliptic curve E_n has infinitely many rational points. By the above transformation, we have

$$x_{n-2} = \frac{y - 3xn - 9x + 9n^3 + 81n^2 + 27n + 243}{6(-x + 3n^2 + 18n + 27)},$$

$$x_{n-1} = \frac{72n}{3(n+3)^2 - x},$$

$$x_n = \frac{-y - 3xn - 9x + 9n^3 + 81n^2 + 27n + 243}{6(-x + 3n^2 + 18n + 27)}.$$

Since all the solution are positive rational points, we have the following condition:

$$x < 3(n+3)^2, |y| < -3xn - 9x + 9n^3 + 81n^2 + 27n + 243.$$

According to Poincaré–Hurwitz theorem, it only needs to find a rational point satisfying the above condition, then we can get infinitely many rational points. By some calculations, the point $[3]R$ satisfies this condition. Hence, there are infinitely many positive rational solutions satisfying (A).

In our proof the solutions are constructive. For example, (A) has three solutions

$$(1, 4, 2, 1), \left(1, \frac{49}{20}, \frac{128}{35}, \frac{25}{28}\right), \left(1, \frac{103058}{24497}, \frac{34969}{29737}, \frac{68644}{42449}\right).$$

Remark. When $x_1, x_2, \ldots, x_{n-3}$ take positive rational numbers, we can turn (A) into

$$\begin{cases} x_1 + x_2 + \cdots + x_n = a, \\ x_1 x_2 \cdots x_n = b \end{cases}$$

and study it, where a, b are fixed rational numbers, by similar method we can get more general result.

Elliptic curve is one of the most important research objects in algebraic geometry, and it plays important role in the proof of Fermat's Last Theorem, which was proved by verified the Taniyama–Shimura conjecture (1957) named after two Japanese mathematicians. This conjecture says that any elliptic curve over

rational field is modular. The French mathematician Andre Weil (1906–1998) proposed this conjecture again in 1967 and generalized it, so this is also called Taniyama–Shimura–Weil conjecture. It's worth to point out that his sister Simone Weil (1909–1943) was a French philosopher, Christian mystic, and political activist, and she had influence in China.

Besides the above conjecture, there is a famous conjecture on elliptic curve, i.e., BSD conjecture. This was raised by two English mathematicians Birch and Swinnerton-Dyer between 1963–1965 with the help of an old computer in Cambridge University. BSD conjecture is one of the Millennium problems of the Clay Mathematics Institute. Let $L(E, s)$ be the L-function on elliptic curve E, $R(E)$ denotes the order of zeros at $s = 1$, the first part of BSD conjecture says: $r(E) = R(E)$. In 1977, Coates and Wiles proved that if $r(E) \geq 1$, then $R(E) \geq 1$. In 1986, Gross and Zagier showed that if $R(E) = 1$, then $r(E) \geq 1$.

28. Lehmer's Congruence

In this section, we first give a congruence of Eisenstein, then we show an elegant congruence established by E. Lehmer.

For any coprime positive integers r, n, Fermat's quotient is defined as follows:

$$q_r(n) = \frac{r^{\phi(n)} - 1}{n}.$$

For any given positive integer r, $q_r(n)$ is a completely multiplicative function because

$$(ab)^{p-1} = a^{p-1}b^{p-1} = (1 + pq_r(a))(1 + pq_r(b))$$

$$\equiv 1 + p(q_r(a) + q_r(b))(\bmod p^2).$$

In 1850, Eisenstein showed that for any odd prime

$$\sum_{i=1}^{(p-1)/2} \frac{1}{i} \equiv -2q_2(p)(\bmod p). \tag{6.9}$$

Portraits of E. Lehmer, she lived to 100 years of age

In 1938, the Russian-born American female mathematician E. Lehmer (1906–2007, her husband D. H. Lehmer and father-in-law D. N. Lehmer formed a rare number theory family) improved Eisenstein's congruence by using some properties of Bernoulli polynomials. She proved the following theorem.

Theorem 6.4 (E. Lehmer). *For any odd prime* p,

$$\sum_{i=1}^{(p-1)/2} \frac{1}{i} \equiv -2q_2(p) + pq_2^2(p) \pmod{p^2}. \tag{6.10}$$

We need two lemmas to prove (9.6) and 6.10.

Lemma 6.1. *If* $1 \le a \le p-1$, *then* $\frac{a^p-a}{p} \equiv \sum_{j=1}^{p-1}[\frac{ja}{p}]\frac{1}{j} \pmod{p}$.

Proof. Let $a\bar{a} \equiv 1 \pmod{p}$, Then

$$(a + pt)^{p-1} \equiv a^{p-1} + p(p-1)ta^{p-2} \equiv 1 + pq_r(a) + p(p-1)ta^{p-1}\bar{a}$$

$$\equiv 1 + p_rq(a) + p(p-1)t\bar{a}(1 + pq_r(a))$$

$$\equiv 1 + p(q_r(a) - t\bar{a}) \pmod{p^2},$$

so $q_r(a + pt) \equiv q_r(a) - \bar{a}t(\bmod\, p)$. Let $aj = pq_j + r_j, 1 \le r_j \le p - 1$, then

$$q_r(aj) = q_r(r_j + pq_j) \equiv q_r(r_j) - \bar{r}_j q_j (\bmod\, p).$$

By the multiplicative property of $q_r(a)$, we have

$$q_r(a) \equiv \sum_{j=1}^{p-1} q_r(j) - \sum_{j=1}^{p-1} q_r(aj) \,(\bmod\, p).$$

When j runs through the reduced residue classes modulo p, so does r_j. Combining the above two congruences, we obtain

$$q_r(a) \equiv \sum_{j=1}^{p-1} \frac{q_j}{r_j} \equiv \sum_{j=1}^{p-1} \left[\frac{ja}{p}\right] \frac{1}{ja} (\bmod\, p).$$

This completes the proof of Lemma 6.1. $\qquad\square$

In particular, when $a = 2$

$$\frac{2^p - 2}{p} \equiv \sum_{p/2 < j < p} \frac{1}{j} \,(\bmod\, p).$$

By Wolstenholme's theorem, we have

$$\sum_{j=1}^{(p-1)/2} \frac{1}{j} \equiv -\frac{2^p - 2}{p} \equiv -2q_2(p) \,(\bmod\, p).$$

Lemma 6.2. *Let m and k be positive integers, $p > m$ is prime. Then*

$$\sum_{r=1}^{[p/m]} r^{2k-1} \equiv \frac{1}{2k} \left\{ B_{2k}\left(\frac{s}{m}\right) - B_{2k} \right\} - \frac{p}{m} B_{2k-1}\left(\frac{s}{m}\right) (\bmod\, p^3),$$

where $\{x\}$ is the decimal part of x, s is the least positive residue of p modulo m.

Proof. In Theorem 6.2, take $x = \frac{p-rm}{m}, r = 1, 2, \ldots, [\frac{p}{m}]$. Adding and simplifying, then

$$\sum_{r=0}^{[p/m]} (p - rm)^n = \frac{m^n}{n+1} \left\{ B_{n+1}\left(\frac{p}{m}\right) - B_{n+1}\left(\frac{s}{m}\right) \right\}. \tag{6.11}$$

In Theorem 6.1, take $n = 2k$, $x = \frac{p}{m}$, p is odd prime greater than m, then

$$B_{2k}\left(\frac{p}{m}\right) \equiv B_{2k} + \frac{p^2}{m^2}\binom{2k}{2}B_{2k-2}(\text{mod } p^3), \tag{6.12}$$

since $B_{2k-1} = B_{2k-3} = 0$, other terms are multiples of the product of p^4 and Bernoulli numbers, by von Staudt–Clausen theorem, this terms has a factor p^3. Noting that the denominator of B_{2k-2} has a factor p if and only if $2k - 2 \equiv 0 \pmod{p - 1}$, so

$$B_{2k}\left(\frac{p}{m}\right) \equiv B_{2k}(\text{mod } p^2) \text{ if } 2k - 2 \not\equiv 0 \pmod{p - 1}. \tag{6.13}$$

Similarly, in Theorem 6.1, take $n = 2k + 1, x = \frac{p}{m}$, then

$$B_{2k+1}\left(\frac{p}{m}\right) \equiv p\frac{2k+1}{m}B_{2k} \pmod{p^3} \quad \text{if } 2k - 2 \not\equiv 0 \pmod{p - 1}. \tag{6.14}$$

Combining (6.13), (6.14) with (6.11), we have

$$\sum_{r=1}^{[p/m]} (p - rm)^{2k-1} \equiv \frac{m^{2k-1}}{2k}\left\{ B_{2k} - B_{2k}\left(\frac{s}{m}\right) \right\} \pmod{p^2}, \tag{6.15}$$

$$\sum_{r=1}^{[p/m]} (p - rm)^{2k} \equiv \frac{m^{2k}}{2k+1}\left\{ \frac{2k+1}{m}pB_{2k} - B_{2k+1}\left(\frac{s}{m}\right) \right\} \pmod{p^3}, \tag{6.16}$$

where s is the least positive residue of p modulo m, $2k - 2 \not\equiv 0 \pmod{p - 1}$.

Noting that

$$\sum_{r=1}^{[p/m]} (p-rm)^n \equiv (-1)^n \left\{ m^n \sum_{r=1}^{[p/m]} r^n - pnm^{n-1} \sum_{r=1}^{[p/m]} r^{n-1} \right\} \pmod{p^2}.$$

$$(6.17)$$

Combining (6.15), (6.16) and (6.17), we complete the proof of Lemma 6.2. □

Combining (6.12) but not (6.13) with (6.11), and take $m = 1$ in (6.16), we have the following corollary.

Corollary. *If p is odd prime, k is positive integer and $2k - 2 \not\equiv 0 \pmod{p-1}$, then*

$$\sum_{r=1}^{p-1} r^{2k+1} \equiv p^2 \frac{2k+1}{2} B_{2k} \pmod{p^3},$$

$$\sum_{r=1}^{p-1} r^{2k} \equiv p B_{2k} \pmod{p^3}.$$

Proof of Theorem 6.4. Let $(a,p) = 1$, $q_a(p) = \frac{a^{p-1}-1}{p}$ is Fermat's quotient, Wilson's quotient is defined as

$$w_p = \frac{(p-1)! + 1}{p},$$

they are positive integers and $\sum_{a=1}^{p-1} q_a \equiv w_p \pmod{p}$. Using this congruence, and add $p-1$ equations $r^{p-1} = 1 + pq_r(p)$, where r from 1 to $p-1$, we obtain

$$p - 1 + pw_p \equiv pB_{p-1} \pmod{p^2},$$

hence

$$\frac{pB_{p-1}}{p-1} \equiv 1 + \frac{pw_p}{p-1} \equiv 1 - pw_p \pmod{p^2}. \qquad (6.18)$$

By Lemma 6.2 and the values of Bernoulli polynomials at $x = \frac{1}{2}$, i.e.,

$$B_{2k}\left(\frac{1}{2}\right) = (1 - 2^{2k-1})\frac{B_{2k}}{2^{2k-1}}, \quad B_{2k-1}\left(\frac{1}{2}\right) = 0,$$

we have

$$\sum_{r=1}^{(p-1)/2} r^{2k-1} \equiv (1-2^{2k})\frac{B_{2k}}{2^{2k}k} \pmod{p^2}, \quad \text{if } 2k-2 \not\equiv 0 \pmod{p-1}.$$

Take $2k = p - 1$, by (6.16),

$$\sum_{r=1}^{(p-1)/2} r^{p-2} \equiv -2q_2(p)(1 - pw_p) + 2pq_2^2(p) \pmod{p^2}. \tag{6.19}$$

By (6.18) and Eisenstein's congruence, we have

$$\sum_{r=1}^{(p-1)/2} \frac{q_r p}{r} \equiv 2q_2(p)w_p + q_2^2(p) \pmod{p}. \tag{6.20}$$

By (6.18) $-p \times 6.19$, we obtain (6.10). This completes the proof of Theorem 6.4.

Using the similar method, Lehmer also got the other four congruences. ☐

Theorem 6.5 (Lehmer). *If p is odd prime, then*

$$\sum_{r=1}^{(p-1)/2} \frac{1}{p-2r} \equiv q_2(p) - \frac{pq_2^2(p)}{2} \pmod{p^2}, \quad p \geq 3,$$

$$\sum_{r=1}^{[p/3]} \frac{1}{p-3r} \equiv \frac{q_3(p)}{2} - \frac{pq_3^2(p)}{4} \pmod{p^2}, \quad p \geq 5,$$

$$\sum_{r=1}^{[p/4]} \frac{1}{p-4r} \equiv \frac{3q_2}{4} - \frac{3pq_2^2}{8} \pmod{p^2}, \quad p \geq 5,$$

$$\sum_{r=1}^{[p/6]} \frac{1}{p-6r} \equiv \frac{q_3(p)}{4} + \frac{q_2(p)}{3} - p\left(\frac{q_3^2(p)}{8} + \frac{q_2^2(p)}{6}\right)$$
$$\pmod{p^2}, p \geq 7.$$

Using these congruences, Lehmer obtained four criteria for the first case of Fermat's Last Theorem. That is, if the first case of

Fermat's Last Theorem is true for prime exponent p, then for $n = 2, 3, 4, 6$, we have

$$\sum_{r=1}^{[p/n]} \frac{1}{r} \equiv 0 \;(\mathrm{mod}\; p).$$

It must be pointed out that before Wiles announced his proof, propulsion the proof of Fermat's Last Theorem (increasing the range of the exponent) relies heavily on this class of congruence.

It is well known that, Euler's theorem generalizes Fermat's Little Theorem, Gauss's theorem generalizes Wilson's theorem, they both generalize a congruence from modulo prime number to modulo an arbitrary positive integer. The former ones were separated more than a hundred years, while the latter ones separated by 30 years. Today there are many congruences modulo prime numbers, but congruences modulo integral powers are very few. Like the Irish mathematician Cosgrave and Canadian mathematician Dilcher points out, our result is the first time since 1906.

Firs of all, the author (*Acta Arith.* **103**, (2002), 313–320) generalizes Lehmer's congruence to a congruence modulo arbitrary positive integer. Let $\chi_n(r)$ denote the trivial Dirichlet character modulo n, i.e.,

$$\chi_n(r) = \begin{cases} 1, & \text{if } (r, n) = 1, \\ 0, & \text{if } (r, n) > 1. \end{cases}$$

We obtain the following theorem.

Theorem 6.6. *For odd $n>1$, we have*

$$\sum_{r=1}^{(n-1)/2} \frac{\chi_n(r)}{r} \equiv -2q_2(n) + n q_2^2(n) \;(\mathrm{mod}\; n^2). \tag{6.21}$$

When n is prime, (6.21) is Lehmer's congruence.

Elementary methods can be used to prove Theorem 6.6, but it is rather complicated and lengthy. Here, we intend to use a known identity, which is associated with estimates of character sums, to give a simple proof. First, we introduce the generalized Bernoulli

numbers and Bernoulli polynomials. Let χ denote the Dirichlet character modulo n, the generalized Bernoulli number is the complex number $B_{k,\chi}$ satisfying the following identity:

$$\sum_{a=1}^{n} \chi(a)\frac{ze^{az}}{e^{nz}-1} = \sum_{k=1}^{\infty} B_{k,\chi}\frac{z^k}{k!}.$$

Also, the generalized Bernoulli polynomial is defined as follows:

$$B_{n,\chi}(x) = \sum_{k=0}^{n} \binom{n}{k} B_{k,\chi} x^{n-k}.$$

We have

$$\sum_{n=0}^{N-1} \chi(n)n^m = \frac{1}{m+1}(B_{m+1,\chi}(N) - B_{m+1,\chi}).$$

If the character χ modulo M is derived from character χ_1 modulo a factor of M, then

$$B_{s,\chi} = B_{s,\chi_1} \prod_{p|M}(1 - \chi_1(p)p^{s-1}). \tag{6.22}$$

We need the following result (*Acta. Arith.* **71** (1995) 273–278).

Lemma 6.3 (Szmidt–Urbanowicz–Zagier). *Let χ be Dirichlet character modulo M, N is a multiple of M, $r > 1$ is an integer coprime to N, then for any non-negative integer m, we have*

$$(m+1)r^m \sum_{0<n<N/r} \chi(n)n^m$$

$$= -B_{m+1,\chi}r^m + \frac{\overline{\chi}(r)}{\phi(r)}\sum_{\psi}\overline{\psi}(-N)B_{m+1,\chi\psi}(N),$$

where ψ runs all Dirichlet characters modulo r.

German mathematician Zagier

Proof of Theorem 6.6. Let

$$S_r(n) = \sum_{0 < i < n/r} \chi_n(i) i^{n\phi(n)-1},$$

by Euler's theorem, it is easy to know

$$\sum_{i=1}^{(n-1)/2} \frac{\chi_n(i)}{i} \equiv S_2(n) (\text{mod } n^2). \tag{6.23}$$

In Lemma 6.3, take $N = M = n, m = n\phi(n) - 1, \chi = \chi_n, r = 2$, then there is only one character $\chi_{0,2}$ modulo r, by (6.22) we have

$$S_2(n) = -\frac{B_{n\phi(n)}}{n\phi(n)} \prod_{p|n} (1 - p^{n\phi(n)-1}) + \frac{2^{1-n\phi(n)}}{n\phi(n)} B_{n\phi(n), \chi_{0,2n}}(n).$$

$$\tag{6.24}$$

Noting that i is odd, then $B_{n\phi(n)-i} = 0$ except $n\phi(n) - i = 1$, so $i \geq 3$. Hence

$$\frac{2^{1-n\phi(n)}}{n\phi(n)} \sum_{i=1}^{n\phi(n)} \binom{n\phi(n)}{i} n^i B_{n\phi(n)-i} \prod_{p|2n} (1 - p^{n\phi(n)-i-1}) \equiv 0 \ (\text{mod } n^2).$$

$$\tag{6.25}$$

Appling 6.23, 6.25 and von Staudt–Clausen theorem to (6.24), we obtain

$$S_2(n) \equiv \frac{(1 - 2^{n\phi(n)})}{n\phi(n)} B_{n\phi(n)} \prod_{p|n}(1 - p^{n\phi(n)-1}) \pmod{n^2}. \quad (6.26)$$

Finally, from $pB_{n\phi(n)} \equiv p - 1 \pmod{p^r}, p^r||n$ we know that

$$\frac{n}{\phi(n)} B_{n\phi(n)} \prod_{p|n}(1 - p^{n\phi(n)-1}) \equiv 1 \pmod{n^2}. \quad (6.27)$$

By the definition $2^{\phi(n)} = nq_2(n) + 1$, combining (6.23), (6.26) with (6.27), it deduce (6.21). We complete the proof of Theorem 6.6. $\quad\square$

Similar methods may be used for other generalization. For example, in 2007, we (Tianxin Cai, Xudan Fu and Xia Zhou, *Acta Arith.* **130**, 203–214) generalized the four congruences of Theorem 6.4, a congruence modulo a power of arbitrary positive integer.

Theorem 6.7. *Let n be odd. Then*

$$\sum_{r=1}^{(n-1)/2} \frac{\chi_n(r)}{n - 2r} \equiv q_2(n) - \frac{1}{2}nq_2^2(n) \pmod{n^2}, \quad \text{if } (3, n) = 1,$$

$$\sum_{r=1}^{[n/3]} \frac{\chi_n(r)}{n - 3r} \equiv \frac{1}{2}q_3(n) - \frac{1}{4}nq_3^2(n) \pmod{n^2}, \quad \text{if } (3, n) = 1,$$

$$\sum_{r=1}^{[n/4]} \frac{\chi_n(r)}{n - 4r} \equiv \frac{3}{4}q_2(n) - \frac{3}{8}nq_2^2(n) \pmod{n^2}, \quad \text{if } (3, n) = 1,$$

$$\sum_{r=1}^{[n/6]} \frac{\chi_n(r)}{n - 6r} \equiv \frac{1}{3}q_2(n) + \frac{1}{4}q_3(n)$$

$$- \frac{1}{6}nq_2^2(n) - \frac{1}{8}nq_3^2(n) \pmod{n^2}, \quad \text{if } (15, n) = 1.$$

Remark. In 2009, Hui-Qin Cao and Hao Pan (*J. Number Theory* **129**, 1813–1819) also got the last three congruences. However, some recent work has shown that the fourth congruence is true for $(3,n)=1$; the first and third congruences are also true for $3|n$ if and only if n

has a factor of form $6k + 1$, otherwise modulo n^2 must be changed to modulo $n^2/3$.

It is worth mentioning that the edge number of regular polygon which fills up the plane also happens to be 3, 4 or 6. (In Section 2, the example of Mosaic geometry.) This indicates that, in the geometry and number theory, there are some things in common.

The abc conjecture

After Fermat's Last Theorem was proved, number theory seems to lack a driving force, which has been described as a golden chicken be slaughtered. Fortunately, in the recent over 20 years, there is a conjecture whose importance is clearly increasingly, this is abc conjecture. It is comparable to BSD conjecture, which is mentioned in the previous section, they are two wonderful flowers in the field of number theory.

For a positive integer n, the radical of n, denoted $\text{rad}(n)$, is the product of the distinct prime factors of n. For example, if $n = 12 = 2^2 \times 3$, then $r(n) = 6$.

In 1985, the French mathematician Joseph Oesterlé (1954–) and the British mathematician David Masser (1948–) independently proposed the following conjecture.

French mathematician Oesterlé

English mathematician Masser

abc Conjecture I. *For every ε > 0, there exist only finitely many triples (a, b, c) of positive coprime integers, with a + b = c, such that*

$$c > \operatorname{rad}(abc)^{1+c}.$$

An equivalent form states the following conjecture.

abc Conjecture II. *For every ε > 0, there exists a constant K_ε such that for all triples (a, b, c) of coprime positive integers, with a + b = c, the inequality*

$$c < K_\varepsilon \cdot \operatorname{rad}(abc)^{1+c} \tag{A}$$

holds.

From the form (A), abc conjecture seems the inherent property of addition and multiplication in the domain of Z under certain conditions, or the inherent nature of the integers. For $1 \le a \le b \le 50$, checked by the computer, only four triples dissatisfy $c \le \operatorname{rad}(abc)$, i.e., $\{1, 8, 9; 6\}, \{5, 27, 32; 30\}, \{1, 48, 49; 42\}, \{32, 49, 81; 42\}$, where the last number is the radical of abc.

A weak form of abc conjecture is replace the left-hand side of inequality (A) by $(abc)^{\frac{1}{3}}$. Another weak form is more concise, that is,

for all triples (a, b, c), with $a + b = c, (a, b) = 1$, we have

$$c \le (\operatorname{rad}(abc))^2. \tag{B}$$

Let $q(a, b, c) = \frac{\log c}{\log(\operatorname{rad}(abc))}$, we have another form of abc conjecture.

abc Conjecture III. For every $\varepsilon > 0$, there exist only finitely many triples $(a, \ b, \ c)$ of coprime positive integers with $a + b = c$ such that

$$q(a, b, c) > 1 + \varepsilon. \tag{C}$$

abc conjecture or its weak form can lead to solve a number of important conjectures in number theory. At the same time, some well-known theorems and results are consequence of abc conjecture or its weak form. The latter include Roth theorem (Fields prize in 1958), Baker's theorem (Fields prize in 1970), Mordell conjecture (Fields prize in 1974), Fermat's Last Theorem (Fields prize in 1998). Take Fermat's Last Theorem as an example: If there exist positive integers x, y, z for some $n \ge 3$ such that

$$x^n + y^n = z^n.$$

Take $a = x^n, b = y^n, c = z^n$, by the weak form of abc conjecture (B),

$$z^n < (\operatorname{rad}(xyz)^n)^2 < (xyz)^2 < z^6.$$

So $n = 3, 4$ or 5, which can be ruled out by elementary methods.

If abc conjecture is true, we can also prove the following Fermat–Catalan conjecture, which is to the stronger form of Fermat's Last Theorem, proposed by the French–Canadian mathematician Henri Darmon and the British mathematician Andrew Granville (*Bull. London Math. Soc.* **27**, (1995), 513–543).

Fermat–Catalan Conjecture. There are only finitely many triples $\{x^p, y^q, z^r\}$ satisfying the equation

$$x^p + y^q = z^r, \tag{F-C}$$

where positive integers p, q, r satisfy $\frac{1}{p} + \frac{1}{q} + \frac{1}{r} < 1$.

So far, people only found 10 groups (kind of) solutions, i.e.,
$1 + 2^3 = 3^2$, $2^5 + 7^2 = 3^4$, $13^2 + 7^3 = 2^9$, $2^7 + 17^3 = 71^2$, $3^5 + 11^4 = 122^2$, $33^8 + 1549034^2 = 15613^3$, $1414^3 + 2213459^2 = 65^7$, $9262^3 + 15312283^2 = 113^7$, $17^7 + 76271^3 = 21063928^2$, $43^8 + 96222^3 = 30042907^2$. It is worth noting that, each set of solutions contain exactly one square.

After Gerd Faltings proved Mordell conjecture, we know that for any given p, q, r, (F–C) has only finitely many solutions.

If abc conjecture is true, we can deduce that (F–C) has only finitely many solutions, i.e., Fermat–Catalan conjecture is true. In fact, we may assume that $p \le q \le r$, if $r \ge 8$, then $\frac{1}{p} + \frac{1}{q} + \frac{1}{r} \le \frac{1}{2} + \frac{1}{3} + \frac{1}{8} = \frac{23}{24}$; if $r \le 7$, then $\frac{1}{p} + \frac{1}{q} + \frac{1}{r}$ has only finitely many values, the largest one is $\frac{1}{2} + \frac{1}{3} + \frac{1}{7} = \frac{41}{42} > \frac{23}{24}$. Therefore, $\frac{1}{p} + \frac{1}{q} + \frac{1}{r} \le \frac{41}{42}$ for any p, q, r. If there exist triples $\{a, b, c\}$ of coprime positive integers satisfying (F–C), by $x < z^{\frac{r}{p}}, y < z^{\frac{r}{q}}$, we have

$$q(x^p, y^q, z^r) = \frac{\log z^r}{\log(\mathrm{rad}(x^p y^q z^r))} = \frac{r \log z}{\log(\mathrm{rad}(xyz))} > \frac{r \log z}{\log(z^{r/p + q/p + 1})}$$

$$= \frac{r}{r/p + r/q + 1} = \frac{1}{1/p + 1/q + 1/r} \ge \frac{42}{41}.$$

By abc conjecture (C), (F–C) has only finitely many solutions.

We must point out that, Fermat–Catalan conjecture is a combination of Fermat's Last Theorem and the following so-called Catalan's conjecture.

Theorem (Catalan's conjecture). *The only case of two consecutive powers $\{x^m, y^n\}$ is 8 and 9, where m and n are any integer greater than one.*

This conjecture originated from the 14th century French mathematician Gersonides (1288–1344), who proved a special case of the conjecture in 1343 where (x, y) was restricted to be $(2, 3)$ or $(3, 2)$. It is said that Gersonides's father, Gerson ben Solomon Catalan is a Jewish writer, he is one of the earliest French mathematician. In 2002, the Romanian mathematician Mihăilescu (1955–) use the theory of cyclotomic fields and Galois modules proved Catalan's conjecture. Therefore, this conjecture is also called Mihăilescu's theorem.

Portrait of mathematician Catalan, the only son of a jeweler

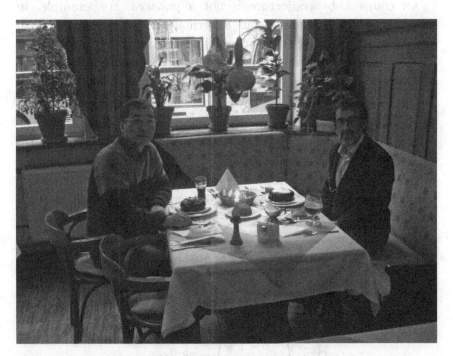

The author having lunch with Mihăilescu in Göttingen in 2010

It's interesting to mention, his proof made use of Wieferich pair of primes.

We consider the weak form of Catalan's equation

$$a - b = 1, \ a \text{ and } b \text{ are square-full.}$$

Here, square-full number is a positive integer m such that for every prime number p dividing m, p^2 also divides m. Obviously, the above equation is equivalent to

$$a - b = 1, \ ab \text{ are square-full.}$$

By using the properties of Pell's equation we might obtain that there are infinitely many solutions to the above equation, for example, (8, 9), (288, 289), (675, 676), etc. However, we have found and raise the following conjecture by calculating.

For any positive integer s, there is only one solution $(a, b) = (2^{s+1}, 2^s)$ satisfies $a - b = 2^s$, where ab is cube-full.

Of course, abc conjecture is not a panacea. For example, in 1988, American mathematician Silverman (1955–) proved that, if abc conjecture is true, there are infinitely many primes p such that $2^{p-1} \not\equiv 1 \pmod{p^2}$. That is, there are infinitely many prime numbers of non-Wieferich prime, but whether the existence of infinitely many Wieferich primes, is still a mystery.

Remark. Several people announced the proof of abc conjecture, the most recent and famous one was in August 2012, the Japanese mathematician Schinichi Mochizuki (1969–) announced and released the proof of abc conjecture on the website of Kyoto University. Later two number theorists find gaps in it. Right now, Schinichi is still working on it, rejecting all interviews.

29. Morley's Theorem and Jacobstahl's Theorem

In 1892, the British mathematician G. B. Mathews (1861–1922) showed a congruence in his *A Course in Number Theory* , i.e., for any prime $p \geq 3$,

$$(-1)^{\frac{p-1}{2}} \binom{p-1}{(p-1)/2} \equiv 4^{p-1} \pmod{p^2}.$$

Three years later, the British-born American mathematician Frank Morley (1860–1937) obtained a more beautiful congruence, called Morley's theorem. Morley graduated from King's College, Cambridge, when he was 27-year-old he moved to the United States, trained more than 50 mathematics Ph.D. for the new world. It can be said that he made a great contribution to change the state of the American Mathematical obscurity. His professional and major achievements were in algebra and geometry, from the proof of Morley's theorem we can see his particular methods. It is worth mentioning that, Morley was the chairman of the American Mathematical Society, one of his sons became a mathematician, another son were the reporter of "Washington Post" who won the Pulitzer Prize (1936), the first time for Washington Post.

Theorem 6.8 (Morley). *If $p \geq 5$ is prime, then*

$$(-1)^{\frac{p-1}{2}} \binom{p-1}{(p-1)/2} \equiv 4^{p-1} \ (\text{mod } p^3).$$

English-American mathematician Morley

Proof. Considering the integration of trigonometric function $\cos^{2n+1} x$, by using the Fourier series expansion,

$$2^{2n} \cos^{2n+1} x = \cos(2n+1)x + (2n+1)\cos(2n-1)x$$
$$+ \frac{(2n+1)\cdot 2n}{1\cdot 2}\cos(2n-3)x + \cdots$$
$$+ \frac{(2n+1)2n...(n+2)}{n!}\cos x$$

Taking the integral from $x = 0$ to $x = \frac{\pi}{2}$ on both sides, we have

$$2^{2n}\frac{2n(2n-2)...2}{(2n+1)(2n-1)...3} = (-1)^n\left\{\frac{1}{2n+1} - \frac{2n+1}{2n-1} + \frac{(2n+1)2n}{1\cdot 2(2n-3)}\right.$$
$$\left. - \cdots + (-1)^n\frac{(2n+1)2n...(n+2)}{n!}\right\}.$$

Let $2n+1 = p$ be prime, the above equation turns to

$$\frac{1}{p}\left\{4^{p-1}\binom{p-1}{(p-1)/2} - (-1)^{\frac{p-1}{2}}\right\}$$
$$= (-1)^{\frac{p-1}{2}}p\left\{-\frac{1}{2n-1}\right.$$
$$\left. + \frac{2n}{1\cdot 2(2n-3)} - \cdots + (-1)^n\frac{2n...(n+2)}{n!}\right\}.$$

Denote the right-hand side of the summation in brackets by \sum, it's easy to see

$$\sum \equiv \frac{1}{1^2} + \frac{1}{3^2} + \cdots + \frac{1}{(p-2)^2}(\text{mod } p).$$

We only need to prove that the right-hand side of the above formula is divisible by p. In fact,

$$\frac{1}{1^2} - \frac{1}{(2n)^2} + \frac{1}{2^2} - \frac{1}{(2n-1)^2} + \cdots + \frac{1}{n^2} - \frac{1}{(n+1)^2}$$
$$= \frac{(2n)^2 - 1}{1^2(2n)^2} + \frac{(2n-1)^2 - 2^2}{2^2(2n-1)^2} + \cdots + \frac{(n+1)^2 - n^2}{n^2(n+1)^2} \equiv 0 \ (\text{mod } p).$$

By Wolstenholme's theorem,

$$\frac{1}{1^2} + \frac{1}{(2n)^2} + \frac{1}{2^2} + \frac{1}{(2n-1)^2} + \cdots + \frac{1}{n^2} + \frac{1}{(n+1)^2} \equiv 0 \pmod{p},$$

hence

$$\frac{1}{1^2} + \frac{1}{2^2} + \cdots + \frac{1}{n^2} \equiv \frac{1}{(n+1)^2} + \cdots + \frac{1}{(2n-1)^2} + \frac{1}{(2n)^2} \equiv 0 \pmod{p},$$

therefore

$$\frac{1}{2^2} + \frac{1}{4^2} + \cdots + \frac{1}{(2n)^2} \equiv 0 \pmod{p}.$$

By (6.6), we obtain $\sum \equiv 0 \pmod{p}$. We complete the proof of Theorem 6.8. $\qquad\square$

Similarly, we can prove, for any prime $p \geq 3$,

$$(-1)^{[\frac{p}{4}]} \binom{p-1}{[p/4]} \equiv 8^{p-1} \pmod{p^2}.$$

In 2002, the author (*Acta Arith.* **103** (2002), 313–320) generalize Morley's theorem to modulo positive integer by using Theorem 6.6.

Theorem 6.9. *If n is odd, then*

$$(-1)^{\phi(n)/2} \prod_{d|n} \left(\frac{d-1}{(d-1)/2} \right)^{\mu(n/d)} \equiv 4^{\phi(n)} \begin{cases} \pmod{n^3}, & if \ 3 \nmid n, \\ \pmod{n^3/3}, & if \ 3 \mid n, \end{cases}$$

where $\mu(n)$ is Möbius function and $\phi(n)$ is Euler function.

To prove Theorem 6.9, we need the following lemma.

Lemma 6.4. *Let $n>1$ be an integer. Then*

$$\sum_{i=1}^{n-1} \frac{\chi_n(i)}{i^2} \equiv 0 \begin{cases} \pmod{n}, & if \ 3 \nmid nn \neq 2^a, \\ \pmod{n/3}, & if \ 3 \mid n, \\ \pmod{n/2}, & if \ n = 2^a. \end{cases}$$

German mathematician von Staudt

Proof. By Euler's theorem, we have

$$\sum_{i=1}^{n-1} \frac{\chi_n(i)}{i^2} \equiv \sum_{i=1}^{n-1} \chi_n(i) i^{\phi(n)-2} \pmod{n}.$$

Denote the right-hand side of the above congruence by $T(n)$. From Lemma 6.1, we have

$$T(n) = \frac{1}{\phi(n-1)} \left(B_{\phi(n)-1,\chi_n}(n) - B_{\phi(n)-1,\chi_n} \right).$$

Meanwhile, by von Staudt–Clausen theorem, we obtain

$$T(n) = \frac{1}{\phi(n-1)} \sum_{k=1}^{\phi(n)-2} \binom{\phi(n)-1}{k} B_{\phi(n)-1,\chi_n} n^k$$

$$\equiv B_{\phi(n)-1-k,\chi_n} n \pmod{n}.$$

Since

$$B_{s,\chi_n} = \prod_{p|n} (1 - p^{s-1}) B_s,$$

again, by von Staudt–Clausen theorem, we see that $B_{\phi(n)-2,\chi_n}$ is a p-integral rational number for $p|n$ if $p - 1 \nmid \phi(n) - 2$. If $p|n$ we have $p - 1 | \phi(n)$ and so $p - 1 | 2$, i.e., $p = 2$ or 3. Hence Lemma 6.4 follows. $\qquad\square$

Proof of Theorem 6.9. Denote

$$A_n = \binom{n-1}{(n-1)/2}.$$

Then

$$A_n = \prod_{r=1}^{(n-1)/2} \frac{n-r}{r} = \prod_{d|n} \prod_{\substack{r=1 \\ (r,n)=d}}^{(n-1)/2} \frac{n-r}{r} = \prod_{d|n} T_{n/d} = \prod_{d|n} T_d,$$

where

$$T_d = \prod_{\substack{r=1 \\ (r,d)=1}}^{(d-1)/2} \frac{d-r}{r}.$$

By using the inverse formula for the Möbius function (Section 14), we have

$$T_n = \prod_{d|n} A_d^{\mu(n/d)} = \prod_{d|n} \binom{d-1}{(d-1)/2}^{\mu(n/d)}.$$

On the other hand,

$$T_n = \prod_{\substack{r=1 \\ (r,n)=1}}^{(n-1)/2} \frac{n-r}{r} = (-1)^{\phi(n)/2} \prod_{\substack{r=1 \\ (r,n)=1}}^{(n-1)/2} \left(1 - \frac{n}{r}\right)$$

$$= (-1)^{\phi(n)/2} \left\{ 1 - n \sum_{\substack{r=1 \\ (r,n)=1}}^{(n-1)/2} \frac{1}{r} \right. \tag{6.28}$$

$$\left. + \frac{n^2}{2} \left\{ \left(\sum_{\substack{r=1 \\ (r,n)=1}}^{(n-1)/2} \frac{1}{r} \right)^2 - \sum_{\substack{r=1 \\ (r,n)=1}}^{(n-1)/2} \frac{1}{r^2} \right\} \right\} \pmod{n^3}.$$

Noting that

$$\sum_{\substack{r=1 \\ (r,n)=1}}^{(n-1)/2} \frac{1}{r^2} = \frac{1}{2} \sum_{\substack{r=1 \\ (r,n)=1}}^{(n-1)/2} \left\{ \frac{1}{r^2} + \frac{1}{(n-r)^2} \right\} \equiv \sum_{\substack{r=1 \\ (r,n)=1}}^{n-1} \frac{1}{r^2} \pmod{n}. \tag{6.29}$$

Applying Theorem 6.6 and Lemma 6.4 to (6.28), we have

$$\prod_{d|n} \binom{d-1}{(d-1)/2}^{\mu(n/d)} \equiv (-1)^{\phi(n)/2}(1 + nq_2(n))^2$$

$$\equiv (-1)^{\phi(n)/2}4^{\phi(n)} \pmod{n^3}$$

for $3 \nmid n$; if $3|n$ the modulus must be replaced by $n^3/3$. This completes the proof of Theorem 6.9. $\qquad\square$

Corollary 6.3. *If $p \geq 5$ is prime, then and for any $l \geq 1$,*

$$(-1)^{(p-1)/2} \binom{p^l-1}{(p^l-1)/2} \bigg/ \binom{p^{l-1}-1}{(p^{l-1}-1)/2} \equiv 4^{\phi(p^l)} \pmod{p^{3l}},$$

$$(-1)^{(p-1)l/2} \binom{p^l-1}{(p^l-1)/2} \equiv 4^{p^l-1} \pmod{p^3}.$$

Meanwhile, the author also get a deep congruence.

Theorem 6.10. *If p and $q \geq 5$ are distinct odd primes, then*

$$\binom{pq-1}{(pq-1)/2} \equiv 4^{(p-1)(q-1)} \binom{p-1}{(p-1)/2}\binom{q-1}{(q-1)/2} \pmod{p^3q^3}.$$

$$(6.30)$$

This congruence of double primes form reminds us of Gauss' quadratic reciprocity law. It could be deduced from Theorem 6.6, in fact, we only need to deal with the case $p = 3, q > 3$. It takes two cases. If $q \equiv 2 \pmod 3$, then

$$\binom{q-1}{(q-1)/2} \equiv 0 \pmod 3,$$

since $[(q-1)/3] - 2[(q-1)/6] = 1$, hence (6.30) is true. If $q \equiv 1$ (mod 3), by using Fermat's Little Theorem, we have

$$\sum_{\substack{r=1 \\ (r,6q)=1}}^{3q-1} \frac{1}{r^3} \equiv \sum_{\substack{r=1 \\ (r,2q)=1}}^{3q-1} r \equiv \sum_{\substack{r=1 \\ 2\nmid r}}^{3q-1} r - q = \frac{(3q-1)^2}{4} - q \equiv 0 \pmod 3.$$

This improves Lemma 6.4 (the case $n = 3q$) to

$$\sum_{r=1}^{3q-1} \frac{\chi_{3q}(r)}{r^2} \equiv 0 \,(\text{mod } 3q).$$

Therefore, (6.30) follows immediately from the proof of Theorem 6.9. This completes the proof of Theorem 6.10.

From Theorem 6.10, the corollary of Theorem 3.16 follows immediately

Remark. In 2019, we (Tianxin Cai, Hao Zhong, Shane Chern, A congruence involving the quotient of Euler and its application III, *Acta Math. Sinica* **62**(4), 529–540) generalize Theorem 6.8 and its note, Theorem 6.9 and Theorem 6.10. For example, for any positive integer and any positive odd n,

$$\prod_{d|n} \left(\frac{kd-1}{(d-1)/2} \right)^{\mu(\frac{n}{d})} \equiv (-1)^{\phi(n)/2} 4^{k\phi(n)} \begin{cases} (\text{mod } n^3), & \text{if } 3 \nmid n, \\ (\text{mod } n^3/3), & \text{if } 3 \mid n. \end{cases}$$

For any positive integer n and two distinct odd primes p, q,

$$\left(\frac{kpq-1}{(pq-1)/2} \right) \equiv 4^{k(p-1)(q-1)} \left(\frac{kp-1}{\frac{p-1}{2}} \right) \left(\frac{kq-1}{\frac{q-1}{2}} \right) (\text{mod } p^3 q^3)$$

In 1952, the German mathematician Jacobstahl (1882–1965) proved the following theorem.

Theorem 6.11 (Jacobstahl). *If $p \geq 5$ is prime, then*

$$\left(\frac{up}{vp} \right) \Big/ \left(\frac{u}{v} \right) \equiv 1 \,(\text{mod } p^3),$$

where $u > v > 0$.

The results improved Wolstenholme theorem. In fact, Jacobstahl got stronger results: If $q(q \geq 3)$ is the maximum power of p which divisible $p^3 ab(a-b)$, then the modulus in Theorem 6.11 can be replaced by p^q.

Jacobstahl also defined the so-called Jacobstahl number, it is a variable of Fibonacci number,

$$J_0 = 0, J_1 = 1, J_n = J_{n-1} + 2J_{n-2}.$$

Then, J_n is $0, 1, 1, 3, 5, 11, 21, 43, 85, 171, 341, \ldots$.

Now, we want to generalize Jacobstahl's theorem to modulo positive integer by using Theorem 6.6.

Theorem 6.12. *Let $n \geq 1$ be an integer, then for any integers $u > v > 0$,*

$$\prod_{d|n} \binom{ud}{vd}^{\mu(n/d)} \equiv 1 \begin{cases} (\bmod\ n^3), & \text{if } 3 \nmid n, n \neq 2^a, \\ (\bmod\ n^3/3), & \text{if } 3 \mid n, \\ (\bmod\ n^3/2), & \text{if } n = 2^a, a \geq 2, \\ (\bmod\ n^3/4), & \text{if } n = 2. \end{cases}$$

Proof. We follow the proof of Theorem 6.10. Denote

$$D_n = \binom{un}{vn},$$

then

$$D_n = \frac{u}{v}\binom{un-1}{vn-1} = \frac{u}{v}\prod_{d|n}\prod_{\substack{r=1 \\ (r,n)=d}}^{vn-1} \frac{un-r}{r} = \frac{u}{v}\prod_{d|n} V_{n/d} = \frac{u}{v}\prod_{d|n} V_d,$$

where

$$V_d = \prod_{\substack{r=1 \\ (r,d)=1}}^{vd-1} \frac{ud-r}{r}.$$

By using the inverse formula for Möbius function, we have

$$V_n = \prod_{d|n} D_d^{\mu(n/d)} = \prod_{d|n} \binom{ud}{vd}^{\mu(n/d)}.$$

On the other hand,

$$V_n = \prod_{\substack{r=1 \\ (r,n)=1}}^{vn-1} \frac{un-r}{r} = (-1)^{v\phi(n)} \prod_{\substack{r=1 \\ (r,d)=1}}^{vd-1} \left(1 - \frac{un}{r}\right)$$

$$\equiv (-1)^{v\phi(n)} \left\{ 1 - un \sum_{\substack{r=1 \\ (r,n)=1}}^{vn-1} \frac{1}{r} \right. \tag{6.31}$$

$$\left. + \frac{u^2 n^2}{2} \left\{ \left(\sum_{\substack{r=1 \\ (r,n)=1}}^{vn-1} \frac{1}{r} \right)^2 - \sum_{\substack{r=1 \\ (r,n)=1}}^{vn-1} \frac{1}{r^2} \right\} \right\} \pmod{n^3}.$$

Since

$$\sum_{\substack{r=1 \\ (r,n)=1}}^{vn-1} \frac{1}{r} \equiv v \sum_{\substack{r=1 \\ (r,n)=1}}^{n-1} \frac{1}{r^2} \pmod{n},$$

hence

$$\sum_{\substack{r=1 \\ (r,n)=1}}^{vn-1} \frac{1}{r} = \sum_{i=0}^{v-1} \sum_{\substack{r=1 \\ (r,n)=1}}^{n-1} \frac{1}{r+in}$$

$$= \sum_{i=0}^{v-1} \sum_{\substack{r=1 \\ (r,n)=1}}^{[n/2]} \left(\frac{1}{r+in} + \frac{1}{(i+1)n-r} \right)$$

$$\equiv -n \sum_{i=0}^{v-1} (2i+1) \sum_{\substack{r=1 \\ (r,n)=1}}^{[n/2]} \frac{1}{r^2}$$

$$\equiv -v^2 n \sum_{\substack{r=1 \\ (r,n)=1}}^{[n/2]} \frac{1}{r^2} \pmod{n^2}.$$

Applying the above equation and (6.29) to (6.31), we have

$$V_n \equiv (-1)^{v\phi(n)} \left\{ 1 + \frac{uv(u-v)}{2} \sum_{\substack{r=1 \\ (r,n)=1}}^{n-1} \frac{1}{r^2} \right\} \pmod{n^3}. \qquad (6.32)$$

Noting that $uv(u-v) \equiv 0 \pmod 2$, applying Lemma 6.4 to (6.32), we complete the proof of Theorem 6.12. $\qquad \square$

Theorem 6.13. *If $p, q \geq 5$ are distinct primes, then for any integers $u > v > 0$,*

$$\binom{up}{vp}\binom{uq}{vq} \equiv \binom{u}{v}\binom{upq}{vpq} \pmod{p^3q^3}.$$

Theorem 6.13 is a consequence of Theorem 6.12. We only need to deal with the case $p = 3, q \equiv 2 \pmod 3$. By (6.32), we may assume $3 \nmid uv(u-v)$.

If $u \equiv 1 \pmod 3, v \equiv 2 \pmod 3$, then

$$3 \mid \binom{3u}{3v},$$

since $[3u/3^2] - [3v/3^2] - [3(u-v)3^2] = 1$.

If $u \equiv 2 \pmod 3, v \equiv 1 \pmod 3$, then

$$3 \mid \binom{uq}{vq}.$$

Since $[uq/3] - [vq/3] - [(u-v)q/3] = 1$, we complete the proof of Theorem 6.13.

Automorphic form and modular form

In 1770, Euler prove the following pentagonal number theorem:

$$\prod_{n=1}^{\infty} (1 - q)^n = \sum_{m=-\infty}^{\infty} (-1)^m q^{\frac{3m^2-m}{2}},$$

where $(3m^2 - m)/2 = 1, 5, 12, \ldots$ are called pentagonal numbers, This is considered to be the earliest automorphic forms.

In 1916, Ramanujan defined

$$\Delta(z) = q \prod_{n=1}^{\infty} (1 - q^n)^{24} = \sum_{n=1}^{\infty} \tau(n)q^n,$$

where $q = e^{2\pi i z}$, z is a variable in the upper half-plane H, i.e., the absolute value of q is less than 1, and thus both the infinite product expression of $\Delta(z)$ and its Fourier expansion converge absolutely. Meanwhile, $\Delta(z)$ have the automorphic property, that is

$$\Delta\left(\frac{az + b}{cz + d}\right) = (cz + d)^{12}\Delta(z),$$

for any

$$\begin{pmatrix} a & b \\ c & d \end{pmatrix} \in \mathrm{SL}_2(\mathbb{Z}),$$

where $\mathrm{SL}_2(\mathbb{Z})$ is the special linear group of order 2 over \mathbb{Z}, i.e., for any $a, b, c, d \in Z$ such that $ad - bc = 1$.

We call $\Delta(z)$ the modular form of weight 12. In general, we have the following definition.

Definition 6.2. Let k be a positive integer, an automorphic form of weight k for the modular group $\mathrm{SL}_2(\mathbb{Z})$ is a complex-valued function $f : H \to \mathbb{Z}$, satisfying the following three conditions:

(1) f is a meromorphic function on H (analytic except for some pole points);

(2) For any $z \in H$, $\begin{pmatrix} a & b \\ c & d \end{pmatrix} \in \mathrm{SL}_2(\mathbb{Z})$,

$$f\left(\frac{az + b}{cz + d}\right) = (cz + d)^k f(z);$$

(3) f is meromorphic at $i\infty$, i.e., $f = \sum_{n=-\infty}^{+\infty} a_n(f)q^n$.

Example. When $k > 2$ is even, the following so-called Eisenstein series

$$E_k(z) = \frac{1}{2} \sum_{(c,d)=1} \frac{1}{(cz + d)^k}$$

converge absolutely. It is a automorphic form of weight k.

The new kind of ζ function discovered by Ramanujan is

$$L(s, \Delta) = \sum_{n=1}^{\infty} \tau(n) n^{-s}.$$

He conjectured (it was proved by Mordell in the following year) τ is multiplicative and has the Euler product

$$L(s, \Delta) = \prod_{p} (1 - \tau(p)p^{-s} + p^{11-2s})^{-1}.$$

Compared with the Euler products of Riemann ζ function and Dirichlet ζ function, the local factors of Euler product raise from linear polynomials in p^{-s} to quadratic form. Ramanujan's second conjecture is

$$|\tau(p)| < 2p^{\frac{11}{2}}.$$

Former residence of Riemann in Göttingen. Photo taken by the author

This is equivalent to say that the discriminant of the quadratic polynomial in p^{-s} appearing as the local factor of Euler product is negative. This conjecture was proved by the Belgian mathematician

Portrait of Riemann, on the corridor of Göttingen Mathematical Institute. Photo taken by the author

Ramanujan at Cambridge, with Hardy on the right

Deligne (1944–) with the method of algebraic geometry in 1974. In fact, Deligne used German-born French math genius Grothendieck's (1928–2014, Fields prize in 1966) revolution in algebraic geometry to prove Weil conjecture, which is the algebraic geometric analogs of Riemann hypothesis. He also won Fields Medal in 1978.

By calculation, it is easy to obtain

$$\tau(1) = 1, \tau(2) = -24, \tau(3) = 252, \tau(4) = -1472, \tau(5) = 4830,$$
$$\tau(6) = -6048, \tau(7) = -16744, \tau(8) = 84480, \tau(9) = -113643.$$

Ramanujan found that for any prime p,

$$\tau(p) \equiv 1 + p^{11} (\text{mod } 691),$$

where 691 is prime. This congruence was proved by himself. Moreover

$$\tau(n) \equiv \sigma_{11}(n) (\text{mod } 691),$$

where $\sigma_k(n) = \sum_{d|n} d^k$.

Ramanujan's problems and conjectures, specially those about τ function and Ramanujan ζ function $L(s, \Delta)$ are the beginning of the automorphic form study and the basis of theory of modern automorphic forms. This can be described as a driving force of number theory in 20th century. In particular, when the meromorphic condition of the definition (1) is strengthen to holomorphic, and $i\infty$ is the only cusp point, i.e.,

$$f = \sum_{n=0}^{+\infty} a_n(f) q^n,$$

where f is called modular form. It is a special kind of automorphic forms. Modular forms is useful not only in number theory, but also has important applications in algebraic topology and string theory.

30. Congruence on Harmonic Sum

At the beginning of the new century, Jianqiang Zhao, a China-born mathematician in Florida, announced on the internet that he gives a concise interesting congruence.

Theorem 6.14 (Jianqiang Zhao). *For any odd prime* p,

$$\sum_{\substack{i+j+k=p \\ i,j,k>0}} \frac{1}{ijk} \equiv -2B_{p-3}(\text{mod } p).$$

His method uses partial sums of multiple zeta value series, formally published in 2007 (*J. Number Theory* **123**, 18–26). Two years before that, namely 2005, Chungang Ji published an elementary proof which we will show below. It requires a lemma.

Lemma 6.5. *For any odd prime* $p > 3$,

$$\sum_{1 \le i < j \le p-1} \frac{1}{ij} \equiv -\frac{p}{3} B_{p-3}(\text{mod } p^2).$$

Proof. Let m be a positive integer, using the corollary of Theorem 6.2 and Theorem 6.1,

$$\sum_{i=1}^{p-1} i^n = \frac{B_{n+1}(p) - B_{n+1}}{n+1} = \frac{1}{n+1} \sum_{r=1}^{n+1} \binom{n+1}{r} B_{n+1-r} p^r$$

$$= pB_n + \frac{p^2}{2} n B_{n-1} + \sum_{r=3}^{n+1} \binom{n}{r-1} pB_{n+1-r} \frac{p^{r-3}}{r} p^2 (\text{mod } p^2)$$

by von Staudt–Clausen theorem, when $r \ge 3$, the denominators of pB_{n+1-r} and p^{r-3}/r do not contain p (p-integral), hence

$$\sum_{i=1}^{p-1} i^n \equiv pB_n + \frac{p^2}{2} n B_{n-1} (\text{mod } p^2).$$

Using von Staudt–Clausen theorem and Euler's theorem, we have

$$\sum_{i=1}^{p-1} \frac{1}{i^2} \equiv \sum_{i=1}^{p-1} i^{\phi(p^2)-2} \equiv pB_{\phi(p^2)-2}(\text{mod } p^2), \qquad (6.33)$$

where ϕ is the Euler totient function. From Kummer congruence, we have

$$\frac{B_{\phi(p^2)-2}}{\phi(p^2) - 2} \equiv \frac{B_{p-3}}{p - 3}(\text{mod } p).$$

That is, $B_{\phi(p^2)-2} \equiv \frac{2}{3} B_{p-3} (\mathrm{mod}\ p)$, put it into (6.33), then

$$\sum_{i=1}^{p-1} \frac{1}{i^2} \equiv \frac{2p}{3} B_{p-3} (\mathrm{mod}\ p^2).$$

Finally, from Wolstenholme theorem,

$$\sum_{1 \leq i < j \leq p-1} \frac{1}{ij} = \frac{1}{2} \left(\sum_{i=1}^{p-1} \frac{1}{i} \sum_{j=1}^{p-1} \frac{1}{j} - \sum_{i=1}^{p-1} \frac{1}{i^2} \right) \equiv -\frac{p}{3} B_{p-3} (\mathrm{mod}\ p^2).$$

The proof of Lemma 3 is completed. □

Proof of Theorem 14. It is obvious true when $p = 3$. Assume $p > 3$, it is easy to see that

$$\sum_{\substack{i+j+k=p \\ i,j,k>0}} \frac{1}{ijk} = \frac{1}{p} \sum_{\substack{i+j+k=p \\ i,j,k>0}} \frac{i+j+k}{ijk} = \frac{3}{p} \sum_{\substack{i+j<p \\ i,j>0}} \frac{1}{ij}. \qquad (6.34)$$

Write $l = i + j$ so that $1 \leq j < l \leq p - 1$. Note that

$$\frac{1}{ij} = \frac{i+j}{ijl} = \frac{1}{l} \left(\frac{1}{i} + \frac{1}{j} \right).$$

By Lemma 6.3, we have

$$\sum_{\substack{i+j<p \\ i,j>0}} \frac{1}{ij} = 2 \sum_{1 \leq j < l \leq p-1} \frac{1}{jl} \equiv -\frac{2p}{3} B_{p-3} (\mathrm{mod}\ p^2).$$

Put it into (6.34), this completes the proof of Theorem 6.14. □

In 2007, 2010, 2014 and 2017, we (Xia Zhou, Tianxin Cai, *Proc. Amer. Math. Soc.* **135**, 1329–1333; Binzhou Xia, Tianxin Cai, *Inter. J. of Number Theory* **6**, 849–855; Liuquan Wang, *J. Number Theory* **144**, 15–24; Tianxin Cai, Zhongyan Shen, Lirui Jia, *Inter. J. Number Theory* **13** (2017), 1085–1094) use more elaborate method to improve Theorem 6.14 as follows.

Theorem 6.15. *For any odd prime* $p > 3$ *and positive integer* $n \le p - 2,$

$$\sum_{\substack{l_1+l_2+\dots+l_n=p \\ l_1,l_2,\dots,l_n>0}} \frac{1}{l_1 l_2 \dots l_n} \equiv \begin{cases} -(n-1)!B_{p-n}(\bmod\ p), & \text{if } n \text{ is odd,} \\ -\dfrac{n}{2n+1}n!B_{p-n-1}(\bmod\ p^2), & \text{if } n \text{ is even.} \end{cases}$$

(A mistake, here is the second formula on the right B_{p-n-1} *should be* $B_{p-n-1}p$*).*

Theorem 6.16. *For any odd prime* $p > 5,$

$$\sum_{\substack{i+j+k=p \\ i,j,k>0}} \frac{1}{ijk} \equiv \frac{12B_{p-3}}{p-3} - \frac{3B_{2p-4}}{p-2}(\bmod\ p^2).$$

Theorem 6.17. *For any odd prime* p *and positive integer* $r,$

$$\sum_{\substack{i+j+k=n \\ i,j,k\in P_p}} \frac{1}{ijk} = -2p^{r-1}B_{p-3}(\bmod\ p^r),$$

where P_p *is the set of all positive integers coprime with* $p.$

Theorem 6.18. *Let* p *and* q *be distinct odd primes,* α *and* β *positive integers. Then*

$$\sum_{\substack{i+j+k=p^\alpha q^\beta \\ i,j,k\in P_{pq}}} \frac{1}{ijk} \equiv 2(2-q)\left(1-\frac{1}{q^3}\right)p^{\alpha-1}q^{\beta-1}B_{p-3}(\bmod\ p^\alpha)$$

In general, for any integers $n > 1$, the author raise the following conjecture: let p be odd prime, $p^\alpha \| n$, then

$$\sum_{\substack{i+j+k=n \\ i,j,k\in P_n}} \frac{1}{ijk} \equiv \prod_{\substack{q|n \\ q\neq p}}\left(1-\frac{2}{q}\right)\left(1-\frac{1}{q^3}\right)\left(-\frac{2n}{p}\right)B_{p-3}\,(\bmod\ p^\alpha).$$

By Kummer's Theorem, Theorem 6.14 is a corollary of Theorem 6.16, while both Theorem 6.17 and Theorem 6.18 are corollaries of the above conjecture.

It is worth mentioning that, Jianqiang Zhao also obtain a generalization of Wolstenholme Theorem, i.e., for any positive integer n and α, $n\alpha \leq p - 3$, he obtains

$$\sum_{1 \leq l_1 < l_2 < \ldots < l_n \leq p-1} \frac{1}{l_1^\alpha l_2^\alpha \ldots l_n^\alpha} \equiv 0 \begin{cases} (\bmod\ p^2), & \text{if } 2 \nmid n\alpha, \\ (\bmod\ p), & \text{if } 2 \mid n\alpha. \end{cases}$$

We improve it to

$$\sum_{1 \leq l_1 < l_2 < \ldots < l_n \leq p-1} \frac{1}{l_1^\alpha l_2^\alpha \ldots l_n^\alpha}$$

$$\equiv \begin{cases} (-1)^n \dfrac{\alpha(n\alpha + 1)}{2(n\alpha + 2)} B_{p-n\alpha-2} p^2 (\bmod\ p^3), & \text{if } 2 \nmid n\alpha, \\[2mm] (-1)^{n-1} \dfrac{\alpha}{n\alpha + 1} B_{p-n\alpha-1} p (\bmod\ p^2), & \text{if } 2 \mid n\alpha. \end{cases}$$

In particular, if $n = \alpha = 1$, then, for any prime $p > 3$,

$$\sum_{r=1}^{p-1} \frac{1}{r} \equiv -\frac{p^2}{3} B_{p-3} (\bmod\ p^3).$$

This result is first obtained by Zhihong Sun's (*Discrete Appl. Math.* **105** (2000), 193–223).

Finally, we should point out that, as early as in 1852, the Italian mathematician Genocchi (1817–1889) proved that, for any odd prime number p,

$$\sum_{r=1}^{(p-1)/2} \frac{1}{r^3} \equiv -2 B_{p-3} \ (\bmod\ p).$$

In 2000, Zhihong Sun improved it, he showed for any prime number $p > 3$,

$$\sum_{r=1}^{(p-1)/2} \frac{1}{r^3} \equiv \frac{12 B_{p-3}}{p-3} - \frac{3 B_{2p-4}}{p-2} \ (\bmod\ p^2)$$

This congruence are surprisingly in accord with Theorems 6.16, that is, for any prime number $p > 5$,

$$\sum_{\substack{i+j+k=p \\ i,j,k>0}} \frac{1}{ijk} \equiv \sum_{r=1}^{(p-1)/2} \frac{1}{r^3} \ (\bmod\ p^2).$$

Here the number of unknowns on the left and the exponent of the sum term on the right are both 3. Comparing the results of Zhou and Cai (2010) with Sun (2000), we can see that if the above exponent is greater than 3 and the prime number $p > k + 2$. Then

$$\sum_{\substack{i_1+\cdots+i_k=p \\ i_1,\ldots,i_k>0}} \frac{1}{i_1..i_k} \equiv \frac{k!}{2^k-2} \sum_{r=1}^{\frac{p-1}{2}} \frac{1}{r^k}$$

$$-\frac{k!p}{2^k-2} \sum_{\substack{2\leq i\leq k-3 \\ i\equiv 0(mod\ 2)}} \frac{B_{p-i-1}B_{p-n+i}}{(i+1)(n-i)} \pmod{p^2}.$$

In particular, take $k = 5$, if $p > 7$,

$$\sum_{\substack{i_1+\cdots+i_5=p \\ i_1,\ldots,i_k>0}} \frac{1}{i_1..i_5} \equiv 4\sum_{r=1}^{\frac{p-1}{2}} \frac{1}{r^5} + \frac{20}{3}B_{p-3}^2 p \pmod{p^2}.$$

Multinomial coefficient of non-power

Let $s \geq 2, n \geq k_1 + \cdots + k_s, k_1 \geq 1, \ldots, k_s \geq 1$. We define the multinomial coefficient as follows:

$$\binom{n}{k_1, \cdots, k_s} = \frac{n!}{k_1 \cdots k_s(n - k_1 - \cdots - k_s)!}.$$

In particular, if $n = a, k_1 = \cdots = k_s = b$, denote

$$\binom{a}{b}_s = \underbrace{\binom{a}{b, \ldots, b}}_{s}.$$

It is not difficult to prove the following Theorem A, Theorem B and their corollaries, they are the generalization of Theorem 6.9, Theorem 6.12 and their corollaries on multinomial coefficients. For brevity, we omit the proofs.

Theorem A. *If $n > 0$ is odd and $k \geq 1$, then*

$$(-1)^{k\phi(n)/2} \prod_{d|n} \{ \frac{1}{d^{k-1}} \binom{kd-1}{(d-1)/2}_{2k} \}^{\mu(n/d)}$$

$$\equiv 4^{k\phi(n)} \begin{cases} (\text{mod } n^3), & \text{if } 3 \nmid n, \\ (\text{mod } n^3/3), & \text{if } 3|n. \end{cases}$$

Corollary A. *If prime $p \geq 5$ and $k \geq 1$, then*

$$\frac{(-1)^{k(p-1)/2}}{p^{k-1}} \binom{kp-1}{(p-1)/2}_{2k} \equiv 4^{k(p-1)} (\text{mod } p^3).$$

Theorem B. *If n, r, k_1, \ldots, k_r are positive integers and $k_1 + \cdots + k_r < k$, then*

$$\prod_{d|n} \binom{kd}{k_1 d, \ldots, k_r d}^{\mu(n/d)} \equiv 1 \begin{cases} (\text{mod } n^3), & \text{if } 3 \nmid n, n \neq 2^a, \\ (\text{mod } n^3/3), & \text{if } 3|n, \\ (\text{mod } n^3/2), & \text{if } n = 2^a, a \geq 2, \\ (\text{mod } n^3/4), & \text{if } n = 2. \end{cases}$$

Corollary B. *If prime $p \geq 5$, r, k_1, \ldots, k_r are positive integers and $k_1 + \cdots + k_r < k$, then*

$$\binom{kp}{k_1 p, \ldots, k_r p} \Big/ \binom{k}{k_1, \ldots, k_r} \equiv 1 (\text{mod } p^3).$$

Corollary C. *If $p, q \geq 5$ are distinct primes, then*

$$\binom{kpq-1}{(pq-1)/2}_{2k} \equiv (k-1)! 4^{k\phi(pq)} \binom{kp-1}{(p-1)/2}_{2k}$$

$$\binom{kq-1}{(q-1)/2}_{2k} (\text{mod } p^3 q^3).$$

In particular, we have

$$\binom{kpq-1}{(pq-1)/2}_{2k} \equiv (k-1)! \binom{kp-1}{(p-1)/2}_{2k} \binom{kq-1}{(q-1)/2}_{2k} (\text{mod } pq).$$

However, we have an unresolved problem. For any given $e > 2$, $n > 1$ and every prime factor is congruent to 1 modulo e, we wonder if it possible to get a congruence similar to Theorem 6.5 containing the following product

$$\prod_{d|n} \binom{d-1}{(d-1)/e}_e^{\mu(n/d)}.$$

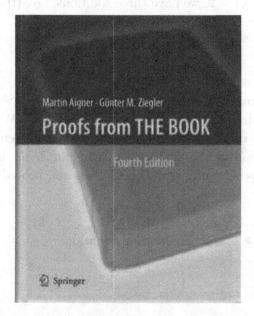

Cover of *Proofs from THE BOOK*, the 4th edition

The following problem comes from *Proofs from THE BOOK*, written by the Austria mathematician Aigner and the German mathematician Ziegler.

Consider a Diophantine equation related with binomial coefficients

$$\binom{n}{k} = x^l, \tag{A}$$

where $n \geq k \geq 2, x, l \geq 2$. When $k = l = 2$, it has infinitely many solutions. Indeed, if $\binom{n}{2}$ is a square, then so is $\binom{(2n-1)^2}{2}$, this is

because, let $\binom{n}{2} = m^2$, then

$$(2n-1)^2((2n-1)^2 - 1) = (2n-1)^2 4n(n-1) = 2(2m(2n-1))^2.$$

Beginning with $\binom{9}{2} = 6^2$ we thus obtain infinitely many solution, the next one is $\binom{289}{2} = 204^2$. However, this does not yield all solutions. For example, $\binom{50}{2} = 35^2$ starts another series, as does $\binom{1682}{2} = 1189^2 \ldots$.

When $k = 3, l = 2$, we have three solutions, i.e., (the first two are trivial)

$$\binom{3}{3} = 1^2, \quad \binom{4}{3} = 2^2, \quad \binom{50}{3} = 140^2.$$

For the remaining cases, including $k = 2, 3, l > 2$, or $k \geq 4$, Erdős conjectured in 1939 that (A) has no solution. With the efforts of himself and other people, the conjecture was proven in 1998.

In 2012, we (Peng Yang, Tianxin Cai, *Acta Arith.* **151**, 7–9) considered the case of multinomial coefficients and proved the following theorem.

Theorem C. For any $n \geq 3, s \geq 3, l \geq 2$, $k_1 + \cdots + k_s = ns$, $k_1 \geq \cdots \geq k_s \geq 1$, the Diophantine equation

$$\binom{n}{k_1, \ldots, k_s} = x^l \tag{B}$$

has no solution.

In order to prove Theorem C, we need some lemmas.

Lemma A (Saradha, Gyory). *For any positive integer b, denote by $P(b)$ the greatest prime factor of b. Apart from $k = b = l = 2$, consider the equation*

$$n(n+1)\ldots(n+k-1) = bx^l,$$

with positive integers n, k, b, x, l satisfying $k \geq 2, l \geq 2, P(b) \leq k$, b is lth power free. When $P(x) > k$, the equation has only one solution $(n, k, b, x, l) = (48, 3, 6, 140, 2)$.

Lemma B. *If $n/2 \geq k_1 \geq \ldots \geq k_s \geq 1$, there exists a prime $p > k_1$, such that*

$$p \left\| \binom{n}{k_1 \ldots k_s} \right. .$$

Proof. First of all, we need to prove that for any integer $n > 1$, there exists a prime $p \leq n$, satisfying $n = p + r, 0 \leq r < p$. It is obviously true when $n = 2$. For $n > 2$ and take the largest prime $p_1 \leq n$. We write $n = t_1 p_1 + r_1$ with $0 \leq r_1 < p_1, t_1 \geq 1$. If $t_1 \geq 2$, by Chebyshev's theorem, we have a prime p_2 satisfying $p_1 < p_2 < 2p_1 \leq n$, which is a contradiction. Hence $t_1 = 1$.

Let $n = p + r, 0 \leq r < p$. Then $k_1 \leq n/2 < p \leq n < 2p$. Since

$$\binom{n}{k_1, \ldots, k_s} = \frac{n(n-1) \ldots (k_1 + 1)}{k_2! \ldots k_s!},$$

and $P(k_2 \ldots k_s) < k_1$, we have $v_p\left(\binom{n}{k_1, \ldots k_s}\right) = 1$, where $v_p(n)$ denotes the exponent of p in the factorization of n. This completes the proof of Lemma B. $\qquad\square$

Proof of Theorem C. By Lemma B, we only need to consider $k_1 > n/2 \geq k_2 \cdots \geq k_s \geq 1$. Suppose that (B) has an integer solution, we rewrite (B) in the form

$$n(n-1) \ldots (n - k_2 - \ldots - k_s + 1) = k_2! \ldots k_s! x^l = bt^l x^l, \qquad (C)$$

where $k_2! \ldots k_s! = bt^l$, b is lth power free. By assumption, we have $P(b) \leq P(k_2! \ldots k_s!) \leq k_2$. Let $k = k_2 + \cdots + k_s, n' = k_1 + 1$, then (C) takes the form

$$n'(n' + 1) \ldots (n' + k - 1) = b(tx)^l.$$

Note that $k = n - k_1 < n/2$ and by Sylvester's Theorem (Section 5), there exists a prime $p > k$, such that $p | \binom{n}{k}$. Since $\binom{n}{k_1, \ldots, k_s} = \binom{n}{k}\binom{k}{k_2, \ldots, k_s}$, we have $p | \binom{k}{k_2, \ldots, k_s}$, whence $p | x$, $P(tx) \geq P(x) \geq p > k$, which contradicts with Lemma A if $k > 3$. Therefore, (B) has no solution if $k > 3$. Now we only need to consider the case $k = 2$ and $k = 3$.

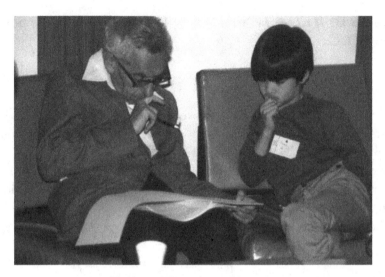

Erdős and Tao, Australia, 1985

For $k = 2$, equation (B) takes the form $\binom{n}{n-2,1,1} = n(n-1) = x^2$, it is easy to see this equation has no solution. For $k = 3$, (B) leads to $\binom{n}{n-3,2,1} = n(n-1)(n-2)/2 = x^2$ or $\binom{n}{n-3,1,1,1} = n(n-1)(n-2) = x^2$. By Lemma A, the above equations have no positive solution. This completes the proof of Theorem C. \square

At this point, we completely solve Erdős' problem.

Chapter 7

Additive and Multiplicative Equations

Additive number theory and multiplicative number theory are both important part of number theory. Additive number theory is also called Dui Lei Su Shu Lun in Chinese by L.K. Hua (1910–1985), and he published a book with the title. As two branches of number theory, there are a few intersections between them. However, inspired by the form of the abc conjecture, we try to combine the addition and multiplication in Waring's problem. We obtain a new class of Diophantine equations and some unexpected results. Later, this idea is gradually expanded, involving several classic number theory problems. Now we select six of them to study, including five classical number theory problems, and the last one is the abcd-equation, which is our creation. Recently, the Romanian-born Germany mathematician Preda Mihailescu published a survey article *Around ABC* (*Newsletter European Math. Soc.* **3** (2014), 29–35), and mentioned two of them, in a note he calls this kind of equations *Yin-Yang equations*.

31. New Waring's Problem

In 1770, Lagrange showed (Section 17) that every positive integer could be expressed as a sum of four squares of non-negative integers. The proposition was probably already known to Diophantus, since he gave some examples in his *Arithmetic* and later it was formally conjectured by French mathematician, poet Bachet (1581–1638), he is the translator of the Latin version of *Arithmetic* (1621).

Portrait of English mathematician Waring

In Fermat's comments, he claimed (as he did to Fermat's Last Theorem) that he had proved this proposition, but no one found it. In the same year that Lagrange gave the proof, the English mathematician Edward Waring (c. 1736–1798) asserted that for every k, there was a number $s = s(k)$ such that every positive integer n could be expressed as a sum of s kth powers of positive integers, i.e.,

$$n = x_1^k + x_2^k + \cdots + x_s^k.$$

During the next 139 years, several special cases of Waring's problem were solved, for example, $k = 3, 4, 5, 6, 7, 8, 10$. It was in 1909 that Hilbert finally solved the problem in the affirmative for all k. So Waring's assertion is also called the Hilbert–Waring theorem. Once we know that the answer to the Hilbert–Waring problem is "yes!", it is natural to ask "How big is $s(k)$?". Taking $g(k)$ as the smallest value of $s(k)$, we have $g(2) = 4$ (1770, Lagrange), $g(3) = 9$ (1909, Wieferich *et al.*), $g(4) = 19$ (1986, R. Balasubramanian *et al.*), $g(5) = 37$ (1964, Chen Jingrun), $g(6) = 73$ (1940, Pillai).

It is worth mentioning that Balasubramanian used the modification of Chen's method. Pillai (1901–1950) was one generation earlier than Balasubramanian and he was one of the most important

Indian mathematician after Ramanujan. Pillai's conjecture is a generalization of Catalan conjecture: for any positive integer k, there exist at most finitely many pairs of positive integers (x^p, y^q), satisfying $x^p - y^q = k$. In 1946, Hua, then a professor of Southwest United University, stopped in Kolkata on his way from Kunming to the Soviet Union visited Pillai. Four years later, Pillai was invited to visit Princeton for one year, he planned to attend the *International Congress of Mathematicians* in Harvard University. Unfortunately, his plane crashed down on the way when he transferred in Cairo.

The tomb of Hilbert in Gottingen. Photo taken by the author

In 1882, J. A. Euler, the son of Leonhard Euler, conjectured that

$$g(k) = 2^k + \left[\left(\frac{3}{2} \right)^k \right] - 2.$$

It is checked to be true for $k \leqslant 471{,}600{,}000$ by L. E. Dickson *et al.*, and Kurt Mahler showed that there are at most finitely exceptional k's.

Meanwhile, from the work of Hardy and Littlewood it became apparent that a more fundamental number than $g(k)$ was $G(k)$,

which is defined to be the least positive integer s such that all sufficiently large integers can be represented as a sum of at most s kth powers of positive integers. Obviously, $G(k) \leq g(k)$. This problem is more complicated, but more meaningful. Using the congruent properties, it is easy to know the integers $\equiv 7 \pmod 8$ can not be expressed as a sum of squares of three integers, so $G(2) = 4$. In 1939, the British mathematician Davenport (1907–1969) proved $G(4) = 16$. The precise value of $G(3)$ is not known, the best result to date being $4 \leqslant G(3) \leqslant 7$. Someone conjecture that $G(3) = 4$. The largest number now known not to be a sum of four cubes is 7, 373, 170, 279, 850. For $k \geqslant 5$, we know little about the precise value of $G(k)$, e.g., $6 \leq G(5) \leq 17, 9 \leq G(6) \leq 24$.

Hardy and Littlewood conjectured that for $k \neq 2^a, a > 1$, $G(k) \leq 2k + 1$. The upper bound estimate of $G(3)$ is the only evidence to date. As one of the central problems of number theory, Waring's problem has not been included in the list of 23 mathematical problems published by German mathematician David Hilbert, can only be said to be a kind of regret. Perhaps in 1900, Hilbert had not been exposed to this problem. But once the Hilbert–Waring theorem was proved, it began to attract more and more mathematicians. Unfortunately, during the long period of time, the progress is very slow.

In April 2011, the author proposed a variant of Waring's problem: let k and s be positive integers, for any $n \in \mathbf{Z}^+$, write

$$n = x_1 + x_2 + \cdots + x_s$$

with all $x \in \mathbf{Z}^+$ such that

$$x_1 x_2 \cdots x_s = x^k.$$

From the Hilbert–Waring theorem, there exists a positive integer $s = s'(k) \leq s(k)$ such that every positive integer n is a sum of s positive integers and the product of the s integers is a kth power. We denote by $g'(k)$(resp., $G'(k)$) the least positive integer such that every integer (resp. all sufficiently large integers) can be represented as a sum of at most $g'(k)$(resp., $G'(k)$) positive integers, and the product of the $g'(k)$(resp., $G'(k)$) integers is a kth power.

We (Tianxin Cai, Deyi Chen, a new variant of the Hilbert–Waring problem, *Math. Comput.* **82**(2) (2013), 2333–2341) made a research on this question and proved the following theorem.

Theorem 7.1. *For every positive integer k, $g'(k) = 2k - 1$.*

Proof. It is obviously true for $k = 1$. Suppose $k > 1$, and for any positive integer n, define i by $n \equiv i \pmod{k}, 0 \leq i \leq k - 1$. Then we have

$$n = \underbrace{1 + \cdots + 1}_{i} + \underbrace{\frac{n-i}{k} + \cdots + \frac{n-i}{k}}_{k}.$$

Therefore $g'(k) \leq 2k - 1$.

On the other hand, if $2k-1$ can be expressed as a sum of $s(s < 2k-1)$ positive integers and the product of the s integers is a kth power, then there exists a positive integer $m > 1$ among the s integers. Let $q = \min_{p \, \text{prime}}\{p | m \equiv 0 \pmod{p}\}$, then $2k - 1 \geq q^{\beta_1} + \cdots + q^{\beta_r}$, where $\beta_1 \geq 1, \ldots, \beta_r \geq 1$ and $\beta_1 + \cdots + \beta_r \equiv 0 \pmod{k}$. Hence

$$2k - 1 \geq \underbrace{q + \cdots + q}_{k} = qk \geq 2k.$$

This contradiction implies that $g'(k) \geq 2k - 1$. We complete the proof of Theorem 7.1.

For $G'(k)$, we have the following upper bound estimate.

Theorem 7.2. *For every prime p, $G'(p) \leq p + 1$.*

Proof. Suppose $n \equiv i \pmod{p}, 0 \leq i \leq p - 1$. If $i = 0$, let $n = mp$, when $n > p^p$,

$$n = m - p^{p-1} + \cdots + m - p^{p-1} + p^p.$$

If $0 < i = p - 1$, by Fermat's little theorem, $n \equiv i \equiv i^p \pmod{p}$. When $n > p^p$,

$$n = i^p + \underbrace{\frac{n - i^p}{p} + \cdots + \frac{n - i^p}{p}}_{p}.$$

This completes the proof of Theorem 7.2.

Remark. In the spring of 2017, Junwei Hu, a junior majoring in finance, generalized this result during the course of our Elementary Number Theory. He proves that if m is a cyclic number, then $G'(m) = m + 1$. Here cyclic number means a positive integer m satisfying $(m, \varphi(m)) = 1$. Obviously, the cyclic number contains all prime numbers, and it has no square factor. Let $n = km + r, 0 \leq r < m, d = (r, m)$. Since m has no square factor, hence $(r, \frac{m}{d}) = 1$, it follows from Euler's theorem that $r^{\phi(\frac{m}{d})} \equiv 1 \pmod{\frac{m}{d}}$. For any integer B, $r^{\phi(\frac{m}{d})B} \equiv 1 \pmod{\frac{m}{d}}$. Noting that d is a factor of r, therefore

$$r^{\phi(\frac{m}{d})+1} \equiv r \pmod{m}, \tag{C}$$

Meanwhile $(m, \varphi(m)) = 1$, hence $(m, \phi(\frac{m}{d})) = 1$. There exist positive integers A_r, B_r such that

$$\phi\left(\frac{m}{d}\right) B_r + 1 = A_r m,$$

apply(C) to the above equation, we have $r \equiv r^{A_r m} \pmod{m}$. Taking $C_r = \frac{r^{A_r m} - r}{r}$, then C_r is a positive integer. Let $C = \max_{0 \leq r \leq m} C_r$. When $n > m(C + 1), k = [\frac{n}{m}] \geq [C + 1] = C + 1 > C \geq C_r$,

$$n = k - C_r + \cdots + k - C_r + mC_r + r$$
$$= k - C_r + \cdots + k - C_r + r^{A_r m}.$$

With the product $(k - C_r)^m (r^{A_r})^m$. Therefore, $G'(m) \leq m + 1$.

In particular, we add that the Carmichael numbers defined in Section 15 are cyclic numbers. This is because, according to Korselt's criterion, the Carmichael number has no square factors, and for any prime $p|n$, it must satisfy $p - 1|n - 1$. Let p and q be any prime factors of n, noting that $(n - 1, q) = 1$, we have $(p - 1, q) = 1$, then $(n, \varphi(n)) = 1$.

Theorem 7.3. *For any composite number k, $G'(k) \leq k + 2$.*

Proof. For every positive integer n, let $n = km + r$, where $0 \leq r \leq k - 1$.

If $r = 0$, when $n > 2k^{2k}$, we have $m = \frac{n}{k} > 2k^{2k-1}$ and

$$n = km = \underbrace{(m - 2k^{2k-1}) + \cdots + (m - 2k^{2k-1})}_{k} + k^{2k} + k^{2k}.$$

If $0 < r \leq k - 1$, when $n > k^{2k-1} + k$, we have $m = \frac{n-r}{k} > \frac{k^{2k-1}}{k} > k^{k-1}r^{k-1}$ and

$$n = km + r = \underbrace{(m - k^{k-1}r^{k-1}) + \cdots + (m - k^{k-1}r^{k-1})}_{k} + k^k r^{k-1} + r.$$

This completes the proof of Theorem 7.3.

Note when $k > 1$, prime $p \equiv 3 \pmod 4$ can't be expressed as a sum of two kth powers, we deduce from Theorems 7.1 and 7.2 that $G'(2) = 3, 3 \leq G'(3) \leq 4, 3 \leq G'(5) \leq 6$, from Theorems 7.1 and 7.3 that $3 \leq G'(4) \leq 6$. In order to discuss $G'(3)$, we set

$$S = \{n \mid n = x + y + z, xyz = m^3, x, y, z \in Z^+\}.$$

It is obvious that if $n \in S$, then $kn \in S$ for any $k \in Z^+$. So we only need to discuss the special case when n is prime. We have the following theorems.

Theorem 7.4. *If p is prime and $p \equiv 1 \pmod 3$ or $p \equiv \pm 1 \pmod 8$, then $p \in S$.*

Theorem 7.5. *If p is prime, $p \equiv 5 \pmod{24}$, and there exists positive integers (x,y) such that*

$$p = 6x^2 - y^2, \tag{7.1}$$

where

$$\frac{x}{y} \in \left(\frac{7}{17}, \frac{19}{46}\right) \cup \left(\frac{11}{26}, \frac{3}{7}\right) \cup \left(\frac{9}{19}, \frac{7}{13}\right) \cup \left(1, \frac{3}{2}\right). \tag{7.2}$$

Then $p \in S$.

In order to prove Theorems 7.4 and 7.5, we need the following lemmas.

Lemma 7.1. *If p is an odd prime, then there exists a pair of positive integers (x,y) such that*

$$x^2 - 2y^2 = p, \tag{7.3}$$

here $x > 2y$.

Lemma 7.2. *If p is prime and $p \equiv 5 \pmod{24}$, then there exists a pair of positive integers (x,z) and a pair of integers (y,m), $(x,y,z)=1$ such that*

$$p = x + y + z,$$

and

$$xyz = m^3.$$

In order to prove Lemmas 7.1 and 7.2, we need to use some basic properties of quadratic form, for example, any prime $p \equiv \pm 1 \pmod 8$ can be expressed as $p = x^2 - 2y^2$ and any prime $p \equiv 5 \pmod{24}$ can be expressed as $p = 6x^2 - y^2$.

Proof of Lemma 7.1. It is obvious that $(u, v) = (3, 2)$ is the fundamental solution of $x^2 - 2y^2 = 1$. Denote $x + y\sqrt{2}$ the fundamental solution of equation (7.3), from the property of quadratic form (cf. T. Nagell, *Introduction to Number Theory*, Theorem 108, Wiley, New York, 1951),

$$0 \leq y \leq \frac{v\sqrt{p}}{\sqrt{2(u+1)}} = \sqrt{\frac{p}{2}}. \tag{7.4}$$

Combining (7.3) and (7.4), we have

$$\left(\frac{x}{y}\right)^2 = 2 + \frac{p}{y^2} \geq 2 + \frac{p}{(\sqrt{\frac{p}{2}})^2} = 4,$$

then

$$\frac{x}{y} \geq 2.$$

But $x \neq 2y$, so the inequality is strict. This complete the proof of Lemma 7.1.

Proof of Lemma 7.2. Making the transformations

$$\begin{cases} x = 3a + b, \\ y = 2a + b, \end{cases} \begin{cases} x = 7a + b, \\ y = 13a + 2b, \end{cases} \begin{cases} x = a + 9b, \\ y = 2a + 19b, \end{cases}$$

$$\begin{cases} x = 3a + 11b, \\ y = 7a + 26b, \end{cases} \begin{cases} x = 19a + 7b, \\ y = 46a + 17b. \end{cases} \tag{7.5}$$

Put them into (7.1), we have

$$p = 50a^2 + 32ab + 5b^2, p = 125a^2 + 32ab + 2b^2,$$

$$p = 2a^2 + 32ab + 125b^2,$$

$$p = 5a^2 + 32ab + 50b^2, p = 50a^2 + 32ab + 5b^2,$$

respectively. The inverse transformations are

$$\begin{cases} a = x - y, \\ b = 3y - 2x, \end{cases} \begin{cases} a = 2x - y, \\ b = 7y - 13x, \end{cases} \begin{cases} a = 19x - 9y, \\ b = y - 2x, \end{cases}$$

$$\begin{cases} a = 26x - 11y, \\ b = 3y - 7x, \end{cases} \begin{cases} a = 17x - 7y, \\ b = 19y - 46x. \end{cases} \tag{7.6}$$

respectively. The required property of coprime is obvious. This completes the proof of Lemma 7.2.

Remark. To prove Lemma 2, we only need one of the five transformations in (7.5). The reason we make five transformations is for the proof of Theorem 7.5.

Proof of Theorem 7.4. If $p \equiv 1 \pmod 3$, then there exists a pair of coprime positive integers x, y such that

$$p = x^2 + xy + y^2.$$

If $p \equiv \pm 1 \pmod 8$, by Lemma 7.1, there exist two positive integers x, y, with $x > 2y$, $(x, y) = 1$, such that $p = x^2 - 2y^2$. Making the transformation

$$\begin{cases} x = 2a + b, \\ y = a, \end{cases}$$

we deduce

$$p = (2a + b)^2 - 2a^2 = 2a^2 + 4ab + b^2,$$

$p \in S$. This completes the proof of Theorem 7.4.

Proof of Theorem 7.5. In the proof of Lemma 7.2, if the inverse transformations of (7.6) satisfy one of the following conditions:

$$\begin{cases} x - y > 0, \\ 3y - 2x > 0, \end{cases} \quad \begin{cases} 2x - y > 0, \\ 7y - 13x > 0, \end{cases} \quad \begin{cases} 19x - 9y > 0, \\ y - 2x > 0, \end{cases}$$

$$\begin{cases} 26x - 11y > 0, \\ 3y - 7x > 0, \end{cases} \quad \begin{cases} 17x - 7y > 0, \\ 19y - 46x > 0, \end{cases}$$

then $a > 0, b > 0, p \in S$. The above condition is equivalent to (7.2). This completes the proof of Theorem 7.5.

By computing, we find that the only primes $p < 10000$ such that $p \notin S$ are 2, 5, 11. Noting that

$$2^6 = 7 + 8 + 49, 7 \times 8 \times 49 = 14^3,$$

$$5^2 = 3 + 4 + 18, 3 \times 4 \times 18 = 6^3,$$

$$11^2 = 1 + 45 + 75, 1 \times 45 \times 75 = 15^3,$$

we deduce that there are only finitely many numbers of the type $2^\alpha 5^\beta 11^\gamma k$ not in S. By computing and checking, we make the following conjectures (it is similar for $k \geq 4$).

Conjecture 1. *Except for* $1, 2, 4, 5, 8, 11, 16, 22, 32, 44, 88, 176$, *every positive integer can be expressed as a sum of three positive integers where the product of those integers is a cube. In particular,* $G'(3) = 3$.

Conjecture 2. *Except for* $1, 2, 3, 5, 6, 7, 11, 13, 14, 15, 17, 22, 23$, *every positive integer can be expressed as a sum of four positive integers where the product of those integers is a fourth power. In particular,* $G'(4) = 4$.

Example 7.1. $3 = 1 + 1 + 1, 6 = 2 + 2 + 2, 7 = 1 + 2 + 4,$ $9 = 3 + 3 + 3, 10 = 1 + 1 + 8, 12 = 4 + 4 + 4, 13 = 1 + 3 + 9, \ldots.$

Example 7.2. $4 = 1+1+1+1, 8 = 2+2+2+2, 9 = 1+2+2+4, 10 = 1+1+4+4, 12 = 3+3+3+3, \ldots$.

Remark. While allowing negative integers in the discussion, we can prove the above two conjectures. Anyway, the new Waring problem leaves many unsolved problems. In general, for any positive integer $m \geq 3$, we conjecture that $G'(m) = m$. Meanwhile, Zhiwei Sun told us a few times that he had checked for $n = 5$ and 6, this result should be true.

Fermat's Last Theorem on French stamp

32. New Fermat's Last Theorem

In 1995, Andrew Wiles finally proved Fermat's Last Theorem, or Fermat's equation,

$$x^n + y^n = z^n, \ n \geq 3$$

has no positive integer solutions.

This is a landmark achievement, and it is regarded as the most important progress in number theory in the 20th century. There are several generalizations of Fermat's Last Theorem, e.g., Fermat–Catalan conjecture, which is mentioned in the section about abc conjecture, and so-called Beal's conjecture. Andrew Beal (1952–) is an American businessman and banker, obsessed with number theory,

he conjectured in 1993 that when $a, b, c \geq 3$, the equation

$$x^a + y^b = z^c, \;\; (x, y, z) = 1$$

has no positive integer solutions.

In the above equation, co-prime condition is necessary, otherwise there are many counterexamples, e.g., $3^3 + 6^3 = 3^5$, $7^6 + 7^7 = 98^3$. Moreover, for any integer $m \geq 3$,

$$[a(a^m + b^m)]^m + [b(a^m + b^m)]^m = (a^m + b^m)^{m+1},$$

where a and b are arbitrary positive integers. In particular, take $a = b = 1$, we have $2^m + 2^m = 2^{m+1}$.

Arithmeticorum Liber II. 61

IN QVÆSTIONEM VII.

QVÆSTIO VIII.

OBSERVATIO DOMINI PETRI DE FERMAT.

QVÆSTIO IX.

The 1670 edition of Diophantus' *Arithmetic* including Fermat's commentary, particulary his "Last Theorem"

In 2013, Beal offered a prize of one million dollars for a proof or a counterexample of his conjecture.

Obviously, Beal's equation is the special case of Fermat–Catalan's equation. Therefore, if the abc conjecture is true, we can deduce that Beal's conjecture has at most a finite number of counterexamples.

In October 2011, inspired by the new Waring's problem, the author propose a new kind of Fermat problem, i.e., considering the positive integer solutions of the equation

$$\begin{cases} A + B = C, \\ ABC = x^n. \end{cases} \tag{7.7}$$

Let $d = (A, B, C)$. When $d = 1$, equation (7.7) is equivalent to Fermat's Last Theorem. That is to say, equation (7.7) has no positive integer solutions.

Let $\omega(n)$ be the number of distinct prime factors of n. When $d = 1$, $\omega(d) = 0$. We consider the case $\omega(d) = 1$, and $n > 3$ is prime p, the data suggests $d = q^\alpha$, where α satisfies $3\alpha + 1 \equiv 0 \pmod{p}$. As an example, let $p = 5$, we study the form of q. Let $a > 0$, a and b be integers, $(a, b) = 1$, $a + b = m^5$, $\frac{a^5 + b^5}{a+b} = q$ be prime. Note that

$$q = (a^3 - b^3)(a - b) + a^2 b^2 \geq \max\{a^2, b^2\},$$

so $(q, a) = (q, b) = 1$. Let

$$A = q^3 a^5, B = q^3 b^5, C = q^4(a + b).$$

Then $\{A, B, C\}$ satisfies (7.7) and $d = (A, B, C) = q^3$.

For example, let $a = n + 1, b = -n$, then $q = (n + 1)^5 - n^5$, $q = 31, 211, 4651, 61051, 371281, \ldots$. On the other hand, let $a_1 + b_1 = c_1^{5r}$ and when $r = 1$, $c_1 = 2$, it has some solutions like

$$\{a_1, b_1, q, c_1\} = \{11, 21, 132661, 2\}, \{9, 23, 202981, 2\}.$$

Here 31 and 132661 are the minimum generalized Mersenne prime and non-generalized Mersenne prime, respectively such that (7.7) has solutions and $d = 31^\alpha, 132661^\alpha$, where $3\alpha + 1 \equiv 0 \pmod 5$.

We (Tianxin Cai, Deyi Chen and Yong Zhang, A New Generalization of Fermat's Last Theorem, *J. Number Theory* **149** (2015), 33–45) study this new kind of Fermat's problem and obtained the following result.

Wiles in front of the monument of Fermat in his hometown

Theorem 7.6. *For any $n \geq 3$, if $n \equiv 0 \,(\mathrm{mod}\ 3)$, then (7.7) has no positive integer solutions; if $n \not\equiv 0 \,(\mathrm{mod}\ 3)$, then (7.7) has infinitely many positive integer solutions.*

Proof. If $n \not\equiv 0 \ (\mathrm{mod}\ 3)$, then there exists a positive integer k such that $3k + 2 \equiv 0(\mathrm{mod}\ n)$. By the property of Pythagorean triple, there exist infinitely many triples of positive integer (a, b, c) such that $a^2 + b^2 = c^2$. Let

$$
\begin{cases}
A = a^{k+2}b^k c^k, \\
B = a^k b^{k+2} c^k, \\
C = a^k b^k c^{k+2}.
\end{cases}
$$

So we have infinitely many positive solutions (A, B, C) for (7.7), where $D = (abc)^{\frac{3k+2}{n}}$.

If $n \equiv 0 \pmod 3$, assume that (A, B, C) is a positive solution of (7.7). Let $d = \gcd(A, B, C)$, then $(\frac{A}{d}, \frac{B}{d}, \frac{C}{d}) = 1$, $d^3 | D^n$. But $3 | n$, so we have $d | D^{\frac{n}{3}}$ and

$$\begin{cases} \dfrac{A}{d} + \dfrac{B}{d} = \dfrac{C}{d}, \\[2mm] \dfrac{A}{d} \cdot \dfrac{B}{d} \cdot \dfrac{C}{d} = \left(\dfrac{D^{n/3}}{d}\right)^3. \end{cases}$$

So $\frac{A}{d} = x^3, \frac{B}{d} = y^3, \frac{C}{d} = z^3$, $x^3 + y^3 = z^3$, a contradiction with Fermat's Last Theorem. This completes the proof of Theorem 7.6.

Theorem 7.7. *Suppose* $(A, B, C) = p^k$, $k \geq 1$. *When* $n = 4$, *if prime* $p \equiv 3 \pmod 8$, *then* (7.7) *has no positive integer solutions; when* $n = 5$, *if an odd prime* $p \not\equiv 1 \pmod{10}$, *then* (7.7) *has no positive integer solutions.*

Remark If $p = 2$ or $p \equiv 1, 5, 7 \pmod 8$, it is possible for (7.7) to have a positive integer solution for $(A, B, C) = p$. For example

$$\begin{cases} 2 + 2 = 4, \\ 2 \times 2 \times 4 = 2^4, \end{cases} \quad \begin{cases} 17 + 272 = 289, \\ 17 \times 272 \times 289 = 34^4, \end{cases} \quad \begin{cases} 5 + 400 = 405, \\ 5 \times 400 \times 405 = 30^4, \end{cases}$$

$$\begin{cases} 47927607119 + 1631432881 = 49559040000, \\ 47927607119 \times 1631432881 \times 49559040000 = 44367960^4. \end{cases}$$

In order to prove Theorem 7.7, we need the following two lemmas.

Lemma 7.3. *For any prime* p, $n \geq 2$ *is an integer, if* $(A, B, C) = p^k$, $k \geq 1$ *and* $k \equiv 0 \pmod n$ *then* (7.7) *has no non-zero integer solutions.*

Proof. Let $A_1 = \frac{A}{p^k}, B_1 = \frac{B}{p^k}, C_1 = \frac{C}{p^k}$, then (7.7) is transformed into

$$\begin{cases} A_1 + B_1 = C_1, \\ p^{3k} A_1 B_1 C_1 = D^n. \end{cases}$$

But $k \equiv 0 \pmod{n}$, A_1, B_1, C_1 are pairwise coprime, so $A_1 = x^n$, $B_1 = y^n$, $C_1 = z^n$, and $x^n + y^n = z^n$. By Fermat's Last Theorem, we deduce that $xyz = 0$, hence $ABC = 0$. A contradiction! This completes the proof of Lemma 7.3.

Lemma 7.4. *Let $A > 2$ be an integer and it has no prime divisor of the form $10k + 1$. Then*

$$\begin{cases} x^5 + y^5 = Az^5 \\ \gcd(x, y) = 1 \end{cases}$$

has no non-zero integer solution. If $A = 2$, the solutions are $(x, y, z) = \pm(1, 1, 1)$.

Lemma 2.2 was first raised by V. A. Lebesgue (earlier than the founder of real analysis H.-L. Lebesgue) as a conjecture in 1843 and proved by E. Halberstadt and A. Kraus (*J. London Math. Soc.*, **69**(2), (2004), 291–302).

Proof of Theorem 7.7. First of all, we prove the first half of the theorem. In view of Lemma 7.3, we only need to discuss the case $k \not\equiv 0 \pmod{4}$. Suppose $p \equiv 3 \pmod{8}$ is a prime and (7.7) has a positive integer solution (A, B, C), note that $(A, B, C) = p^k$, equation (7.7) can be changed into

$$\begin{cases} A_1 + B_1 = C_1, \\ p^{3k} A_1 B_1 C_1 = D^4, \end{cases} \tag{7.8}$$

where $A_1 = \frac{A}{p^k}, B_1 = \frac{B}{p^k}, C_1 = \frac{C}{p^k}$ are pairwise coprime. From $k \not\equiv 0 \pmod{4}$, we deduce that one and only one of A_1, B_1, C_1 is a multiple of p. Let $3k \equiv r \pmod{4}, 1 \leq r \leq 3$. In view of (7.8) and reordering A_1, B_1 if necessary, we obtain

$$\begin{cases} A_1 = x^4, \\ B_1 = y^4, \\ C_1 = p^{4-r} z^4, \end{cases} \tag{7.9}$$

or

$$\begin{cases} A_1 = p^{4-r}z^4, \\ B_1 = y^4, \\ C_1 = x^4, \end{cases} \tag{7.10}$$

where x, y, z are pairwise coprime.

If A_1, B_1 and C_1 satisfy (7.9), then

$$x^4 + y^4 = p^{4-r}z^4. \tag{7.11}$$

But $(y, p) = 1$, we have integer $s \not\equiv 0 \pmod{p}$ such that $sy \equiv 1 \pmod{p}$. By (7.11), we have

$$(xs)^4 \equiv -1 \pmod{p},$$

which implies that -1 is a quadratic residue modulo p, hence $p = 2$ or $p \equiv 1 \pmod 4$. A contradiction!

If A_1, B_1 and C_1 satisfy (7.10), then

$$x^4 - y^4 = p^{4-r}z^4. \tag{7.12}$$

When $r = 2$, we have $x^4 - y^4 = (pz^2)^2$. On the other hand, it is well known that the equation $X^4 - Y^4 = Z^2$ has no non-zero integer solutions. So $r = 1$ or 3 and (7.12) can be changed into

$$x^4 - y^4 = p(pz^2)^2,$$

or

$$x^4 - y^4 = p(z^2)^2,$$

respectively.

Finally, we use two properties of congruent numbers (cf. the definition in Section 35):

(1) Any prime $p \equiv 3 \pmod 8$ is not a congruent number.
(2) If $x^4 - y^4 = cz^2$ $(c \neq 0)$ has a rational solution with $xyz \neq 0$, then $|c|$ is a congruent number. This completes the proof of the first half of Theorem 7.7.

Now we prove the second half of the theorem. In view of Lemma 7.3, we only need to discuss $k \not\equiv 0 \pmod 5$. Note that $(A, B, C) = p^k$, equation (7.7) can be changed into

$$\begin{cases} A_1 + B_1 = C_1, \\ p^{3k} A_1 B_1 C_1 = D^5, \end{cases} \tag{7.13}$$

where $A_1 = \frac{A}{p^k}, B_1 = \frac{B}{p^k}, C_1 = \frac{C}{p^k}$ are pairwise coprime. Since $k \not\equiv 0 \pmod 5$, we deduce that one and only one of A_1, B_1, C_1 is a multiple of p. Let $3k \equiv r \pmod 5, 1 \leq r \leq 4$, by (7.13) and reordering A_1, B_1, C_1 if necessary, we obtain

$$\begin{cases} A_1 = x^5, \\ B_1 = y^5, \\ C_1 = p^{5-r} z^5, \end{cases} \tag{7.14}$$

where x, y and pz are pairwise coprime. From (7.13) and (7.14), one has

$$x^5 + y^5 = p^{5-r} z^5. \tag{7.15}$$

But odd prime $p \not\equiv 1 \pmod{10}$, by Lemma 7.3, (7.15) has no positive integer solutions. This completes the proof of Theorem 7.7.

Moreover, we have the following conjectures. Let p be odd prime.

Conjecture 3. *If $(A, B, C) = p$ and n is odd prime, then (7.7) has no non-negative integer solutions.*

Conjecture 4. *If $(A, B, C) = p^k$, $k \geq 1$, n is odd prime and $p \not\equiv 1 \pmod{2n}$, then (7.7) has no non-negative integer solutions.*

If $n = 3$, Conjecture 3 is true by Theorem 7.6.

Remark. If abc conjecture is true, then Conjecture 3 should be true for fixed prime p and sufficiently large n where n need not be prime.

Proof. Since $(A, B, C) = p$, by (7.7) we have

$$\begin{cases} \frac{A}{p} + \frac{B}{p} = \frac{C}{p}, \\ ABC = D^n. \end{cases}$$

So

$$\text{rad}\left(\frac{ABC}{p^3}\right) = \text{rad}\left(\frac{D^n}{p^3}\right) = \text{rad}\left(\left(\frac{D}{p}\right)^3 D^{n-3}\right)$$

$$\leq \text{rad}(D), \quad C > D^{n/3}.$$

For any $n \geq 7, 0 < \varepsilon < \frac{1}{3}$, we deduce that

$$p \leq D = D^{\frac{7}{3}-1-\frac{1}{3}} \leq D^{\frac{7}{3}-1-\varepsilon}$$

and

$$q\left(-\frac{A}{p}, -\frac{B}{p}, \frac{C}{p}\right) = \frac{\log\left(\frac{C}{p}\right)}{\log\left(\text{rad}\left(\frac{ABC}{p^3}\right)\right)}$$

$$\geq \frac{\frac{n}{3}\log D - \left(\frac{3}{7}-1-\varepsilon\right)\log D}{\log D}$$

By the third form of abc conjecture, there exist only finitely many triples $(-\frac{A}{p}, -\frac{B}{p}, \frac{C}{p})$. Let $A_1 = \frac{A}{p}, B_1 = \frac{B}{p}, C_1 = \frac{C}{p}$, then $(A_1, B_1, C_1) = 1, p^3 A_1 B_1 C_1 = D^n$. Let M be the greatest m such that there is prime q satisfying $q^m | A_1 B_1 C_1$. Then, if $n > M+3$, there is no solution for $p^3 A_1 B_1 C_1 = D^n$. Therefore, when $(A, B, C) = p$, equation (7.7) has no positive integer solutions for sufficiently large n.

Since Fermat equation has no trivial solution in \mathbb{Q}, people looks for solution in other fields, for example, there are a lot of results in the quadratic fields. As early as in 1915, W. Burnside found the following solutions in quadratic field:

$$\begin{cases} x = -3 + \sqrt{-3(1 + 4k^3)}, \\ y = -3 - \sqrt{-3(1 + 4k^3)}, \\ z = 6k. \end{cases}$$

Here $k \neq 0, -1$ is rational. When $k = 0$, Fermat's equation has no trivial solution in $\mathbb{Q}(\sqrt{-3})$. Later there are many results, after nearly a century, in 2013, under BSD conjecture, M. Jones and J. Rouse gave the necessary and sufficient condition for non-trivial solution of Fermat equation in $\mathbb{Q}(\sqrt{t})$, here t is an integer without square factor.

Let's consider the case $n = 3$, and we have the following result.

Proposition *If t is a square-free integer, not 0 or -1, and the elliptic function*

$$tu^2 = 1 + 4k^3$$

has non-zero rational solution (u, k), then when $n = 3$, equation (7.7) has infinitely many non-zero solutions (A, B, C, x) in the quadratic field $Q(\sqrt{t})$.

Proof. Let $A = a + b\sqrt{t}, B = c + d\sqrt{t}, x = e + f\sqrt{t}$. When $n = 3$, it follows from (7.7) that

$$a^2c + ac^2 + 2adtb + ad^2t + b^2tc + 2btcd$$
$$+(2acb + 2acd + a^2d + bc^2 + tb^2d + tbd^2)\sqrt{t}$$
$$= e^3 + 3ef^2t + f(3e^2 + f^2t)\sqrt{t},$$

hence

$$\begin{cases} a^2c + ac^2 + 2adtb + ad^2t + b^2tc + 2btcd = e^3 + 3ef^2t, \\ 2acb + 2acd + a^2d + bc^2 + tb^2d + tbd^2 = f(3e^2 + f^2t). \end{cases}$$

Solve the above two solutions, we have

$$t = -\frac{a^2c + ac^2 - e^3}{ad^2 + b^2c + 2bcd + 2adb - 3ef^2}$$
$$= -\frac{2acb + 2acd + a^2d + bc^2 - 3fe^2}{b^2d + bd^2 - f^3}.$$

Take $e = kc, f = kd$, from the expression of t we can get the following equation:

$$(ad + 2ab + cb - 2ck^3d)$$
$$\times (d^2a^2 + c^2b^2 + 2cbda + 2c^2db + 2cd^2a - 4c^2d^2k^3) = 0.$$

Let $ad + 2ab + cb - 2ck^3d = 0$, we get

$$d = -\frac{b(c + 2a)}{a - 2ck^3}.$$

Hence,

$$t = \frac{(a - 2ck^3)^2}{(1 + 4k^3)b^2}.$$

If $a - 2ck^3 = (1 + 4k^3)b$, then

$$t = 1 + 4k^3.$$

Therefore, we get the solution

$$\begin{cases} A = (1 + 4k^3)a + (a - 2k^3c)\sqrt{1 + 4k^3}, \\ B = (1 + 4k^3)c - (2a + c)\sqrt{1 + 4k^3}, \\ C = (1 + 4k^3)(a + c) - (a + (2k^3 + 1)c)\sqrt{1 + 4k^3}, \\ x = (1 + 4k^3)kc - k(2a + c)\sqrt{1 + 4k^3}, \end{cases}$$

where a, c are parameters, $\mathbb{Q}(\sqrt{t})$ $k \in \mathbb{Q}-\{0\}$.

Let $tu^2 = 1 + 4k^3$, where t is a square-free integer, not 0 or -1. When $n = 3$, equation (7.7) has non-zero solutions in the quadratic field $\mathbb{Q}(\sqrt{t})$:

$$\begin{cases} A = uta + (a - 2k^3c)\sqrt{t}, \\ B = utc - (2a + c)\sqrt{t}, \\ C = ut(a + c) - (a + (2k^3 + 1)c)\sqrt{t}, \\ x = utkc - k(2a + c)\sqrt{t}. \end{cases}$$

Example. When $k = -1$, then $t = -3$. For $n = 3$, equation (7.7) has non-zero solutions in quadratic field $\mathbb{Q}(\sqrt{-3})$.

$$\begin{cases} A = -3ua + (a + 2c)\sqrt{-3}, \\ B = -3uc - (2a + c)\sqrt{-3}, \\ C = -3u(a + c) - (a - c)\sqrt{-3}, \\ x = 3uc + (2a + c)\sqrt{-3}. \end{cases}$$

Problem. Except $t = -3$, does it exist other t, such that when $n = 3$, equation (7.7) has non-zero solutions on the quadratic field $\mathbb{Q}(\sqrt{t})$?

Remark. After my talk on this subject in Princeton on November 2, 2018, I have got several emails from Shai Evra, a member of the Institute for Advanced Study, he believes that the similar proof of Fermat's last theorem might also work in our new Fermat equation when the following two mild conditions holds: A is even and B = 1 (mod 4). Later he said he will try to remove the condition B = 1 (mod 4), but we haven't yet found a counterexample for $n = 9$.

33.　Euler's Conjecture

In 1769, Euler returned to Petersburg from Berlin, then he proposed a conjecture and tried to generalize Fermat's last theorem to multi-variable.

Stamp issued by the former Democratic Republic of Germany in 1983, to commemorate the 200th anniversary of the death of Euler

Euler's Conjecture. For any positive integers $s \geq 3$, the Diophantine equation

$$a_1^s + a_2^s + \cdots + a_{s-1}^s = a_s^s \tag{7.16}$$

has no positive integer solutions.

For $s = 3$, this is a special case of Fermat's Last Theorem, so Euler's conjecture holds. But for $s > 3$, the case is different.

In 1911, Norrie found the Diophantine equation

$$a^4 + b^4 + c^4 + d^4 = e^4$$

has a solution $(a, b, c, d, e) = (30, 120, 272, 315, 353)$, but this is not the counterexample of Euler's conjecture.

In 1966, Lander and Parkin gave the first counterexample of Euler's conjecture for $s = 5$, i.e.,

$$27^5 + 84^5 + 110^5 + 133^5 = 144^5.$$

In 1988, the American mathematician Elkies (1966–) used the method of elliptic curve to construct an infinite series of counterexamples for the case $k = 4$, one of them is:

$$2682440^4 + 15365639^4 + 18796760^4 = 20615673^4.$$

At the same year, Roger Frye found a smaller counterexample:

$$95800^4 + 217519^4 + 414560^4 = 422481^4.$$

In 2004, Jim Frye gave a counterexample for $s = 5$:

$$55^5 + 3183^5 + 28969^5 + 85282^5 = 85359^5.$$

It's worth mentioning that these two Frye and the Germany mathematician Gerhard Frey (1944–) who defined the Frye's curve which played a key role in the proof of Fermat's Last Theorem, are not the same person, though their Chinese name might be the same.

Map of Königsberg with seven bridges

For $s \geq 6$, it is unknown whether Euler's conjecture fails or holds.

In 2008, the American physicist Lee W. Jacobi and mathematician Daniel J. Madden studied the non-zero integer solutions of the Diophantine equation

$$a^4 + b^4 + c^4 + d^4 = (a + b + c + d)^4. \tag{J–M}$$

And it can be transformed into the Pythagoras triples
$$(a^2 + ab + b^2)^2 + (c^2 + cd + d^2)^2$$
$$= ((a + b)^2 + (a + b)(c + d) + (c + d)^2)^2.$$

By using the two solutions found independently by Brudno and Wroblewski around 1964,

$$(a, b, c, d) = (955, -2634, 1770, 5400), (7590, -31764, 27385, 48150)$$

and the method of elliptic curve, Jacobi and Madden proved (J–M) has infinitely many non-zero integer solutions. Then (E) has infinitely many non-zero integer solutions.

In April of 2012, the author was inspired by the new variant of the Hilbert–Waring problem and raised the following Diophantine equation:

$$\begin{cases} n = a_1 + a_2 + \cdots + a_{s-1}, \\ a_1 \, a_2 \, \dots \, a_{s-1}(a_1 + a_2 + \cdots + a_{s-1}) = b^s, \end{cases} \tag{7.17}$$

where $s \geq 3$, n, a_i, b are all positive integers.

Obviously, the solutions of (7.16) are a subset of the solutions of (7.17). The motivation of studying (7.17) is that we try to find a counterexample to Euler's conjecture for $s = 6$. Although we can't get anyone, but we (Tianxin Cai and Yong Zhang, A variety of Euler's sum of powers conjecture, May 2021, Czechoslovak Mathematical Journal, DOI: 10.21136/CMJ.2021.0210-20) find some interesting results about this problem by using the theory of elliptic curve.

Theorem 7.8. *For $s = 3$, equation (7.17) has no positive integer solution.*

Proof. For $s = 3$, equation (7.17) is equivalent to

$$\begin{cases} n = a_1 + a_2, \\ a_1 a_2 (a_1 + a_2) = b^3. \end{cases} \tag{7.18}$$

From the second equation, we have

$$\frac{a_1}{b} \frac{a_2}{b} \left(\frac{a_1}{b} + \frac{a_2}{b} \right) = 1.$$

Let $b_i = \frac{a_i}{b} (1 \le i \le 2)$, we get

$$b_1 b_2 (b_1 + b_2) = 1.$$

Then

$$\left(\frac{b_1}{b_2} \right)^2 + \frac{b_1}{b_2} = \frac{1}{b_2^3}.$$

Put $u = \frac{b_1}{b_2}, v = \frac{1}{b_2}$, we have

$$u^2 + u = v^3.$$

Take $y = 16u + 8, x = 4v$, then

$$y^2 = x^3 + 64.$$

Using the package of Magma, we can get the only rational points on it, i.e.,

$$(x, \pm y) = (8, 24), (0, 8), (-4, 0).$$

Tracing back, there is no integer solution of (7.17). This completes the proof of Theorem 7.8.

Theorem 7.9. *For $s = 4$, equation (7.17) has infinitely many positive integer solutions.*

Proof. For $s = 4$, equation (7.17) becomes

$$\begin{cases} n = a_1 + a_2 + a_3, \\ a_1 a_2 a_3 (a_1 + a_2 + a_3) = b^4. \end{cases} \tag{7.19}$$

From the second equation of 7.19, we have

$$\frac{a_1}{b} \frac{a_2}{b} \frac{a_3}{b} \left(\frac{a_1}{b} + \frac{a_2}{b} + \frac{a_3}{b} \right) = 1.$$

Take $b_i = \frac{a_i}{b} (1 \leq i \leq 3)$, then

$$b_1 b_2 b_3 (b_1 + b_2 + b_3) = 1.$$

It's easy to see that $(a_1, a_2, a_3) = (1, 2, 24)$ satisfies (7.19), so

$$(b_1, b_2, b_3) = \left(\frac{1}{6}, \frac{1}{3}, 4 \right).$$

Hence,

$$\begin{cases} b_1 b_2 b_3 = \frac{2}{9}, \\ b_1 + b_2 + b_3 = \frac{9}{2}. \end{cases} \tag{7.20}$$

Next, we consider b_i be unknowns. Eliminating b_3 in (7.20), then

$$18 b_1^2 b_2 + 18 b_1 b_2^2 - 81 b_1 b_2 + 4 = 0.$$

Take $u = \frac{b_1}{b_2}, v = \frac{1}{b_1}$, we have

$$18 u^2 + 18 u - 81 u + 4 v^3 = 0.$$

Let

$$y = 384 u - 864 v + 192, \quad x = -32 v + 243,$$

we get an elliptic curve

$$E : y^2 = x^3 - 166779 x + 26215254.$$

By the Nagell–Lutz theorem (Section 27), to prove that there are infinitely many rational points on E, it is enough to find a rational point on E with x-coordinate not in \mathbb{Z}. Using the package Magma,

we can get the rational point

$$\left(\frac{30507}{121}, -\frac{584592}{1331}\right),$$

where the x-coordinate is not an integer.

From the above transformations, we have

$$\begin{cases} b_1 = \dfrac{32}{243 - x}, \\[2mm] b_2 = \dfrac{+y - 27x + 6369}{12(243 - x)}, \\[2mm] b_3 = \dfrac{-y - 27x + 6369}{12(243 - x)}. \end{cases}$$

Then

$$\begin{cases} a_1 = \dfrac{32}{243 - x}b, \\[2mm] a_2 = \dfrac{y - 27x + 6369}{12(243 - x)}b, \\[2mm] a_3 = \dfrac{-y - 27x + 6369}{12(243 - x)}b \end{cases}$$

is a solution of (7.19).

In order to make $b_i > 0$, the equivalent condition is

$$x < 243, |y| < 27x - 6369.$$

By the Poincaré–Hurwitz theorem (Section 27), the elliptic curve has infinitely many rational points in every neighborhood of any one of them. We should find a point satisfy the above equivalent condition. It's easy to see that the point $P = (235, 8)$ is what we want, then there are infinitely many rational points (x, y) satisfying the above condition.

Therefore, we can find infinitely many rational numbers $b_i > 0$ $(1 \le i \le 3)$ satisfying (7.20), leading to integer numbers $a_i > 0$, by multiplying the least common multiple of b_i's denominator. It means

that (7.17) has infinitely many positive integer solutions for $s = 4$. This completes the proof of Theorem 7.9.

Example 7.3. For $s = 4$, the rational points

$$(x, y) = (235, 8), \left(\frac{60266587}{257049}, \frac{3852230624}{130323843} \right)$$

lie on E, leading to the solution of (7.20):

$$(a_1, a_2, a_3) = (1, 2, 24), (781943058, 138991832, 18609625).$$

Theorem 7.10. *For $s \geq 5$, equation (7.17) has infinitely many positive integer solutions.*

The proof of Theorem 7.10 is similar to that of Theorems 7.8 and 7.9, but is much more complicated, so we omit it.

Example 7.4. For $s = 5$, equation (7.17) has positive integer solutions

$$(2, 28, 49, 49) \ (b = 28), (5, 81, 90, 324) \quad (b = 90).$$

Next, we give another form of Diophantine equation.

In 2005, A, Murthy (http://oeis.org/A109909) raised the following conjecture.

Conjecture 5. *For any $n > 3$, there exist positive integers a and b satisfying*

$$\begin{cases} n = a + b, \\ ab - 1 = p, \end{cases}$$

where p is a prime. For any $n > 1$, when $n \neq 6, 30, 54$, there are positive integers a and b satisfying

$$\begin{cases} n = a + b, \\ ab + 1 = p, \end{cases}$$

where p is a prime.

In 2012, the author ,who didn't notice the work of Murthy, proposed Conjecture 5 and

Conjecture 6. *For any $n > 14$, there exist positive integers a, b and c such that*

$$\begin{cases} n = a + b + c, \\ abc + 1 = p^2. \end{cases}$$

Conjecture 7. *For any $n > 8$, there exist positive integers a, b, c and d such that*

$$\begin{cases} n = a + b + c + d, \\ abcd = p^3 - p. \end{cases}$$

Remark. Here we give (Tianxin Cai and Deyi Chen) some remarks on Conjecture 5. By

$$\begin{cases} n = a + b, \\ ab \pm 1 = p, \end{cases}$$

and Vieta theorem, we know that a and b are two solutions of $x^2 - nx + p \mp 1 = 0$, i.e.,

$$\frac{n \pm \sqrt{n^2 - 4(p \mp 1)}}{2}.$$

Then there is a positive integer m such that

$$n^2 - 4(p \mp 1) = m^2.$$

If $n = 2k$, then $m = 2s$,

$$k^2 \pm 1 = p + s^2. \tag{7.21}$$

If $n = 2k + 1$, then $m = 2s + 1$,

$$k^2 + k \pm 1 = p + s(s + 1). \tag{7.22}$$

Equation (7.21) is a consequence of the famous conjecture.

Hardy–Littlewood conjecture (1923). Every sufficiently large even number is a square number or the sum of a prime and a square number.

For any odd prime p, we call $\frac{p-1}{2}$ half-prime. Then the twin prime conjecture is equivalent to the proposition that there are infinitely many pair of consecutive half-primes. And Goldbach's conjecture is

equivalent to the proposition that every positive integer greater than 1 can be expressed as the sum of two half-primes.

Equation (7.22) is a consequence of the following assertion.

Every positive integer greater than 1 can be expressed as the sum of a half-prime and a triangular number.

This is coincident with a conjecture of Zhiwei Sun (*J. Comb. Number Theory.* **1** (2009) 65–76): Every odd integer greater than 3 can be expressed as $p+x(x+1)$, where p is a prime and x is a positive integer.

34. *F*-Perfect Number Problem

In Section 1 of this book, we have introduced perfect numbers, i.e., a natural number which is equal to the sum of its proper positive divisors. There is a bijection between even perfect numbers and Mersenne primes, so it is a remarkable and famous problem in computing. The infinity or not of these two kinds of numbers is an immortal riddle, and it can be considered as the oldest and most difficult problem in the history of mathematics. Meanwhile, some great mathematicians tried to find the generalization of perfect numbers, for example,

$$\sum_{d|n,d<n} d = kn,$$

where k is a fixed positive integer, it's called k-perfect number. When $k = 1$, it is ordinary perfect number. These great mathematicians include Mersenne, Descartes and Fermat, as well as Lehmer and Carmichael, etc. However, so far they only found a few isolated solutions to these equations, not related with any infinite property as Mersenne primes.

The first person who found k-perfect number ($k > 1$) is the Welsh physician and mathematician Robert Recorde (1512–1558) and he found 120 is 2-perfect number in 1557, in that year he also invented "=" as equal sign. Fermat found the second 2-perfect number 672 in 1637, in that year he raised Fermat's Last Theorem. In 1644, Fermat gave another 2-perfect number with 11 digits. Before Mersenne

A P P E N D I X,
Continens

O B I E C T I O N E S

QVINTAS & SEPTIMAS

In

R E N A T I D E S - C A R T E S

M E D I T A T I O N E S

De Primâ Philofophiâ,

Cum ejafdem ad illas Refponfionibus & duabus Epiftolis,

Vna ad Patrem D I N E T Societatis Iefu Præpofitum
Provincialem per Franciam,

Altera ad celeberrimum Virum

D. G I S B E R T V M V O E T I V M.

AMSTELODAMI.
Apud Ludovicum Elzevirium,
cɪɔ ɪɔc ʟɪv.

Title pages of Descartes Meditationes, he was an amateur number theorist

and Descartes found two 2-perfect numbers with 9 digits and 10 digits, respectively, these three French mathematicians also found other k-perfect numbers.

Let $\sigma(n) = \sum_{d|n} d$. Then perfect number problem is equivalent with $\sigma(n) = 2n$. In 1969,the Indian mathematician Suryanarayana proposed the definition of superperfect number, i.e., n is a natural number satisfying $\sigma^2(n) = \sigma(\sigma(n)) = 2n$. He proved that all the even superperfect numbers in the form of 2^{p-1}, where $2^p - 1$ is Mersenne primes. Later (m, k) is also called superperfect number, if n is a natural number satisfying

$$\sigma^m(n) = kn.$$

In the spring of 2012,the author considered the case of square sum, i.e, the following equation

$$\sum_{d|n, d<n} d^2 = 3n, \qquad (7.23)$$

Former residence of Descartes in Utrecht. Photo taken by the author

and got some surprising results. We (Tianxin Cai, Deyi Chen and Yong Zhang, Perfect numbers and Fibonacci primes (I), *Inter. J. Number Theory*, **11** (2015), 159–169) proved the following theorem.

Theorem 7.11. *All the positive integer solutions of (7.23) are* $n = F_{2k-1}F_{2k+1}(k \geq 1)$, *where* F_{2k-1} *and* F_{2k+1} *are Fibonacci twin primes.*

We might call the original perfect numbers perhaps defined by Pythagoras M-perfect numbers because of the relation with Mersenne primes. And we call those numbers satisfying (7.23) F-perfect numbers. It is worth mentioning that during the China–Japan number theory conference hold in Fukuoka in 2013, the Japanese mathematician Kohji Matsumoto suggested to call them Ying perfect numbers and Yang perfect numbers, respectively. So far, the largest known Fibonacci prime is F_{81839} and the largest known probable Fibonacci prime is $F_{1968721}$ (it has 411439 digits). By the list of Fibonacci primes, we deduce that n $= F_3F_5, F_5F_7,$ $F_{11}F_{13}, F_{431}F_{433}, F_{569}F_{571}$ are the only 5 F-perfect numbers we know.

There are (it is easy to see that every pair of footnote number must be twin primes too)

10,

65,

20737,

73510803816922669761033626642123533261948011970405233919814585711917444519057612261963528801744523093107269516305744106136707871525711296518385628509088429445930772087319647420825725 7,

35232209573904449595952790620404802458842537915400184965695897596126849742246390276402878432136154463286879043721897517251836590479716000271118557285532827829382383900100646042179787559935516043180579182691829284567616114036685 7

7116737601.

It is easy to see that 10 is the unique even F-perfect number, as there is only one even prime 2. The next probable odd F-perfect number has at least 822878 digits. Anyway, neither we know that there is the sixth F-perfect number nor if there are infinitely many F-perfect numbers.

Next, we study a more general case, for any positive integers a and b, we consider the equation

$$\sum_{d|n,d<n} d^a = bn. \tag{7.24}$$

What we get is the following theorem.

Theorem 7.12. *If $a = 2, b \neq 3$, or $a \geq 3, b \geq 1$, then* (7.24) *has only finitely many solutions. In particular, when $a = 2, b = 1$ or 2, equation* (7.24) *has no solution.*

In other word, except M-perfect numbers and F-perfect numbers, there is no other interesting perfect numbers.

To prove Theorems 7.11 and 7.12, we need the following lemmas.

Lemma 7.5. *Let $d > 0$ be an odd number. Then $x^2 - dy^2 = -4$ has integer solutions if and only if $u^2 - dv^2 = -1$ has integer solutions.*

Proof. The sufficiency is obvious. Now we prove the necessity. From the condition, x and y have the same parity. Make the transformation

$$
\begin{cases}
u = \dfrac{x(x^2 + 3)}{2}, \\
v = \dfrac{(x^2 + 1)y}{2}
\end{cases}
$$

then

$$
u^2 - dv^2 = \left(\frac{x(x^2 + 3)}{2} \right)^2 - d \left(\frac{(x^2 + 1)y}{2} \right)^2 = -1. \qquad \square
$$

Lemma 7.6. *If $N > 0$ is not a perfect square, then $x^2 - Ny^2 = -1$ has integer solutions if and only if $l(\sqrt{N})$ is odd, where $l(\sqrt{N})$ denotes the period of the continued fraction of \sqrt{N}.*

We can refer to P. Kaplan and K. S. Williams (*J. Number Theory* **23** (1986) 169–182) for the proof.

Lemma 7.7. *All solutions of the equation $x^2 + y^2 + 1 = 3xy$ ($1 \leq x < y$) are*

$$
\begin{cases}
x = F_{2k-1}, \\
y = F_{2k+1},
\end{cases}
\tag{7.25}
$$

where $k \leq 1$ and F_n is the nth Fibonacci number.

Proof. First of all, we prove that x and y given by (7.25) satisfy the equation $x^2 + y^2 + 1 = 3xy$. By Cassini's identity

$$
F_n^2 - F_{n+1}F_{n-1} = (-1)^{n-1},
$$

we have

$$
F_{2k}^2 - F_{2k+1}F_{2k-1} = -1.
$$

Then

$$(F_{2k+1} - F_{2k-1})^2 - F_{2k+1}F_{2k-1} = -1,$$

that is to say

$$1 + F_{2k-1}^2 + F_{2k+1}^2 = 3F_{2k-1}F_{2k+1}.$$

Now we prove that (7.25)gives all the solutions. We only need to prove that if $x^2 + y^2 + 1 = 3xy$ $(1 \le x < y)$, then $x = F_{2k-1}$.

Note that this equation is equivalent to

$$5x^2 - 4 = (3x - 2y)^2. \tag{7.26}$$

By I. Gessel (Fibonacci is a square, *Fibonacci Quarterly* **10**(4) (1972) 417–419), we know that x is Fibonacci sequence. If $x = F_{2k-1}$, then we are done. If $x = F_{2k}$, substitute it into (7.26), then $5F_{2k}^2 - 4 = (3x - 2y)^2$. By the result in Gessel's paper, $5F_{2k}^2 + 4 = L_{2k}^2$, where $L_0 = 2, L_1 = 1, L_{n+1} = L_n + L_{n-1}$ is Lucas sequence, therefore

$$8 = 5F_{2k}^2 + 4 - (5F_{2k}^2 - 4) = L_{2k}^2 - (3x - 2y)^2. \tag{7.27}$$

From (7.27) we get $|3x - 2y| = 1, L_{2k} = 3$, hence $x = F_{2k} = \sqrt{\frac{L_{2k}^2 - 4}{5}} = 1 = y$. Also $y = 2 = F_3$. A contradiction! Hence, $x = F_{2k-1}$. $\qquad\square$

Lemma 7.8. *If $k \ne 3$ is a positive integer, then $1 + x^2 + y^2 = kxy$ has no solutions.*

Proof. Since $kxy = x^2 + y^2 + 1 > 2xy$, we only need to consider the case $k \ge 4$. If $1 + x^2 + y^2 = kxy$ has integer solution, the discriminant of the above quadratic form in x must be a square, it means there exists an integer z such that

$$k^2y^2 - 4(y^2 + 1) = (k^2 - 4)y^2 - 4 = z^2,$$

or

$$z^2 - (k^2 - 4)y^2 = -4.$$

Now we show that k is odd. Assume that k is even, then x and y can't be both even. If only one of x and y is odd, then

$$1 + x^2 + y^2 = kxy \pmod 4 \Rightarrow 2 \equiv 0 \pmod 4.$$

If x and y are both odd, then

$$1 + x^2 + y^2 = kxy \pmod 2 \Rightarrow 1 \equiv 0 \pmod 2.$$

Therefore k is odd. Note that $k^2 - 4$ is not a perfect square for odd $k \geq 4$, so by Lemmas 7.5 and 7.6 we obtain

$$z^2 - (k^2 - 4)y^2 = -4 \text{ has integer solution}$$

$$\Leftrightarrow z^2 - (k^2 - 4)y^2 = -1 \text{ has integer solution}$$

$$\Leftrightarrow l(k^2 - 4) \text{ is odd}.$$

On the other hand, for odd $k \geq 4$, by Exercise 11 on p. 503 of K. H. Rosen, *Elementary Number Theory and Its Applications*, 5th edn., Addison-Wesley, 2004, we have

$$\sqrt{k^2 - 4} = \left[k - 1; \overline{1, \frac{k-3}{2}, 2, \frac{k-3}{2}, 1, 2k - 2} \right],$$

i.e., $l(k^2 - 4) = 6$. A contradiction! \square

Proof of Theorem 7.12. Let $n = p_1^{\alpha_1} p_2^{\alpha_2} \ldots p_k^{\alpha_k}$ be the standard prime factorization of n, where $p_1 < p_2 < \cdots < p_k, \alpha_i \geq 1, (1 \leq i \leq k)$. First, we consider the case $a = 2$.

If $k = 1$, equation (7.17) becomes $1 + p_1^2 + p_1^4 + \cdots + p_1^{2(\alpha_1 - 1)} = bp_1^{\alpha_1}$, which is impossible. If $k \geq 3$, by using the arithmetic–geometric mean inequality, we get

$$bn = \sum_{d|n, d<n} d^2 \geq \frac{n^2}{p_1^2} + \frac{n^2}{p_2^2} + \cdots + \frac{n^2}{p_k^2}$$

$$\geq k \left(\frac{n^2}{p_1^2} \cdot \frac{n^2}{p_2^2} \cdots \frac{n^2}{p_k^2} \right)^{\frac{1}{k}} = \left(k p_1^{\alpha_1 - \frac{2}{k}} p_2^{\alpha_2 - \frac{2}{k}} \ldots p_k^{\alpha_k - \frac{2}{k}} \right) n$$

Note that $\alpha_i - \frac{2}{k} \geq \frac{\alpha_i}{3} (1 \leq i \leq k)$, we have

$$b \geq k p_1^{\alpha_1 - \frac{2}{k}} p_2^{\alpha_2 - \frac{2}{k}} \ldots p_k^{\alpha_k - \frac{2}{k}} \geq k \prod_{i=1}^{k} p_i^{\frac{\alpha_i}{3}} \geq k n^{\frac{1}{3}}. \tag{7.28}$$

Hence, $n \leq \left(\frac{b}{k} \right)^3$. Note that $\sum_{k=1}^{\infty} \frac{1}{k^3}$ converges, therefore n is bounded.

If $k = 2$, from (7.28) we have

$$b \geq 2p_1^{\alpha_1-1}p_2^{\alpha_2-1}, \qquad (7.29)$$

from (7.29) we know that α_1, α_2 are bounded. If $\alpha_2 > 1$, then we deduce from (7.29) that p_2 is bounded, so n is bounded. If $\alpha_2 = 1, \alpha_1 > 1$, then p_1 is bounded. From (7.23) we have

$$(1 + p_1^2 + p_1^4 + \cdots + p_1^{2\alpha_1})(1 + p_2^2) - p_1^{2\alpha_1}p_2^2 = bp_1^{\alpha_1}p_2,$$

then

$$p_2 | 1 + p_1^2 + p_1^4 + \cdots + p_1^{2\alpha_1}.$$

Hence p_2 is bounded and n is bounded. If $\alpha_1 = \alpha_2 = 1$, from (7.24) we have

$$1 + p_1^2 + p_2^2 = bp_1p_2. \qquad (7.30)$$

If $b \neq 3$, from Lemma 7.8 we get (7.30) has no integer solution.

Therefore, when $a = 2$, $b \neq 3$, (7.24) has only finitely many solutions. In particular, according to (7.28)–(7.30), when $b = 1, 2$, equation (7.24) has no integer solution.

Next we consider the case $a \geq 3$. It's easy to see that, $k \neq 1$. When $k \geq 2$, by a similar argument, we get

$$b \geq kp_1^{(a-1)\alpha_1 - \frac{a}{k}}p_2^{(a-1)\alpha_2 - \frac{a}{k}} ... p_k^{(a-1)\alpha_k - \frac{a}{k}} \geq k \prod_{i=1}^{k} p_i^{\frac{\alpha_i}{2}} \geq kn^{\frac{1}{2}},$$

so $n \leq \left(\frac{b}{k}\right)^2$. Note that $\sum_{k=1}^{\infty} \frac{1}{k^2} = \frac{\pi^2}{6}$, hence n is bounded. \square

Proof of Theorem 7.11. Let $n = p_1^{\alpha_1}p_2^{\alpha_2}...p_k^{\alpha_k}$ be the standard prime factorization of n, where $p_1 < p_2 < \cdots < p_k, \alpha_i \geq 1, (1 \leq i \leq k)$. Now we want to prove that if n is an F-perfect number, then $k = 2$, $\alpha_1 = \alpha_2 = 1$.

If $k = 1$, equation (7.23) is $1 + p_1^2 + p_1^4 + \cdots + p_1^{2(\alpha_1-1)} = 3p_1^{\alpha_1}$, which is impossible.

If $k \geq 3$, from (7.28) we get $n = 1$, this is impossible.

If $k = 2$, assume that $a_1 + \alpha_2 \geq 3$, note that $\alpha_1^2 - \alpha_1 + 2\alpha_1\alpha_2 \geq \alpha_1(\alpha_1+\alpha_2)$, $\alpha_2^2 - \alpha_2 + 2\alpha_1\alpha_2 \geq \alpha_2(\alpha_1+\alpha_2)$. By using the arithmetic–geometric mean inequality, we get

$$
\begin{aligned}
3n &> (1 + p_1^2 + \cdots + p_1^{2(\alpha_1-1)})p_2^{2\alpha_2} + (1 + p_2^2 + \cdots + p_2^{2(\alpha_2-1)})p_1^{2\alpha_1} \\
&> (\alpha_1 + \alpha_2)(p_2^{2\alpha_2} \times p_1^2 p_2^{2\alpha_2} \times \cdots \times p_1^{2(2\alpha_1-1)} p_2^{2\alpha_2} \times p_1^{2\alpha_1} \times p_2^2 p_1^{2\alpha_1} \\
&\qquad \times \cdots \times p_2^{2(2\alpha_2-1)} p_1^{2\alpha_1})^{\frac{1}{\alpha_1+\alpha_2}} \\
&\geq 3p_1^{\frac{\alpha_1^2-\alpha_1+2\alpha_1\alpha_2}{\alpha_1+\alpha_2}} \; p_2^{\frac{\alpha_2^2-\alpha_2+2\alpha_1\alpha_2}{\alpha_1+\alpha_2}} \geq 3p_1^{\alpha_1} p_2^{\alpha_2} = 3n.
\end{aligned}
$$

A contradiction! Hence, $k = 2, \alpha_1 = \alpha_2 = 1$. Theorem 7.11 follows from Lemma 7.7.

Remark. For $a = 1$, $b > 1$, we are not sure whether (7.24) has only finite solutions.

In the spring of 2014, the author raised a more general form about F-perfect numbers, that is

$$
\sum_{d \mid n, d < n} d^2 = L_{2s}n - F_{2s}^2 + 1,
$$

where s is a positive integer, F_{2s} and L_{2s} are Fibonacci sequence and Lucas sequence. We (Tianxin Cai, Liuquan Wang and Yong Zhang, Perfect Numbers and Fibonacci Primes (II), *Integers* **19** (2019) #A21 1–10) proved that except for finitely many computable solutions, all the solutions of the above equation are

$$
n = F_{2k+1}F_{2k+2s+1} \text{ or } n = F_{2k+1}F_{2s-2k+1},
$$

where k is any positive integer, F_{2k+1} and $F_{2k+2s+1}$ ($F_{2k-2s-1}$) are Fibonacci primes.

In particular, when $s = 1$, it is (7.23) and Theorem 7.11.

Meanwhile, we also have that for any positive integer k, the equation

$$
\sum_{d \mid n, d < n} d^2 = 2n + 4k^2 + 1
$$

has infinitely many solutions if and only if Polignac's conjecture holds. In particular, when $k = 1$, the equation

$$\sum_{d|n, d<n} d^2 = 2n + 5$$

has infinitely many solutions if and only if the twin prime conjecture holds.

Recently, we discover that for any positive integer k, the equation

$$\sum_{\substack{d|n \\ d<n}} 2kd^2 - (2k-1)d = (4k^2 + 1)n + 2$$

has infinitely many solutions if and only if there exist infinitely many primes p such that $2kp + 1$ is prime. In particular, if $k = 1$, i.e., there are infinitely many Sophie Germain primes. While there exist infinitely many primes p such that $2kp - 1$ is prime, if and only if the equation

$$\sum_{\substack{d|n \\ d<n}} 2kd^2 + (2k-1)d = (4k^2 + 1)n + 4k$$

has infinitely many solutions.

Recently, we (Hao Zhong and Tianxin Cai) have also considered the corresponding problems of generalized Lucas sequences. However, there is still some problem left over: is there a polynomial $f(n) = an^2 + bn + c$ with integer coefficients such that the solutions (except for finitely many computable solutions) of

$$\sum_{d|n, d<n} d^2 = f(n)$$

has a representation about the primes?

Remark. The German-born physicist Albert Einstein (1879–1955) wrote in his autobiography: "The really true theory could not be linear and cannot be deduced from linear form." These words might be some extreme, but it had been verified in this book.

35.　New Congruent Number Problem

A congruent number is a positive integer that is the area of a right triangle with three rational number sides. The simplest example is 6 because it is the area of a 3-, 4-, 5- triangle. The so-called congruent number problem is "to find a simple criterion to determine whether a given natural number is a congruent number". Obviously, for any positive integers m and n, $m^2 n$ is a congruent number if and only if n is a congruent number. Hence, we just need to study the square-free positive integers.

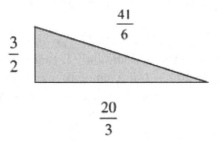

The least congruent number $n = 5$

It is said that the Arab mathematician Al-Karaji who lived in Baghdad studied the congruent number problem in the 10th century. The Italian Fibonacci, who is considered "the most talented western mathematician of the Middle Ages", proved that $n = 5$ is a congruent number, the corresponding rational sides are $(\frac{3}{2}, \frac{20}{3}, \frac{41}{6})$, see the left picture. Fibonacci asserted that $n = 1$ is not a congruent number, but his proof is wrong. After four centuries, Fermat gave a right proof, this reduced that Fermat's Last Theorem holds for $n = 4$, i.e., the equation $x^4 + y^4 = z^4$ has no positive integer solutions. That is a proof by Fermat hardly be seen before his death. Therefore, all square numbers are not congruent numbers. In the 18th century, Euler found that 7 is a congruent number.

The congruent numbers less than 50 are 5, 6, 7, 13, 14, 15, 20, 21, 22, 23, 24, 28, 29, 30, 31, 34, 37, 38, 39, 41, 45, 46, 47, total 23. There are many special congruent numbers, such as the product of three consecutive integers $n^3 - n$ is a congruent number, $4n^3 + n$, $n^4 - m^4, n^4 + 4m^4, 2n^4 + 2m^4$ are congruent numbers and so on.

In the second half of the 20th century, mathematicians found that the congruent number problem has a close relationship with elliptic curves as Fermat's Last Theorem. In fact, n is a congruent number if and only if

$$\begin{cases} a^2 + b^2 = c^2 \\ \dfrac{1}{2}ab = n \end{cases} \qquad (7.31)$$

has positive rational solutions (a, b, c).

Plus and subtract 4 times the second equation of (7.31) to the first equation, we get

$$(a \pm b)^2 = c^2 \pm 4n.$$

Multiplying 16 on both sides in the above equation, we have

$$\left(\frac{a^2 - b^2}{4}\right)^2 = \left(\frac{c}{2}\right)^4 - n^2.$$

This means that if n is a congruent number, then $u^4 - n^2 = v^2$ has rational solutions $u = \frac{c}{2}, v = \frac{a^2 - b^2}{4}$. Multiplying u^2 on both sides, taking

$$x = u^2 = \left(\frac{c}{2}\right)^2, y = uv = \frac{(a^2 - b^2)c}{8},$$

then there exists a pair of rational numbers (x, y) satisfying

$$y^2 = x^3 - n^2 x. \qquad (7.32)$$

Conversely, if x and y satisfying (7.32) and $y \neq 0$, put

$$a = \frac{x^2 - n^2}{y}, b = \frac{2nx}{y}, c = \frac{x^2 + n^2}{y}.$$

It's easy to verify that they satisfy (7.31). Then there exists a bijection between (a, b, c) and (x, y). By the theory of elliptic curve, we find that the torsion points on (7.32) are those satisfying $y = 0$. Hence the existence of rational points with y not zero is equivalent to positive rank of the elliptic curve.

By the property of elliptic curve, we have

If $p \equiv 3 \pmod 8$ is a prime, then p is not a congruent number, while $2p$ is r.

If $p \equiv 5 \pmod 8$ is a prime, then p is a congruent number.

If $p \equiv 7 \pmod 8$ is a prime, then both p and $2p$ are congruent numbers.

The second result above is got by the German radio engineer Kurt Heegner (1893–1965) (*Math. Z.* **56**, 227–253), and he was the first person to prove that there are infinitely many square-free congruent numbers. In 2012, Ye Tian (Congruent Numbers and Heegner Points, *Cambridge J. Math.*, **2**(1), (2014), 117–161) proved that for any positive integers k, there are infinitely many square-free congruent numbers with exactly $k + 1$ odd prime divisors in each residue class of 5, 6, and 7 modulo 8.

According to Goldfeld conjecture, the number of congruent numbers and non-congruent numbers are both half of all. More exactly, almost all positive integers congruent to 5, 6, 7 modulo 8 are congruent numbers, and almost all positive integer congruent to 1, 2, 3 modulo 8 are non-congruent numbers. While under BSD conjecture, it has been proved that each $n \equiv 5, 6, 7 \pmod 8$ is a congruent number. The least congruent number in other form is 34, its sides are $(225/30, 272/30, 353/30)$; the least congruent numbers $n \equiv 1 \pmod 8$ and $n \equiv 3 \pmod 8$ are 41 with sides $(40/3, 123/20, 881/60)$ and 219 with sides $(55/4, 1752/55, 7633/220)$, respectively. For any positive integer $n \equiv 5, 6, 7 \pmod 8$.

Recently, Ye Tian, Xinyi Yuan, and Shouwu Zhang(Genus periods, genus points and congruent number problem, *Asian J. Math.* **21**(4), (2017) 721–774) showed some necessary and sufficient conditions for $n \equiv 5, 6, 7 \pmod 8$ to be a congruent number. They believe that these results can lead to positive density for congruent numbers. Later, A. Smith (The congruent numbers have positive natural density, preprint (2016); arXiv:1603.08479v2) announced and he proved:in the residue $n \equiv 5$ or $7 \pmod 8$, the proportion of congruent numbers is at least 62.9%, while in the residue $n \equiv 6 \pmod 8$, the proportion of congruent numbers is at least 41.9%.

Tunnell Theorem (1983). *If n is an odd congruent number, then*

$$\#\{n = 2x^2 + y^2 + 8z^2\}\# = 2 \times \#\{n = 2x^2 + y^2 + 32z^2\}\#.$$

If n is an even congruent number, then

$$\#\{n = 2x^2 + y^2 + 16z^2\}\# = 2 \times \#\{n = 2x^2 + y^2 + 64z^2\}\#.$$

where $\#\{\ \}\#$ denotes the number of integer solutions of this equation. Under the BSD conjecture, the inverse also holds.

The criterion of Tunnell theorem is practical, if combined with the method of exhaustion. For example, when $n = 1$, the solution numbers of the first two equations are both 2, i.e., $(0,\pm1,0)$, hence 1 is not a congruent number.

It's worth mentioning that in 2013 two American mathematicians Jones and Rouse proved under BSD conjecture: There exist non-trivial solutions of Fermat equation $x^3 + y^3 = z^3$ over quadratic fields if and only if the numbers of integer solutions of two ternary equations with degree 2 are the same.

Recently, Hourong Qin (Congruent numbers, quadratic forms and K_2) declared a few times at number theory conferences, that he proved the following results: if n is an odd congruent number, then

$$\#\{n = x^2 + 2y^2 + 32z^2\} = \#\{n = 2x^2 + 4y^2 + 9z^2 - 4yz\};$$

if n is an even congruent number, then

$$\#\{\frac{n}{2} = x^2 + 4y^2 + 32z^2\}\# = \#\{\frac{n}{2} = 4x^2 + 4y^2 + 9z^2 - 4yz\}\#.$$

Under the assumption of BSD conjecture, the inverse proposition of the above conclusion is also true.

When we know that n is a congruent number, to find the rational sides is still not easy. The Germany-born American mathematician Zagier (1951–) computed the rational sides of right triangle for $n = 157$, with the numerator and denominator of the hypotenuse 45 and 47 digits, respectively.

Like other problems in number theory, congruent number problem is also generalized. For example, there are t-congruent numbers and θ-congruent numbers. Let n be a positive integer and t be a positive

rational number, if there are positive rational numbers a, b, c such that

$$a^2 = b^2 + c^2 - 2bc\frac{t^2 - 1}{t^2 + 1}, \; bc\frac{2t}{t^2 + 1} = 2n,$$

then n is called a t-congruent number.

The case $t = 1$ corresponds to the classical congruent number. Its corresponding elliptic curve is

$$E_{n,t} : y^2 = x\left(x - \frac{n}{t}\right)(x + nt).$$

Let $0 < \theta < \pi$ be a real number, and $\cos\theta = \frac{s}{r}$ be a proper fraction, if $n\sqrt{r^2 - s^2}$ is the area of a triangle with rational sides and an intersection angle θ, then n is called a θ-congruent number. When $\theta = \frac{\pi}{2}$, it is the classical congruent number.

In October of 2012, the author observed when the length of base is zero, right trapezoid becomes right triangle. Then we have a new kind of congruent number.

Definition 7.1. A positive integer n is called congruent integers, if it is the area of the right trapezoid (as in Congruent number diagram 1), here a, b and c are positive integers, d is a non-negative integer, and $(b, c) = 1$.

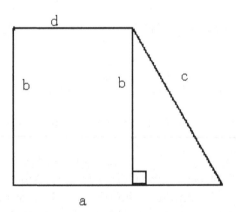

Congruent number diagram 1

Definition 7.2. A positive integer n is called k-congruent numbers, if it is the area of the right trapezoid (as in Congruent number diagram 1), here a, b, c and d are positive rational numbers, $k \geq 2$ is an integer, and $a = kd$.

From Definition 7.2, we have

$$n = \frac{(a+d)b}{2}, (a-d)^2 + b^2 = c^2. \tag{7.33}$$

Definition 7.3. A positive integer n is called d-congruent numbers, if it is the area of the right trapezoid (as in Congruent number diagram 2), here a, b and c are positive rational numbers and d is a non-negative integer.

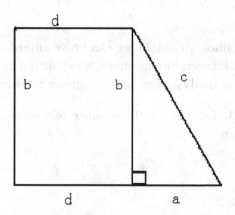

Congruent number diagram 2

From Definition 7.3, we have

$$n = \frac{(a+2d)b}{2}, a^2 + b^2 = c^2. \tag{7.34}$$

Now we (Tianxin Cai and Yong Zhang, Congruent numbers on the right trapezoid) study the above three kind of congruent numbers.

From Definition 7.1, we have

$$n = \frac{(a+d)b}{2}, \quad (a-d)^2 + b^2 = c^2, \quad (b,c) = 1.$$

By the property of Pythagoras triples, we get $(a-d, b) = 1$, $a-d$ and b have opposite parity,

$$(a-d, b) = (2xy, x^2 - y^2) \text{ or } (x^2 - y^2, 2xy),$$

where $x > y, (x,y) = 1$, x and y have opposite parity. Then we have the following condition.

A positive integer n is congruent integer if and only if $n = pk$, where p is any odd prime and $k \geq \frac{p^2-1}{4}$ or $n = 2^i k, k \geq 2^{2i} - 1, i \geq 1$, where k is any odd integer.

Conversely, $n > 1$ is non-congruent integer if and only if n has one of the following forms:

$$p, \quad p^2(p \neq 3), \quad pq\left(5 < p < q < \frac{p^2-1}{4}\right),$$

$$2^i(i \geq 0), \quad 2^i p(i \geq 2, 2^{1+i/2} < p < 2^{2i} - 1),$$

where p, q are primes.

By the Pigeonhole principle, we can prove almost all positive integer is congruent integer. Furthermore, we show the following theorem by the method of analysis and prime number theorem.

Theorem 7.13. *Let $f(x)$ be the number of non-congruent integers less than x. Then*

$$f(x) \sim \frac{cx}{\log x},$$

where $c = 1 + \ln 2$.

Remark. For the classical congruent integer, we also can define congruent integer number, which is the area of right triangle with integer sides. Let $g(x)$ be the number of non-congruent integers less than x, we don't find its approximate evaluation, but we can prove

$$\frac{\sqrt{x}}{2} + O(1) < g(x) \leq \frac{1}{2\sqrt[3]{4}} x^{\frac{2}{3}} + O(x^{\frac{5}{9}}).$$

Next we study the k-congruent number and d-congruent number by using the theory of elliptic curve.

Theorem 7.14. *For every positive integer n, there exists k such that n is a k-congruent number.*

Proof. In (7.33), take

$$b = \left|\frac{x^2 - (k^2 - 1)^2 n^2}{(k+1)y}\right|, \quad d = \left|\frac{2nx}{y}\right|,$$

we get a family of elliptic curves

$$E_{n,k} : y^2 = x^3 - (k^2 - 1)n^2 x,$$

and

$$a = \left| \frac{2knx}{y} \right|, \quad c = \left| \frac{x^2 + (k^2 - 1)^2 n^2}{(k+1)y} \right|.$$

For $k \geq 2$, $E_{n,k}$ is the special kind of congruent number curves, we call it k-congruent number curve. Note that $n^3 - n$ is a congruent number, let $p = n + 1, q = n - 1$, then

$$4(n^3 - n) = pq(p^2 - q^2)$$

is a congruent number (by Euclid's formula). Put $k = n$, then the rank of $E_{n,n}$ is positive, hence $n > 1$ is a k-congruent number.

When $n = 1$, take $k = k_1^2, k_1 > 1$, then $E_{1,k_1^2} : y^2 = x^3 - (k_1^4 - 1)^2 x$. As $k_1^4 - 1$ is a congruent number, then E_{1,k_1^2} has infinitely many solutions. Hence, 1 is a k-congruent number. \square

Example 7.5. Take $n = 1$, $k = 4$, we have

$$(a, b, c, d) = \left(2, \frac{4}{5}, \frac{17}{10}, \frac{1}{2} \right), \left(\frac{16}{15}, \frac{3}{2}, \frac{17}{10}, \frac{4}{15} \right),$$

$$\left(\frac{544}{161}, \frac{161}{340}, \frac{141121}{54740}, \frac{136}{161} \right).$$

Take $k = n = 2$, from E_{22}, we have

$$(a, b, c, d) = \left(\frac{8}{3}, 1, \frac{5}{3}, \frac{4}{3} \right), \left(\frac{80}{7}, \frac{7}{30}, \frac{1201}{210}, \frac{40}{7} \right),$$

$$\left(\frac{6808}{4653}, \frac{1551}{851}, \frac{7776485}{3959703}, \frac{3404}{4653} \right).$$

Take $k = n = 3$, from $E_{3,3}$, we have

$$(a, b, c, d) = \left(\frac{9}{4}, 2, \frac{5}{2}, \frac{3}{4} \right), \left(\frac{21}{40}, \frac{60}{7}, \frac{1201}{140}, \frac{7}{40} \right),$$

$$\left(\frac{851}{517}, \frac{4653}{1702}, \frac{7776485}{2639802}, \frac{851}{1551} \right).$$

Theorem 7.15. *Every positive integer is a d-congruent number.*

Proof. In (7.34), take

$$
\begin{cases}
a = \dfrac{(3x - d^2 - 3n)(3x - d^2 + 3n)}{3(-3y + 3dx - d^3)}, \\[2ex]
b = \dfrac{2n(3x - d^2)}{-3y + 3dx - d^3}, \\[2ex]
c = \dfrac{(9 - 6d^2)x^2 + 9n^2 + d^4}{3(-3y + 3dx - d^3)},
\end{cases}
$$

we get a family of elliptic curves

$$
E_{n,d} : y^2 = x^3 - \frac{3n^2 + d^4}{3}x + \frac{(9n^2 + 2d^4)d^2}{27}.
$$

We call it d-congruent number curve. Let $d = 3n$, we have

$$
E_{n,3n} : y^2 = x^3 - (1 + 27n^2)n^2 x + 3n^4(1 + 18n^2).
$$

The discriminant of $E_{n,3n}$ is $\Delta = (4 + 81n^2)n^6 > 0$, this means that $E_{n,3n}$ is non-singular.

We want to prove that for each positive integer n, there are infinitely many rational points on $E_{n,3n}$, select a suitable to lead a solution a, b, c, d of (7.34). Noting that the point $P = (-6n^2, 3n^2)$ lies on $E_{n,3n}$, using the group law on the elliptic curve, we get the point

$$
[2]P = \left(\frac{(27n^2 + 1)(243n^2 + 1)}{36} - \frac{(81n^2 + 1)(6561n^4 + 324n^2 - 1)}{216} \right).
$$

It's easy to see that for any positive integer n, the x-coordinate of the point $[2]P$ is not an integer. By the Nagell–Lutz theorem (Section 27), for all $n \geq 4$ the point $[2]P$ is of infinite order. Then there are infinitely many rational points on $E_{n,3n}$, moreover, the point $[2]P$ is the exact point such that $a, b, c, d \, (= 3n)$ satisfying (7.34), which

leads to

$$
\begin{cases}
a = \dfrac{(729n^3 - 81n^2 + 27n + 1)(9n - 1)}{6(1 + 81n^2)}, \\[3mm]
b = \dfrac{12n(1 + 81n^2)}{(1 + 9n)(729n^3 + 81n^2 + 27n - 1)}, \\[3mm]
c = \dfrac{43046721n^8 + 2125764n^6 + 39366n^4 + 1620n^2 + 1}{6(1 + 81n^2)(1 + 9n)(729n^3 + 81n^2 + 27n - 1)}.
\end{cases}
$$

This completes the proof of Theorem 7.15. □

Example 7.6. Taking $n = 1, 2, 3$, respectively, from $E_{n,3n}$ we have

$$
(a, b, c, d) = \left(\frac{1352}{123}, \frac{123}{1045}, \frac{1412921}{128535}, 3 \right),
$$

$$
(a, b, c, d) = \left(\frac{94571}{1950}, \frac{7800}{117971}, \frac{11156645809}{230043450}, 6 \right),
$$

$$
(a, b, c, d) = \left(\frac{123734}{1095}, \frac{3285}{71722}, \frac{8874450677}{78535590}, 9 \right).
$$

Furthermore, we have the following results and conjecture.

For any positive integer $k \geq 2$, there are infinitely many positive integers n which are k-congruent numbers.

By the proof of Theorem 7.13, for $n = 1$, there are infinitely many positive integers k such that 1 is a k-congruent number. For $n \geq 2$, the general result is hard to get. But we have the following condition:

If n is a square, then there are infinitely many positive integers k such that n is a k-congruent number.

With the help of computer, we raise a conjecture.

Conjecture 8. *For any positive integer n, there are infinitely many positive integers k such that n is a k-congruent numbers.*

For d-congruent numbers, we have the following result:

For any positive integer d, if $n \neq d^2$, then n is a d-congruent number. For any positive integer n, except $d^2 = n$, n is a d-congruent number.

For any positive integer d, d^2 is not a d-congruent number.

36. The ABCD Equation

This section will describe the abcd equation, which was proposed by the author in January 2013, maybe it is the most original problem in the book, its research method is both elegant and rich, and the difficulty is incalculable. Whether its judge of solvability, or the number and structure of the solution when the equation is solvable, is very worthy of discussion. We (Tianxin Cai, Deyi Chen, On abcd problem) made the preliminary discussion.

Definition 7.4. Let n be a positive integer, a, b, c, d are positive rational numbers, the so-called abcd equation is

$$n = (a + b)(c + d), \qquad (7.35)$$

where

$$abcd = 1.$$

By the arithmetic–geometric inequality, $(a + b)(c + d) \geq 2\sqrt{ab} \times 2\sqrt{cd} = 4$, so equation (7.35) has no solution when $n = 1, 2$, or 3. On the other hand, $4 = (1 + 1)(1 + 1), 5 = (1 + 1)(2 + \frac{1}{2})$.

It is easy to see that if (7.35) has positive rational solutions, so does the following equation:

$$n = x + \frac{1}{x} + y + \frac{1}{y}. \qquad (7.36)$$

It's true vice versa, since

$$x + \frac{1}{x} + y + \frac{1}{y} = (x + y)\left(1 + \frac{1}{xy}\right).$$

It's not hard to see that each solution of (7.36) corresponding to infinitely many solutions $(ka, kb, \frac{c}{k}, \frac{d}{k})$ of (7.35). In particular, (7.36)

has unique solution $(x, y) = (1, 1)(n = 4), (2, 2)(n = 5)$. The proof of the former is obvious, the latter need to use the theory of elliptic curve.

Theorem 7.16. *When $8|n$ or $2||n$, abcd equation has no rational solution; when n is odd or $4||n$ and n has a prime factor congruent to 3 modulo 4, abcd equation also has no rational solution.*

Proof. If (7.36) has a rational solution, then there exist positive integers a_1, b_1, c_1, d_1 such that

$$n = \frac{a_1}{b_1} + \frac{b_1}{a_1} + \frac{c_1}{d_1} + \frac{d_1}{c_1}, \quad (a_1, b_1) = (c_1, d_1) = 1.$$

Multiplying on both sides by $a_1 b_1$, we have

$$a_1 b_1 n - (a_1^2 + b_1^2) = \frac{a_1 b_1}{c_1 d_1}(c_1^2 + d_1^2).$$

Since $(a_1 b_1, a_1^2 + b_1^2) = 1$, we deduce that $c_1 d_1 | a_1 b_1$, otherwise, the right-hand side is not an integer. Similarly we have $a_1 b_1 | c_1 d_1$. So $a_1 b_1 = c_1 d_1$.

On the other hand, we have

$$a_1 b_1 n = a_1^2 + b_1^2 + c_1^2 + d_1^2 = (a_1 \pm b_1)^2 + (c_1 \mp d_1)^2. \tag{7.37}$$

By parity, symmetry and $a_1 b_1 = c_1 d_1$, (a_1, b_1, c_1, d_1) can be divided into two cases: (odd, odd, odd, odd) and (odd, even, odd, even). But the square of an even number is congruent to 0 modulo 4 and the square of an odd number is congruent to 1 modulo 8, then we can deduce that (7.36) has no rational solution when $8|n$ or $2||n$. Therefore, abcd equation has no rational solutions.

When n is odd or $4||n$, by the theory of quadratic residue, the left-hand side of (7.37) does not have a prime factor congruent to 3 modulo 4, so is n. Otherwise, let prime $p \equiv 3 \pmod 4$, if $p|c_1 + d_1$ and $p|c_1 - d_1$, then $p|(c_1, d_1)$. A contradiction. If $p \nmid c_1 \pm d_1$, then $(\frac{-(c_1 \pm d_1)^2}{p}) = -1$, where () is Legendre symbol. Again a contradiction. This completes the proof of Theorem 7.16.

Next, we consider the equation

$$n = \left(a + \frac{1}{a}\right)\left(b + \frac{1}{b}\right), \tag{7.38}$$

where a and b are positive integers. Obviously, if (7.38) has an integer solution, then abcd equation is also solvable. Equation (7.38) has integer solution if and only if

$$a|b^2 + 1, b|a^2 + 1.$$

By the property of F-perfect number (Section 34), the solutions of above relation are $a = F_{2k-1}, b = F_{2k+1}$. By Cassini's identity of Fibonacci number, we have the following theorem.

Theorem 7.17. *Equation* (7.38) *has integer solution if and only if* $n = F_{2k-3}F_{2k+3}(k \geq 0)$. *The solutions are* $(a, b) = (F_{2k-1}, F_{2k+1})$.

By Theorem 7.17, there are infinitely many n $(4, 5, 13, 68, 445, 3029, 20740, \ldots)$ such that abcd equation has rational solutions.

Moreover, by the theory of Pisano period, we have the following theorem.

Theorem 7.18. *If* n *is odd and* (7.38) *has integer solution, then* $n \equiv 5 \pmod 8$. *If* n *is even and* (7.38) *has integer solution, then* $n = 4m, m \equiv 1 \pmod{16}$.

Coin used in Bejaia, North Africa, at the time of Pisano

Proof. If (7.38) has integer solution, by Theorem 7.17, $n = F_{2k-3}F_{2k+3}$. From Section 11, we know that Pisano cycle of Fibonacci number modulo 8 is $\{1, 1, 2, 3, 5, 0, 5, 5, 2, 7, 1, 0\}$, its length is 12, so the cycle length of n is 6. When $k = 3$ or 6, n is even. When $k = 1, 2, 4$ or 5, $n \equiv 5 \pmod 8$. Therefore, when n is even and (7.38) has integer solutions, $n = F_{6k-3}F_{6k+3}$. Because the terms $\equiv 3 \pmod 6$ in Pisano cycle of Fibonacci numbers are all 2, i.e., have the form $32k+2$, so product of them $n \equiv 4 \pmod{64}$. This completes the proof of Theorem 7.18.

Now we consider the equation

$$n = \left(\frac{a}{b} + \frac{b}{a}\right)\left(\frac{c}{d} + \frac{d}{c}\right), \tag{7.39}$$

where a, b, c, d are positive integers and $(a, b) = (c, d) = 1$.

Obviously, (7.39) is the special form of (7.38). If (7.39) has a solution, then (7.35) and (7.36) are also solvable, vice versa. In fact, if (7.36) is solvable, by Theorem 7.16, let $x = \frac{a_1}{b_1}, y = \frac{c_1}{d_1}$, noting that $a_1 b_1 = c_1 d_1$, we have $xy = (\frac{c_1}{d_1})^2$. Substitute it into (7.36), we have

$$n = (x + y)\left(1 + \frac{1}{xy}\right) = \left(\frac{a_1}{c_1} + \frac{c_1}{a_1}\right)\left(\frac{c_1}{b_1} + \frac{b_1}{c_1}\right).$$

So (7.39) is solvable. Therefore, equations (7.35), (7.36) and (7.39) are all called abcd equation.

For (7.39), we have only obtained partial results. For example, when $b = 1$, it has the following solutions:

$[a, 1, c, 2]$, where $2c|a^2 + 1, a|c^2 + 4$, it has solutions:

$$[a, c; n] = [17, 145; 1237], [337, 41; 6925], \text{ where } 1237 \text{ is prime;}$$

$[a, 1, c, d]$, where a, c, d are odd, $cd|a^2 + 1, a|c^2 + d^2$, it has solutions:

$$[a, c, d; n] = [157, 5, 17; 580], [73, 13, 205; 1156], [697, 5, 37; 5252];$$

$[a, 1, c, (a^2 + 1)/c]$, where $a|c^4 + 1, c|a^2 + 1$, it has solutions:

$$[a, c; n] = [17, 2; 1237], [73, 10; 3893], [697, 202; 8357],$$

$$[137, 10; 25717],$$

$$[4097, 1657; 25717], [313, 5; 1226597], [3778, 85; 7463597],$$

$$[13817, 85; 365092933];$$

$[c^2 + d^2, 1, c, d]$, where $c|a^4 + 1, d|c^4 + 1$, it has solutions:

$$[a, c; n] = [293, 2; 2525], [41, 137; 74453], [386, 35521; 116137055245],$$

$$[41, 20626; 214025054125], [386, 624977; 632419324106333].$$

And $[a, b, c, (a^2 + b^2)/c]$, where $c|a^2 + b^2, a|b^4 + c^4, b|a^4 + c^4$, it has solutions:

$$[a, b, c; n] = [2, 641, 5; 5267597], [17, 3394, 29; 2734717].$$

Therefore, we find seven primes such that abcd equation is solvable, i.e., 5, 13, 1237, 25717, 74453, 2734717, 365092933.

The more interesting thing is, we can get infinitely many n such that (7.38) or the abcd equation has solutions. For example, $(c, d) = (1, 1), (41, 137)$ or $(386, 35521)$, each generates a sequence, in which each pair of the neighboring numbers produces a solution of (7.38) for $n = \left\{ \left(c^2 + d^2\right)^2 + 1 \right\} / cd$, i.e.,

$$\cdots, 41761, 17, 2, 1, 1, 2, 17, 41761 \cdots,$$

$$\cdots, 20626, 41, 137, 8592082, \cdots$$

$$\cdots, 624977, 386, 35531, \cdots$$

with each neighboring triple $\{a, b, c\}$ satisfying $ac = b^4 + 1$.

Theorem 7.19. *Suppose abcd equation has solutions, n is odd or satisfies $4\|n$, n has no prime factor congruent to 3 modulo 4, if $n \pm 4$ has a prime factor p congruent to 3 modulo 4, then $p^{2k}\|n \pm 4$, where k is a positive integer.*

Proof. By Theorem 7.16, if (7.36) is solvable, then there exist positive integers a, b, c, d, $(a, b) = (c, d) = 1$, $ab = cd$, such that

$$nab = (a \pm b)^2 + (c \mp d)^2.$$

By the theory of quadratic residue, a and b have no prime factor congruent to 3 modulo 4. The above equation can be changed to

$$(n \pm 4)ab = (a \pm b)^2 + (c \pm d)^2. \qquad (7.40)$$

Suppose there is a prime $p \equiv 3 \pmod 4$ satisfying $p|n + 4$. If $p \nmid c + d$, by the theory of quadratic residue, (7.40) is impossible; if $p|c + d$, then $p|a + b$, so $p^2|n + 4$. If $p^k|n + 4, k > 2$, divide by p^2 on both sides of (7.40), continue it and we can prove that $p^{2k}|n + 4$. Similarly, we can prove that if prime $p \equiv 3 \pmod 4$ satisfying $p|n-4$, then $p^{2k}||n - 4$. This completes the proof of Theorem 7.19.

Corollary 7.1. *For any non-negative integer k, if $n = F_{2k-3}F_{2k+3}$, then n is odd or $4||n$, n has no prime factor congruent to 3 modulo 4; moreover, if $n \pm 4$ has a prime factor p congruent to 3 modulo 4, then $p^{2k}||n \pm 4$ where k is a positive integer.*

Corollary 7.2. *If $n = 4m$ and abcd equation has solution, then $m \equiv 1 \pmod 8$.*

Proof. By Theorem 7.16, $m \equiv 1 \pmod 4$. If $m = 8k + 5$, then $n + 4 = 8(4k + 3)$. Therefore, $n + 4$ has a prime factor congruent to 3 modulo 4. By Theorem 7.19, abcd equation has no solution. This completes the proof of Corollary 7.2. ·

By Theorems 7.16, 7.17 and 7.19, among the positive integers not exceeding 1000, the abcd equation has solution when $n = 4, 5, 13, 68, 445$, and 580. It may also has solutions when $n = 41, 85, 113, 149, 229, 265, 292, 365, 373, 401, 481, 545, 761, 769, 797, 905, 932$.

Conjecture 9. *If positive integer $n \equiv 1 \pmod 8$, then abcd equation has no solution.*

Conjecture 10. *If $n = 4m$ and abcd equation has solution, then $m \equiv 1 \pmod{16}$.*

Remark. If the above two conjectures are true, and positive integer n not exceeding 1000, then the abcd equation may have solutions when $n = 85, 149, 229, 365, 373, 797$.

In addition, 149 has the following solution with positive a, b and negative c, d, i.e.,

$$149 = \frac{14640}{91} + \frac{91}{14640} - \frac{3965}{336} - \frac{336}{3965}$$
$$= \left(\frac{14640}{91} - \frac{336}{3965}\right)\left(1 - \frac{91 \times 3965}{14640 \times 336}\right).$$

Problem 7.1. Is it possible to have infinitely many positive integers n such that (7.38) has no solution and abcd equation has solution?

Problem 7.2. Is it possible to have infinitely many primes n such that abcd equation has solution?

Problem 7.3. Does there exist a positive integer n satisfying $n = (a + b)(c + d), abcd = 1, ab \neq k^2$?

Remark. If the other assumptions in Definition 7.4 remain, and let the condition $abcd = 1$ be changed to $abcd = k^2$. Then some sufficient and necessary conditions above for the abcd equation still remain, we will have a new problem worth studying.

Now we discuss the number of solutions of (7.36). By the arithmetic–geometric inequality, it is easy to deduce that (7.36) has only one solution when $n = 4$. When $n > 4$, we may change (7.36) into an elliptic curve.

Theorem 7.20. *If $n > 4$,*

$$E_n : Y^2 = X^3 + (n^2 - 8)X^2 + 16X$$

is a class of elliptic curves, then (7.36) has solution if and only if E_n has rational solution for $X < 0$.

Proof. Let $x \geq 1, y \geq 1, x + y > 2$, making the transformation

$$\begin{cases} x = \dfrac{s + nt}{2(t + t^2)}, \\[2mm] y = \dfrac{s + nt}{2(1 + t)}, \end{cases} \quad s, t > 0. \tag{7.41}$$

Its inverse transformation is

$$\begin{cases} t = \dfrac{y}{x}, \\ s = \dfrac{2y^2 + (2x - n)y}{x}. \end{cases}$$

This correspondence is a bijection. Substitute it into (7.36), we have

$$n = \frac{(s + nt)^2 + 4t(1 + t^2)^2}{2t(s + nt)}.$$

Simplify it and let

$$\begin{cases} X = -4t, \\ Y = 4s. \end{cases}$$

We get E_n. This completes the proof of Theorem 7.20.

Here,

$$\begin{cases} x = \dfrac{2Y - 2nX}{X^2 - 4X}, \\ y = \dfrac{Y - nX}{24 - X}. \end{cases} \tag{7.42}$$

Example 7.7. The equation $5 = x + \frac{1}{x} + y + \frac{1}{y}$ has unique positive rational solution $x = y = 2$.

By Theorem 7.20, we only need to consider the following elliptic curve:

$$E_5 : Y^2 = X^3 + 17X^2 + 16X, \ X < 0.$$

By Magma package, we can calculate that rank $E_5 = 0$. By Mordell's theorem:

$$E_5(\mathbb{Q}) \cong T \oplus \mathbb{Z}^{\mathrm{rank}(E_5)},$$

here T is the torsion part of E_5. By calculation, all rational points on E_5 are

$$E_5(Q)_{\text{Torsion}} = \{(-16, 0), (-4, -12), (-4, 12), (-1, 0), (0, 1),$$

$$(4, -20), (4, 10), \infty\},$$

where the last one is the point at infinity. Substitute into (7.42), we obtain the unique solution $x = y = 2$.

Now we consider for general n, there are infinitely many solutions when there exists a rational solution. We need the following variant of Nagell–Lutz theorem (Silverman J. Tate J. *Rational Points on Elliptic Curves*, Springer, 2004, P. 104, Ex.3.7).

Lemma 7.9. *For non-singular elliptic curve of Weierstrass form*

$$y^2 = x^3 + ax^2 + bx,$$

where a and b are integers. If (x,y) is a point with limited upper bound, $y \neq 0$, then $x|b$, and

$$x + a + \frac{b}{x}$$

is perfect square.

Theorem 7.21. *When $n \geq 6$, the torsion points of $E_n : Y^2 = X^3 + (n^2 - 8)X^2 + 16X$ are*

$$(0,0), (4, -4n), (4, 4n), \infty.$$

Proof. It's easy to verify that $(0,0), (4, -4n), (4, 4n), \infty$ are the torsion points of E_n. Now we want to prove that they are all the torsion points. When $Y = 0$, note that $n \geq 6$, hence $X^3 + (n^2 - 8)X^2 + 16X = 0$ has unique rational solution $X = 0$. When $Y \neq 0$, let (X, Y) be the torsion point of E_n. By Lemma 7.9,

$$X|16 \text{ and } X + n^2 - 8 + \frac{16}{X} \text{ is perfect square.}$$

From $X|16$, we have $X \in \{1, 2, 4, 8, 16\}$. If $X = 4$, then $X + n^2 - 8 + \frac{16}{X} = n^2$ is perfect square. If $X \in \{1, 2, 8, 16\}$, then $X + n^2 - 8 + \frac{16}{X} = n^2 + 2$ or $n^2 + 9$; but $n^2 + 2$ is not perfect square; $n^2 + 9$ is perfect square if and only if $n = 0, 4$. Hence, when $Y \neq 0$, if (X, Y) is a torsion point of E_n, then $X = 4$, therefore $Y = \pm 4n$. This proves Theorem 7.21.

Theorem 7.22. *When $n \geq 6$, if abcd equation has positive rational solution, then there exist infinitely many rational solutions.*

Proof. We only need to show that when $n \geq 6$, if $E_n : Y^2 = X^3 + (n^2 - 8)X^2 + 16X$ has a rational point $P_0(X_0, Y_0)$ satisfying $X_0 < 0$, then there exist infinitely many rational points satisfying $X < 0$.

In fact, by Theorem 7.21, $P_0(X_0, Y_0)$ is not a torsion point, hence $[n]P_0$ are distinct. We need to prove that $[3]P_0$ satisfies $X < 0$, then $[2k + 1]P_0$ satisfying $X < 0$ by recursion.

First of all, if the straight line $Y = kX + b$, $b \neq 0$ has intersection with $Y^2 = X^3 + (n^2 - 8)X^2 + 16X$, then

$$X^3 + (n^2 - 8 - k^2)X^2 + (16 - 2kb)X - b^2 = 0,$$

hence the product of all the roots is positive. Hence, the intersection of the tangent of E_n at P_0 and E_n satisfying $X > 0$ (since it has double roots $X_0 < 0$), by the symmetry we get that $[2]P_0$ satisfying $X > 0$. Moreover, the straight line connecting P_0 and $[2]P_0$ has an intersection with E_n which satisfies $X < 0$ (since $X_0 < 0$, $[2]P_0$ satisfies $X > 0$). Therefore,it follows from symmetry that $[3]P_0$ satisfies $X < 0$. This completes the proof of Theorem 7.22.

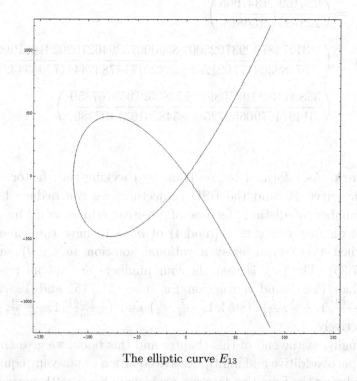

The elliptic curve E_{13}

Example 7.8. $13 = x + \frac{1}{x} + y + \frac{1}{y}$ has infinitely many positive rational solutions.

By Theorem 7.20, we only need to find the solutions of the following elliptic curve:

$$E_{13} : Y^2 = X^3 + 161X^2 + 16X, X < 0.$$

By Magma, we can calculate that the rank of E_{13} is 1. By Mordell theorem, we can get the generator of E_{13} is $P(X, Y) = (-100, 780)$. By the group theory of elliptic curve, all the rational points in E_{13} with $X < 0$ are $[2k + 1]P$, here k is arbitrary non-negative integer. Substitute it into (7.41), we have

$$[1]P = (-100, 780)(k = 0), (x, y) = \left(\frac{2}{5}, 10 \right),$$

$$[3]P = \left(-\frac{6604900}{776161}, \frac{71411669940}{683797841} \right) (k = 1), (x, y)$$

$$= \left(\frac{924169}{228730}, \frac{1347965}{156818} \right),$$

$$[5]p = \left(-\frac{31274879093702500}{57589364171021281}, \frac{85900073394621020231661900}{138201712783244441779434321} \right),$$

$$(x, y) = \left(\frac{33896240819350898}{3149745790659725}, \frac{12489591059767450}{8548281631402489} \right).$$

$$\vdots$$

Remark. As informed by Ye Tian, by checking the ε factor of the elliptic curve E_n and the BSD conjecture, we can deduce that if the number of distinct factors of positive odd n and the number of distinct factor $\equiv 1 \pmod 4$ of $n^2 - 16$ have the same parity, then there must exist a rational solution (a, b, c, d) satisfying (7.35). The pity is that the four numbers are not all positive. Jingzhao Pan found a solution for $n = 11$, 15, and 19, which are $\left(-\frac{64}{9}, 1, -\frac{15}{8}, \frac{3}{40} \right)$, $\left(-64, 1, -\frac{7}{24}, \frac{3}{56} \right)$ and $\left(-\frac{14161}{576}, 1, -\frac{1320}{1547}, \frac{312}{6545} \right)$, respectively

Finally, at the end of this chapter and this book, we give another example of additive and multiplicative equation or yin-yang equation. Considering the Egyptian fraction problem in Section 21, noting that $5 = 4 + \frac{1}{2} + \frac{1}{2}$, $4 \times \frac{1}{2} \times \frac{1}{2} = 1$. Then, for any positive integer n, the

yin-yang equation of Sierpinski's type

$$\begin{cases} \dfrac{5}{n} = x + y + z, \\ xyz = \dfrac{1}{A} \end{cases}$$

always has a positive solution. Here x, y, z are rational and A is an integer. However, for the following equation

$$\begin{cases} 4 = x + y + z, \\ xyz = \dfrac{1}{A} \end{cases}$$

so far we neither find any positive rational solution, nor conclude that there is no such solution. What we have found is some rational solution, e.g., $(x, y, z) = (-\frac{1}{6}, -\frac{1}{3}, \frac{9}{2})$. Therefore, in general we cannot judge the solvability of the yin-yang equation of Erdős–Strauss' type

$$\begin{cases} \dfrac{4}{n} = x + y + z, \\ xyz = \dfrac{1}{A}. \end{cases}$$

The latter is a week Erdős–Strauss conjecture. Meanwhile, we guess that the original Erdős–Strauss conjecture might be more difficult than the Sierpinski conjecture.

On the other hand, for any positive integer n, let $A(n)$ denote the least positive integer A, so that the following yin-yang equation has a positive rational solution

$$\begin{cases} n = x + y + z, \\ xyz = \dfrac{1}{A}. \end{cases} \tag{E}$$

Noting $1 = \frac{1}{3} + \frac{1}{3} + \frac{1}{3}, 2 = 1 + \frac{1}{2} + \frac{1}{2}$, by using the inequality of arithmetic–geometric means, we have $A(1) = 27$, $A(2) = 4$. Since

$$3 = 1 + 1 + 1, \; 5 = 4 + \frac{1}{2} + \frac{1}{2}, \; 6 = \frac{9}{2} + \frac{4}{3} + \frac{1}{6},$$

$$9 = \frac{49}{6} + \frac{9}{14} + \frac{4}{21}, \; 10 = \frac{324}{35} + \frac{25}{126} + \frac{49}{90},$$

we deduce that $A(3) = A(5) = A(6) = A(9) = A(10) = 1$. As a matter of fact, when $A = 1$, (E) is equivalent to

$$x^3 + y^3 + z^3 = nxyz$$

or

$$n = \frac{x}{y} + \frac{y}{z} + \frac{z}{x}.$$

Here x, y, z are positive integers. According to http://oeis.org/ A072716, there are solutions for 29's $n \leq 100$, i.e., 3, 5, 6, 9, 10, 13, 14, 17, 18, 19, 21, 26, 29, 30, 38, 41, 51, 53, 54, 57, 66, 67, 69, 73, 74, 77, 83, 86, 94. For the rest of n, if it is not 4 or multiple of 4, we can count $A(n)$ by using the Magma package and the theory of elliptic curves. For example, $A(7) = 6$, here is a solution $(\frac{4232}{825}, \frac{1875}{1012}, \frac{121}{6900})$. However, for $n = 4$ or multiple of 4, it is difficult to determine the value of $A(n)$.

Looking at the six problems we have discussed, there are still much work to be done. They are the most valuable part of this book, and we expect them to bloom beautiful flowers and rich fruits. After Fermat's Last Theorem was proved in 1995, there have been triumphant pieces of news in number theory one after another. At first, it was the Catalan conjecture (2002), later the abc conjecture (2012, it seems there exists some gap), and the odd Goldbach conjecture (2013). The twin prime conjecture (2013) also made important progress. Meanwhile, this is a double-edged sword, which rings the alarm bells for number theorists: The gold chicken is getting fewer and fewer. I hope that the publication of this book will be refreshing blood, and it can help to produce one or two chicken.

Solutions to Exercises

Exercises 1

1. Prove that each odd number can be the difference of two squares.

Proof. Any odd number can be written as $2k + 1$. Note that $2k + 1 = (k + 1)^2 - k^2$, we prove it. \square

2. Prove that for any integer n, $3 \mid n^3 - n$, $5 \mid n^5 - n$, $7 \mid n^7 - n$, and how about $9 \mid n^9 - n$?

Proof. (1) Since $n^3 - n = (n - 1)n(n + 1)$, therefore it must be divided by 3.

(2) Let $n = 5k + 1$, $0 \leq i \leq 4$, then $n^5 - n = (5k+i)^5 - (5k+i) = 5A + (i^5 - i)$. It is easy to see that $5 \mid (i^5 - i)$, $0 \leq i \leq 4$, we prove it.

(3) It can be proved as (2).

(4) Note that $9 \nmid (2^9 - 2)$, therefore it is not true. \square

3. Prove that if both p and $p + 2$ are primes greater than 3, then $6 \mid p + 1$.

Proof. Since $p > 3$, p is odd prime, $p + 1$ is even, i.e., $2 \mid p + 1$; note that $p, p + 1, p + 2$ are three consecutive integers and $3 \nmid p, 3 \nmid p + 2$, it must be $3 \mid p + 1$. Therefore $6 \mid p + 1$. \square

4. Show that $\sqrt{2}$ and $\sqrt{6}$ are irrational.

Proof. If $\sqrt{2}$ is rational, let $\sqrt{2} = p/q$, with $(p, q) = 1$, then $2q^2 = p^2$, p^2 and hence p is even. Let $p = 2m$, we have $q^2 = 2m^2$, hence q is also even. This contradicts $(p, q) = 1$.

Similarly, we can prove that $\sqrt{6}$ is irrational. □

5. Show that, for any positive integer k, there are k consecutive integers which are composite.

Proof. Let $M = (k + 1)! + 2$, note that

$$M + i = (k + 1)! + 2 + i = (i + 2) \left(\frac{(k + 1)!}{i + 2} + 1 \right),$$

$$i = 0, 1, 2, \ldots, k - 1$$

and $i + 2 \mid (k + 1)!$. Hence, $M + i$ are all composite. □

6. Determine all the positive integer n such that $7 \mid 2^n - 1$.

Proof. Let $n = 3k + i, k \in Z, i = 0, 1, 2$. Then

if $i = 0$, $2^n - 1 = 2^{3k} - 1 = (7 + 1)^k - 1 = 7A$, hence $7 \mid (2^n - 1)$;
if $i = 1$, $2^n - 1 = 2^{3k+1} - 1 = 2 \cdot (7 + 1)^k - 1 = 2 \cdot 7A + 1$, hence $7 \nmid (2^n - 1)$;
if $i = 2$, $2^n - 1 = 2^{3k+2} - 1 = 4 \cdot (7 + 1)^k - 1 = 4 \cdot 7A + 3$, hence $7 \nmid (2^n - 1)$.

Therefore, if and only if n is a multiple of 3, $7 \mid 2^n - 1$. □

7. Let n be odd, find the number of ways that n can be the difference of two square numbers.

Solution. Since n is odd, we have

$$n = x^2 - y^2 = (x - y)(x + y) = ab \Leftrightarrow \begin{cases} x = \pm \dfrac{a + b}{2}, \\ y = \pm \dfrac{a - b}{2}, \end{cases}$$

where $\frac{a \pm b}{2}$ are both integers.

That is to say, we only need to consider the number of the decomposition of n into the product of two odd numbers. Note that $n = ab = (-a)(-b)$, therefore the number is $2d(n)$, here $d(n)$ is the number of divisors of n.

8. Let a, b, c be positive integers, $(a, c) = 1$, and $\frac{1}{a} + \frac{1}{b} = \frac{1}{c}$. Show that $a + b$ is square.

Proof. Since $\frac{1}{a} + \frac{1}{b} = \frac{1}{c}$, we have $c(a + b) = ab$, hence

$$a^2 = a^2 + ab - c(a + b) = (a - c)(a + b).$$

Note that $(a, c) = 1$, then $(a - c, a) = 1$. Therefore $a - c = 1$, $a + b = a^2$. \square

9. Show that there are infinitely many primes in the form of $4n + 3$.

Proof. Suppose there are only finitely many primes of the form $4n + 3$, denoted them by p_1, p_2, \ldots, p_k. Let $q = 4p_1p_2 \ldots p_k - 1$, obviously $p_i \nmid 1$. Note that $q = 4(p_1p_2 \ldots p_k - 1) + 3$, q is also of the form $4n + 3$. Since the product of any two integers of the form $4n + 1$ keeps the same form, therefore q must have at least one prime factor of the form $4n + 3$, denoted by p. Then, p must be different from other p_i. A contradiction! \square

10. Prove that there are no consecutive odds which are both the sum of two non-zero squares.

Proof. Let $m = a^2 + b^2, n = c^2 + d^2$ be two consecutive odd numbers, $m > n$.

Then both a, b and c, d have different parity. Suppose that $a = 2k + 1, c = 2t + 1, b = 2p, d = 2q$, then

$$2 = m - n = a^2 + b^2 - c^2 - d^2 = (a - c)(a + c) + (b - d)(b + d)$$

$$= 4((k - t)(k + t + 1) + (p - q)(p + q)).$$

A contradiction! \square

11. Prove that for any positive integer n, $\frac{1}{3} + \frac{1}{5} + \cdots + \frac{1}{2n+1}$ cannot be an integer.

Proof. Let $3^k \leq 2n + 1 < 3^{k+1}$. Then only one of $3, 5, 7, \ldots, 2n + 1$ is divided by 3^k, the other is divided at most by 3^{k-1}. Hence

$$\sum_{i=1}^{n} (2i + 1) = \frac{1}{3^k} + \frac{A}{3^{k-1}t} = \frac{3A + t}{3^k t}, \quad (t, 3) = 1.$$

Note that $3 \mid 3^k t, 3 \nmid (3A + t)$, we prove it. \square

12. Pigeonhole principle or Dirichlet principle states that if n items are put into m containers, with $n > m$, then at least one container must contain more than one item. Using this principle, show, if $1 \leq a_1 < a_2 < \cdots < a_{n+1} \leq 2n$, there must exist $1 \leq i < j \leq n + 1$, with $a_i \mid a_j$.

Proof. For $1 \leq i \leq n$, let $A_i = \{\leq (2i - 1)2^l \leq 2n, l \geq 0\}$, then A_i are disjoint whose sum set covers all the integers from 1 to $2n$. By Pigeonhole principle, among $n + 1$ a_i, there must be two belong to the same A_i. This completes the proof. \square

Exercises 2

1. Prove that one can choose three from any five integers such that their sum is a multiple of 3.

Proof. The remainder of any integer divided by 3 can only be 0, 1 or 2. For given five integers, consider the remainders divided by 3.

If there is only one or two remainders of $\{0, 1, 2\}$, one of them must appear at least three times. Then the sum of the three integers with the same remainder can be divided by 3.

If the remainders takes each of $\{0, 1, 2\}$. Then the sum of any three integers with different remainders can also be divided by 3. \square

2. Prove that one can choose m $(\leq n)$ from any n integers such that their sum is a multiple of n.

Proof. Let a_1, a_2, \ldots, a_n be n integers, consider the set of n integers $\{a_1, a_2 + a_2 \ldots, a_1 + a_2 + \cdots + a_n\}$. If they are all incongruent modulo n, then there must be one divided by n. Otherwise, by Pigeonhole principle, there are at least two of them having the same remainder modulo n. Therefore, their difference is a multiple of n, and it is a sum of a_i. \square

3. A palindromic number or numeral palindrome is a number that remains the same when its digits are reversed, for example, 3, 11, 242, which are "symmetric". Show that each palindromic number with four digits is divisible by 11. Try to generalize this result.

Proof. The generalization is as follows: if n is a positive integer, then each palindromic number with $2n$ digits is a multiple of 11.

In fact, a palindromic number with $2n$ digits can be written as $x_1 \ldots x_n x_n \ldots x_1$, or

$$(10^{2n-1} + 1)x_1 + \cdots + (10^n + 10^{n-1})x_n.$$

Noting that

$$10^{2n-1} + 1 \equiv \cdots \equiv 10^n + 10^{n-1} \equiv 0 \pmod{11},$$

we complete the proof. \square

4. Let x be real and n a positive integer. Show that

$$[x] + \left[x + \frac{1}{n}\right] + \cdots + \left[x + \frac{n-1}{n}\right] = [nx].$$

Proof. Let $k/n \leq \{x\} < (k+1)/n$, $0 \leq k \leq n-1$. Then

$$[nx] = [n[x] + n\{x\}] = n[x] + k.$$

Meanwhile,

$$[x] = [x + 1/n] = \cdots = [x + (n - k - 1)/n],$$
$$[x + (n - k)/n] = [x + (n - k + 1)/n]$$
$$= \cdots = [x + (n - 1)/n] = [x] + 1.$$

Hence

$$[x] + [x + 1/n] + \cdots + [x + (n-1)/n]$$
$$= (n-k)[x] + k([x] + 1) = n[x] + k = [nx].$$ \square

5. Let x be real and n a positive integer. Show that $\left[\frac{[x]}{n}\right] = \left[\frac{x}{n}\right]$.

Proof. Since $[x] \le x < [x] + 1$, we have $\left[\frac{[x]}{n}\right] \le \frac{[x]}{n} \le \frac{x}{n} < \frac{[x]}{n} + 1$.

Note $\frac{a}{b} \le \left[\frac{a}{b}\right] + 1 - \frac{1}{b}, a, b \in N$, then $\frac{[x]}{n} \le \left[\frac{[x]}{n}\right] + 1 - \frac{1}{n}$.

Hence $\left[\frac{[x]}{n}\right] \le \frac{x}{n} < \left[\frac{[x]}{n}\right] + 1$, i.e., $\left[\frac{[x]}{n}\right] = \left[\frac{x}{n}\right]$. \square

6. Let n be a positive integer. Show that $\frac{(2n)!}{n!(n+1)!}$ is an integer.

Proof. Since $C_{2n}^n = \frac{(2n)!}{n!n!}, C_{2n}^{n+1} = \frac{(2n)!}{(n+1)!(n-1)!}$, then $C_{2n}^n - C_{2n}^{n+1} = \frac{(2n)!}{(n+1)!n!}$. \square

7. Show that $\phi(n)$ is 1 or even.

Proof. When $n = 1, 2$, $\phi(n) = 1$.

When $n > 2$, if $2^2 \mid n$, then $2 \mid \phi(n)$; if $2^2 \nmid n$, then n must have an odd prime factor p. Hence $2 \mid (p-1), 2 \mid \phi(n)$. \square

8. Prove that

(1) for any positive integers m and n, the number of integers $m, m+1, \ldots, m+n-1$ relatively prime to n is $\phi(n)$.
(2) for any positive integer n we have $\phi(n^2) = n\phi(n)$, find all the positive integers such that $\phi(n) \mid n$.

Proof. (1) It is known that $(r, n) = 1 \Leftrightarrow (qn + r, n) = 1$, moreover the remainders of n consecutive numbers divided by n exactly consist of a complete residue modulo n.

(2) Let $n = p_1^{r_1} \cdots p_k^{r_k}$, then

$$\phi(n) = n\left(1 - \frac{1}{p_1}\right) \cdots \left(1 - \frac{1}{p_k}\right)$$
$$= p_1^{r_1-1} \cdots p_k^{r_k-1}(p_1 - 1) \cdots (p_k - 1).$$

Hence

$$\phi(n^2) = n^2 \left(1 - \frac{1}{p_1}\right) \cdots \left(1 - \frac{1}{p_k}\right) = n\phi(n).$$

When $\phi(n) \mid n$, then $(p_1 - 1) \cdots (p_k - 1) \mid p_1 \cdots p_k$. It's easy to see that only if $n = 2^{r_0}, r_0 \geq 0$ or $n = 2^{r_1}3^{r_2}, r_1, r_2 \geq 1$, $\phi(n) \mid n$. \square

9. Let $(m, n) = 1$. Show that $m^{\phi(n)} + n^{\phi(m)} \equiv 1 \pmod{mn}$.

Proof. It follows from Euler's theorem, $m^{\phi(n)} \equiv 1 \pmod{n}$, then $m^{\phi(n)} + n^{\phi(m)} \equiv 1 \pmod{n}$. Similarly, we have $m^{\phi(n)} + n^{\phi(m)} \equiv 1 \pmod{m}$. Therefore, $m^{\phi(n)} + n^{\phi(m)} \equiv 1 \pmod{mn}$. \square

10. Prove that

(1) $\sum_{d^2 \mid n} \mu(d) = \mu^2(n)$.
(2) $\sum_{d \mid n} | \mu(d) | = 2^k$.

Proof. (1) When $n = 1$ or n doesn't contain a square factor greater than 1, then $\sum_{d^2 \mid n} \mu(d) = \mu(1) = 1$.

When n contains a square factor greater than 1, let $n = n_0^2 m$, $n_0 > 1$, m doesn't contain a square factor greater than 1. If $d^2 \mid n$, then $d^2 \mid n_0$, $\sum_{d^2 \mid n} \mu(d) = \sum_{d^2 \mid n_0} \mu(d) = 0$. Anyway, we have $\sum_{d^2 \mid n} \mu(d) = \mu^2(n)$.

(2) Let $n = p_1^{\alpha_1} \dots p_k^{\alpha_k}$, we have $\sum_{d \mid n} |\mu(d)| = \sum_{d \mid p_1 \cdots p_k} |\mu(d)|$.
We prove by induction on k. If $k = 1$,

$$\sum_{d \mid p} |\mu(d)| = |\mu(1)| + |\mu(p)| = 1 + 1 = 2.$$

Suppose it holds for the number of distinct prime factors less than k, now we consider n with distinct k prime factors,

$$\sum_{d \mid p_1 \cdots p_k} |\mu(d)| = \sum_{d_1 \mid p_1 \cdots p_{k-1}} \sum_{d_2 \mid p_k} |\mu(d_1 d_2)|$$

$$= \sum_{d_1 \mid p_1 \cdots p_{k-1}} |\mu(d_1)| \sum_{d_2 \mid p_k} |\mu(d_2)| = 2^{k-1} \cdot 2 = 2^k.$$

This completes the proof. \square

11. Liouville's function is defined as follows:

$$\lambda(n) = \begin{cases} 1 & \text{if } n = 1 \text{ or the product of even primes,} \\ -1 & \text{if } n \text{ is the product of odd primes.} \end{cases}$$

Proof. (1) $\lambda(n)$ is complete multiplicative function.

(2) If $f(n)$ is defined as

$$f(n) = \begin{cases} 1 & \text{if } n \text{ is a square,} \\ 0 & \text{if } n \text{ is not a square.} \end{cases}$$

Then $f(n)$ is multiplicative function, show that $\lambda(n) = \sum_{d|n} \mu(d) f(n/d)$. □

Proof. (1) Note that even number + even number = odd number + odd numbers = even number, even number + odd number = odd number. Then, it's easy to deduce, $\lambda(mn) = \lambda(m)\lambda(n)$.

(2) Let $n = p_1^{r_1} \ldots p_k^{r_k}$. If n is a square number, then r_1, \ldots, r_k are all even. Hence

$$\sum_{d|n} \lambda(d) = \sum_{d_1|p_1^{r_1}} \lambda(d_1) \sum_{d_k|p_k^{r_k}} \lambda(d_k)$$

$$= (\lambda(1) + \lambda(p_1) + \cdots + \lambda(p_1^{r_1})) \cdots (\lambda(1) + \lambda(p_k)$$

$$+ \cdots + \lambda(p_k^{r_k})) = 1 = f(n).$$

If n is not a square, similarly we can show that one term of the above product must be 0, i.e., $f(n) = \sum_{d|n} \lambda(d)$. By using Möbius' inversive formula in Section 9, we have $\lambda(n) = \sum_{d|n} \mu(d) f(n/d)$. □

12. Let $\{0, 1, 2\}$ and $\{0, 2, 1\}$ be two least residue systems modulo 3. The difference of corresponding residues forms a least absolute residue system. Discuss the generalization for modulo $n(\geq 3)$.

Proof. When $n = 2k$, the sum of elements of least absolute residue system modulo n is congruent to $\pm k \pmod{n}$. Let $\{a_1, \ldots, a_n\}$, $\{b_1, \ldots, b_n\}$ be two sets of least residue system modulo n, then

$$(a_1 - b_1) + \cdots + (a_n - b_n) \equiv 0 \pmod{n}.$$

Hence $\{a_1 - b_1, \ldots, a_n - b_n\}$ can't be a least absolute residue system.

When $n = 2k + 1$, then

$$\{0, 1, 2, \ldots, k, k+1, k+2, \ldots, 2k\}, \{0, 2, 4, \ldots, 2k, 1, 3, \ldots, 2k-1\}$$

are two least residue systems modulo n, the difference of the corresponding residues forms a least absolute residue system modulo n. \square

Exercises 3

1. Find the smallest positive integer $n > 2$ such that $2 \mid n$, $3 \mid n + 1$, $4 \mid n + 2$, $5 \mid n + 3$, $6 \mid n + 4$.

Solution. Note that $n \equiv -2 \pmod{4} \Rightarrow n \equiv 0 \pmod{2}$, and

$$n \equiv -4 \pmod{6} \Leftrightarrow \begin{cases} n \equiv -4 \pmod{2} \\ n \equiv -4 \pmod{3} \end{cases} \Leftrightarrow \begin{cases} n \equiv 0 \pmod{2}, \\ n \equiv -1 \pmod{3}. \end{cases}$$

Hence the following congruence groups are equivalent:

$$\begin{cases} n \equiv 0 \pmod{2} \\ n \equiv -1 \pmod{3} \\ n \equiv -2 \pmod{4} \\ n \equiv -3 \pmod{5} \\ n \equiv -4 \pmod{6} \end{cases} \Leftrightarrow \begin{cases} n \equiv -1 \pmod{3} \\ n \equiv -2 \pmod{4} \\ n \equiv -3 \pmod{5} \end{cases}$$

$$\Leftrightarrow \begin{cases} n \equiv 2 \pmod{3} \\ n \equiv 2 \pmod{4} \\ n \equiv 2 \pmod{5} \end{cases} \Leftrightarrow n \equiv 2 \pmod{60}$$

with which we find the solution $n = 62$.

2. Prove that

(1) for arbitrary distinct prime numbers p_1, p_2, \ldots, p_n, there exist n consecutive integers such that the kth one is divisible by p_k for $k = 1, \ldots, n$.
(2) there exist n consecutive integers such that each one of them has a square factor.

Proof. (1) Consider the congruences $x \equiv -i \pmod{p_i}$, $i = 1, 2, \ldots, n$, here $(p_i, p_j) = 1, i \neq j$. By Qin Jiushao Theorem (Chinese Remainder Theorem), we deduce that it has a solution x_0, hence $x_0 + 1, x_0 + 2, \ldots, x_0 + n$ are n integers satisfying the condition.

(2) Similarly, we can prove it if we consider the congruences $x \equiv -i \pmod{p_i^2}$, $i = 1, 2, \ldots, n$. $\qquad\square$

3. Let F_n be the nth Fibonacci number. Show that for any integer $n > 0$,

$$F_{n-1}F_{n+1} - F_n^2 = (-1)^n.$$

Proof. By using the recursion formula of F_n, we have

$$F_{n-1}F_{n+1} - F_n^2$$

$$= \begin{vmatrix} F_{n+1} & F_n \\ F_n & F_{n-1} \end{vmatrix} = \begin{vmatrix} F_n & F_{n-1} \\ F_n & F_{n-1} \end{vmatrix} + \begin{vmatrix} F_{n-1} & F_{n-2} \\ F_n & F_{n-1} \end{vmatrix}$$

$$= (-1) \begin{vmatrix} F_n & F_{n-1} \\ F_{n-1} & F_{n-2} \end{vmatrix} = \cdots = (-1)^n \begin{vmatrix} F_2 & F_1 \\ F_1 & F_0 \end{vmatrix} = (-1)^n.$$

$\qquad\square$

4. Show that the system

$$\begin{cases} x \equiv a \pmod{m} \\ x \equiv b \pmod{n} \end{cases}$$

has a solution if and only if $a \equiv b \pmod{(m, n)}$. If a solution exists, it is unique modulo $[m, n]$.

Proof. We first prove the necessity. Let x_0 be a solution, i.e.,

$$\begin{cases} x_0 \equiv a \pmod{m}, \\ x_0 \equiv b \pmod{n}. \end{cases}$$

Then

$$\begin{cases} x_0 \equiv a \pmod{(m, n)}, \\ x_0 \equiv b \pmod{(m, n)}. \end{cases}$$

Hence $a \equiv b \pmod{(m, n)}$.

Next we prove the sufficiency: Note that $a \equiv b \pmod{(m, n)} \Leftrightarrow (m, n) \mid b - a$.

Hence there exist k, μ, ν such that $(m, n) = \mu m + \nu n$, $b = a + k(\mu m + \nu n)$. Taking $x_0 = a + k\mu m$, then x_0 is a solution.

If y_0 is another solution, then

$$\begin{cases} x_0 \equiv y_0 \pmod{m}, \\ x_0 \equiv y_0 \pmod{n}, \end{cases}$$

hence $[m, n] \mid x_0 - y_0$, which means $x_0 \equiv y_0 \pmod{[m, n]}$. □

5. Prove that for any odd prime p,

(1) $1^2 3^2 \cdots (p-2)^2 \equiv (-1)^{\frac{p+1}{2}} \pmod{p}$;

(2) $2^2 4^2 \cdots (p-1)^2 \equiv (-1)^{\frac{p+1}{2}} \pmod{p}$.

Proof. (1) Note that $i \equiv -(p-i) \pmod{p}$, $i = 1, 2, \ldots, p$, we have

$$1^2 3^2 \cdots (p-2)^2 \equiv (-1)^{(p-1)/2}(p-1)! \equiv (-1)^{(p+1)/2} \pmod{p}.$$

(2) Similarly, we can prove it. □

6. Prove that for any positive integer m and prime $p > 5$, the Diophantine equation $(p-1)! + 1 = p^m$ has no solution.

Proof. Since $p > 5$, $2 < (p-1)/2 < p - 1$, hence $(p-1)^2 \mid (p-1)!$.

If the Diophantine equation has integer solution, then $(p-1)^2 \mid p^m - 1$, i.e., $p - 1 \mid p^{m-1} + \cdots + 1$.

Since $p^k \equiv 1 \pmod{p-1}$, $k = 0, 1, \ldots, m-1$, we have $p^{m-1} + \cdots + 1 \equiv m \pmod{p-1}$.

Hence $p - 1 \mid m$, $p^m \geq p^{p-1} > (p-1)^{p-1} > (p-1)! + 1$. A contradiction! □

7. If p is prime, show that $\left(\left(\frac{p-1}{2}\right)!\right)^2 + (-1)^{\frac{p-1}{2}} \equiv 0 \pmod{p}$.

Proof. Since p is odd prime, $i \equiv -(p-i) \pmod{p}$, $i = 1, 2, \ldots, p$. Then by using Wilson's theorem,

$$(p-1)! \equiv 1 \cdot \ldots \cdot \frac{p-1}{2} \cdot \frac{p+1}{2} \cdot \ldots \cdot (p-1)$$

$$\equiv 1 \cdot \ldots \cdot \frac{p-1}{2} \cdot \left(-\frac{p-1}{2}\right) \cdot \ldots \cdot (-1)$$

$$= (-1)^{(p-1)/2} \left(\frac{p-1}{2}!\right)^2 \equiv -1 \pmod{p}.$$

We have

$$\left(\frac{p-1}{2}!\right)^2 \equiv -(-1)^{(p-1)/2} \pmod{p}. \qquad \square$$

8. If p is prime, $n = 1 + 2 + \cdots + (p-1)$. Show that $(p-1)! \equiv p-1 \pmod{n}$.

Proof. Note that $n = 1 + \cdots + (p-1) = p \cdot \frac{p-1}{2}$.
 When $p = 2$, it is obvious.
 When p is odd prime, $((p-1)/2, p) = 1$. Then

$$(p-1)! \equiv -1 \equiv p-1 \pmod{p}, (p-1)! \equiv 0 \equiv p-1 \pmod{(p-1)/2}.$$

Therefore $(p-1)! \equiv p-1 \left(\bmod\ p \cdot \frac{p-1}{2}\right).$ $\qquad \square$

9. Let $x_1 = 3$, $y_1 = 4$, $z_1 = 5$,

$$\begin{cases} x_{n+1} = 3x_n + 2z_n + 1, \\ y_{n+1} = 3x_n + 2z_n + 2, \\ z_{n+1} = 4x_n + 3z_n + 2. \end{cases}$$

Show that for any positive integer n, $\{x_n, y_n, z_n\}$ is Pythagoras triple, i.e., $x_n^2 + y_n^2 = z_n^2$.

Proof. We prove by induction on n. When $n = 1$, it is obvious. Suppose it is true for $n \leq k$. Note $y_{n+1} = x_{n+1} + 1$, we have

$$x_{k+1}^2 + y_{k+1}^2 = 2x_{k+1}^2 + 2x_{k+1} + 1 = 2(3x_k + 2z_k + 1)^2$$
$$+ 6x_k + 4z_k + 3$$
$$= 18x_k^2 + 24x_k z_k + 8z_k^2 + 18x_k + 12z_k + 5$$
$$= (4x_k + 3z_k + 2)^2 + x_k^2 + (x_k + 1)^2 - z_k^2 = z_{k+1}^2$$

This completes the proof. $\qquad \square$

10. If p is prime, $0 < a < p$, show that the congruence $ax \equiv b \pmod{p}$ has a solution

$$x \equiv b(-1)^{a-1} \frac{(p-1)\cdots(p-a+1)}{a!} \pmod{p}.$$

Proof. Since $p - i \equiv -i \pmod{p}$, we have

$$\prod_{i=1}^{a-1} (p - i) \equiv (-1)^{a-1} \prod_{i=1}^{a-1} i \pmod{p},$$

or

$$(-1)^{a-1} \prod_{i=1}^{a-1} \frac{p - i}{i} \equiv 1 \pmod{p}.$$

Hence

$$ax \equiv ab(-1)^{a-1} \frac{(p - 1) \cdot \ldots \cdot (p - a + 1)}{a!} \equiv b \pmod{p}. \qquad \square$$

11. Prove that the Diophantine equation $x^3 + y^3 + z^3 = x + y + z = 3$ has exactly four solutions: $(x, y, z) = (1, 1, 1), (4, 4, -5), (4, -5, 4), (-5, 4, 4)$.

Proof. Note that $(x + y + z)^3 - (x^3 + y^3 + z^3) = 3(x + y)(x + z)(y + z)$, we have

$$(x + y)(x + z)(y + z) = (3 - x)(3 - y)(3 - z) = 8,$$

$$(3 - x) + (3 - y) + (3 - z) = 6.$$

If $3 - x, 3 - y, 3 - z$ are all even, then $|3 - x| = |3 - y| = |3 - z| = 2$. Hence

$$3 - x = 3 - y = 3 - z = 2, \qquad x = y = z = 1.$$

If only one of $3 - x$, $3 - y$, $3 - z$ is even, let $3 - x$ be that number, then

$$|3 - x| = 8, |3 - y| = |3 - z| = 1.$$

We have $3 - x = 8, 3 - y = 3 - z = -1$, i.e., $x = -5, y = z = 4$. From the symmetry of x, y, z, we can get the other two solutions. So the equation has four solutions. $\qquad \square$

12. If $n = p_1 p_2 \ldots p_k$, with prime p_i distinct, $k > 2$, and for any i, $p_i - 1 \mid n - 1$, show that n is Carmichael number.

Proof. For any $a > 1$, $(a, n) = 1$, we have $(a, p_i) = 1$, $i = 1, 2, \ldots, k$. By Fermat's

Little Theorem, $a^{p_i-1} \equiv 1 \pmod{p_i}$. Since $p_i - 1 \mid n - 1$, then

$$a^{n-1} = (a^{p_i-1})^{\frac{n-1}{p_i-1}} \equiv 1^{\frac{n-1}{p_i-1}} = 1 \pmod{p_i}.$$

Hence $a^{n-1} \equiv 1 \pmod{n}$, n is Carmichael number. □

Exercises 4

1. If p is odd prime, show that the number of solutions of the equation $x^2 \equiv a \pmod{p}$ is $1 + \left(\frac{a}{p}\right)$; moreover, if $p \nmid a$, then the number of solutions of the equation $ax^2 + bx + c \equiv 0 \pmod{p}$ is $1 + \left(\frac{b^2-4ac}{p}\right)$.

Proof. When $p \mid a$, $\left(\frac{a}{p}\right) = 0$, $x \equiv 0 \pmod{p}$ is the unique solution, the number of the solutions is $1 = 1 + \left(\frac{a}{p}\right)$.

When $p \nmid a$, $\left(\frac{a}{p}\right) = \pm 1$. If $\left(\frac{a}{p}\right) = 1$, then a is a quadratic residue modulo p. Let x_0 be one of the solutions, then $-x_0$ is the other one and $x_0 \not\equiv -x_0 \pmod{p}$. The number of the solutions is $2 = 1 + \left(\frac{a}{p}\right)$.

If $\left(\frac{a}{p}\right) = -1$, then a is a non-quadratic residue modulo p, there is no solution for the quadratic congruence. The number of the solutions is $0 = 1 + \left(\frac{a}{p}\right)$.

We prove the first part.

When $p \nmid 4a$, $ax^2 + bx + c \equiv 0 \pmod{p} \Leftrightarrow (2ax + b)^2 \equiv b^2 - 4ac \pmod{p}$.

The second part follows from the first part. □

2. If $p = 4n + 3$ ($n > 1$) and $q = 2p + 1$ are both prime, show that the Mersenne number M_p is not prime.

Proof. If $q = 8n + 7$, $p = \frac{q-1}{2}$, $M_p = 2^p - 1 = 2^{\frac{q-1}{2}} - 1$. Note that when $q = 8n + 7$, 2 is a quadratic residue modulo q, hence $2^{\frac{q-1}{2}} \equiv 1 \pmod{q}$, which means $q \mid M_p$. □

3. Show that there are infinitely many primes of the form $4n + 1$.

Proof. Suppose that there are finitely many primes of the form $4n + 1$, denote them by p_1, \ldots, p_k.

Consider $q = (2p_1 \ldots p_k)^2 + 1$, obviously $p_i \nmid q$. If p is a prime factor of q, then $(2p_1 \ldots p_k)^2 \equiv -1 \ (\mathrm{mod}\ p)$, which means $\left(\frac{-1}{p}\right) = 1$. Therefore p must be $4n + 1$ and $p \neq p_i$, a contradiction! $\qquad\square$

4. Show that there are infinitely many primes of the form $8n + 3$.

Proof. Suppose that there are finitely many primes of the form $8n + 3$, denote them by p_1, \ldots, p_k.

Consider $q = (2^2 p_1 \ldots p_k + 1)^2 + 2$, obviously $p_i \nmid q$ and q is of the form $8n + 3$.

If p is a prime factor of q, then $(2^2 p_1 \ldots p_k + 1)^2 \equiv -2 \ (\mathrm{mod}\ p)$, which means $\left(\frac{-2}{p}\right) = 1$. Hence p must be of the form $8n + 1$ or $8n + 3$, and $p \neq p_i$. Note that a product of two primes of the form $8n + 1$ is still of the form $8n + 1$. Therefore q must contain a prime factor p_0, which is of the form $8n + 3$ and $p_0 \neq p_i$. A contradiction! $\qquad\square$

5. 563 is prime, determine that whether the congruence $x^2 \equiv 429 \ (\mathrm{mod}\ 563)$ has solutions?

Solution. Since 563 is prime, by using the properties of Legendre symbol,

$$\left(\frac{429}{563}\right) = \left(\frac{3}{563}\right)\left(\frac{11}{563}\right)\left(\frac{13}{563}\right) = \left(\frac{563}{3}\right)\left(\frac{563}{11}\right)\left(\frac{563}{13}\right)$$

$$= \left(\frac{2}{3}\right)\left(\frac{2}{11}\right)\left(\frac{4}{13}\right) = (-1)^{\frac{3^2-1}{8}}(-1)^{\frac{11^2-1}{8}} = 1.$$

So the congruence has solutions.

6. By using the properties of Jacobi symbol, determine that whether the congruence in Exercise 5 has solutions?

Solution. Since 563 is prime and 429 is odd, using the properties of Jacobi symbol,

$$\left(\frac{429}{563}\right) = \left(\frac{563}{429}\right) = \left(\frac{134}{429}\right) = \left(\frac{2}{429}\right)\left(\frac{67}{429}\right) = -\left(\frac{67}{429}\right)$$

$$= -\left(\frac{429}{67}\right) = -\left(\frac{3}{67}\right) = \left(\frac{67}{3}\right) = \left(\frac{1}{3}\right) = 1.$$

So the congruence has solutions.

7. If $(x, y) = 1$, find the general form of odd prime factor of $x^2 - 3y^2$.

Solution. Consider $x^2 - 3y^2 \equiv 0 \pmod{p}$, or $x^2 \equiv 3y^2 \pmod{p}$.

When $p \mid 3y^2$, $x^2 \equiv 0 \pmod{p}$. Then $x \equiv 0 \pmod{p}$, $p \mid x$. Since $(x, y) = 1$, hence $p \nmid y$, $p \mid 3$, i.e., $p = 3$.

When $p \nmid 3y^2$, $x^2 \equiv 3y^2 \pmod{p}$. Then $\left(\frac{3y^2}{p}\right) = \left(\frac{3}{p}\right)\left(\frac{y^2}{p}\right) = \left(\frac{3}{p}\right) = 1$.

We divide it into two cases:

(1) If $p = 3k + 1, k \in N$, then $1 = \left(\frac{3}{p}\right) = \left(\frac{3}{3k+1}\right) = (-1)^{\frac{3k}{2}}\left(\frac{1}{3}\right) = (-1)^{\frac{3k}{2}}$, hence $4 \mid p - 1$, p is of the form $12k + 1$.

(2) If $p = 3k + 2, k \in N$, then $1 = \left(\frac{3}{p}\right) = \left(\frac{3}{3k+2}\right) = (-1)^{\frac{3k+1}{2}}\left(\frac{2}{3}\right) = (-1)^{\frac{3k-1}{2}}$, hence $4 \mid p - 3$, p is of the form $12k - 1$. $=$

8. If p, q are two distinct odd primes and $p = q + 4a$, show that $\left(\frac{a}{p}\right) = \left(\frac{a}{q}\right)$.

Proof. Note that $\left(\frac{p}{q}\right) = \left(\frac{q+4a}{q}\right) = \left(\frac{4a}{q}\right) = \left(\frac{a}{q}\right)$, hence

$$\left(\frac{q}{p}\right) = \left(\frac{p - 4a}{p}\right) = \left(\frac{-4a}{p}\right) = \left(\frac{-a}{p}\right) = (-1)^{\frac{p-1}{2}}\left(\frac{a}{p}\right).$$

Since $\frac{p-1}{2} \cdot \frac{q-1}{2} \equiv \left(\frac{p-1}{2}\right)^2 \equiv \frac{p-1}{2} \pmod{2}$, whether $p \equiv q \equiv 1$ or $3 \pmod{4}$, we have

$$\left(\frac{a}{q}\right) = \left(\frac{p}{q}\right) = (-1)^{\frac{p-1}{2} \cdot \frac{q-1}{2}}\left(\frac{q}{p}\right) = (-1)^{\left(\frac{p-1}{2}\right)^2}\left(\frac{q}{p}\right)$$

$$= (-1)^{\frac{p-1}{2}}\left(\frac{q}{p}\right) = \left(\frac{a}{p}\right).$$

\square

9. Find the solutions of $x^2 \equiv 59 \pmod{125}$.

Solution. Note the solutions of $x^2 \equiv 59 \pmod{5^2}$ are $x \equiv \pm 3 \pmod{5^2}$.

Let $x = 25k + 3$. Substituting it into the original congruence,

$$(25k + 3)^2 \equiv 59 \pmod{5^3},$$

we have $6k \equiv 2 \pmod 5$, $k \equiv 2 \pmod 5$. Let $k = 5t + 2$,

$$x = 25(5t + 2) + 3 \equiv 53 \pmod{125},$$

which is one of the two solutions, the other is $x \equiv -53 \pmod{125}$.

10. If p is odd prime, show that

$$|NR| = |NN| = \frac{p - 2 + \left(\frac{-1}{p}\right)}{4}.$$

Proof. If $a \neq p - 1$, then

$$\left\{\left(\frac{a}{p}\right) - 1\right\}\left\{\left(\frac{a+1}{p}\right) - 1\right\} = \begin{cases} 4 & a \in NN, \\ 0 & a \notin NN. \end{cases}$$

Hence

$$|NN| = \frac{1}{4} \sum_{a=1}^{p-2} \left\{\left(\frac{a}{p}\right) - 1\right\}\left\{\left(\frac{a+1}{p}\right) - 1\right\}$$

$$= \frac{1}{4}\left\{\sum_{a=1}^{p-2}\left(\frac{a}{p}\right)\left(\frac{a+1}{p}\right) - \sum_{a=1}^{p-2}\left(\frac{a}{p}\right)\right.$$

$$\left. - \sum_{a=1}^{p-2}\left(\frac{a+1}{p}\right) + p - 2\right\}$$

$$= \frac{1}{4}\left\{\sum_{a=1}^{p-2}\left(\frac{a}{p}\right)^2\left(\frac{a^{-1}+1}{p}\right) + \left(\frac{-1}{p}\right) + p - 1\right\}$$

$$= \frac{p + \left(\frac{-1}{p}\right) - 2}{4}.$$

Note that

$$\left\{1 - \left(\frac{a}{p}\right)\right\}\left\{1 + \left(\frac{a+1}{p}\right)\right\} = \begin{cases} 4 & a \in NR, \\ 0 & a \notin NR, \end{cases}$$

similarly we can get the result for $|NR|$. $\qquad\square$

11. If p is odd prime, show that the number of quadratic residues modulo p is $\frac{p-1}{2}$; if $l \geq 3$, how many is the number of quadratic residues modulo 2^l?

Proof. Let $S = \{x^2 \mid 1 \leq x \leq (p-1)/2, x \in N\}$. For any $1 \leq i \neq j \leq (p-1)/2$, $i^2 \not\equiv j^2 \pmod{p}$. Otherwise $i \equiv \pm j \pmod{p}$, both are impossible. Meanwhile, $i^2 \equiv (p-i)^2 \pmod{p}$, hence the elements in S are exactly those which are quadratic residues modulo p, there are $\frac{p-1}{2}$ all together.

Let $S' = \{(2i-1)^2 \mid 1 \leq i \leq 2^{l-3}, i \in N\}$. For any $1 \leq i \neq j \leq 2^{l-3}$, $(2i-1)^2 \not\equiv (2j-1)^2 \pmod{2^l}$. Otherwise, $i \equiv j \pmod{2^{l-2}}$ or $i+j \equiv 1 \pmod{2^{l-2}}$, both are impossible. Meanwhile,

$$x^2 \equiv (2^{l-1}+x)^2 \equiv (2^l - x)^2 \equiv (2^{l-1} - x)^2 \pmod{2^l}.$$

Hence the elements in S' are exactly those which are quadratic residues modulo 2^l, they are 2^{l-3} all together. □

12. If p is odd prime, show that the congruence

$$1 + x^2 + y^2 \equiv 0 \pmod{p}$$

has a solution which satisfies $0 \leq x, y < \frac{p}{2}$.

Proof. Let $S_1 = \{x^2 \mid 0 \leq x \leq (p-1)/2, x \in Z\}$, then $|S_1| = (p+1)/2$ and for any integer $0 \leq i \neq j \leq (p-1)/2$, $i^2 \not\equiv j^2 \pmod{p}$. Otherwise $i \equiv \pm j \pmod{p}$, it's impossible.

Let $S_2 = \{-1 - y^2 \mid 0 \leq y \leq (p-1)/2, x \in Z\}$, similarly we can get $|S_2| = (p+1)/2$, and for any $0 \leq i \neq j \leq (p-1)/2$, $-1 - i^2 \not\equiv -1 - j^2 \pmod{p}$.

Since there are p elements in each complete congruent system modulo p, according to the pigeonhole principle, there must exist $x_0^2 \in S_1$, $-1 - y_0^2 \in S_2$, so that $x_0 \equiv -1 - y_0^2 \pmod{p}$ and $0 \leq x_0, y_0 \leq (p-1)/2$. □

Exercises 5

1. If $p, q = 2p+1$ are primes, $p \equiv 1 \pmod{4}$, show that 2 is primitive root of q; if $3 < p \equiv 3 \pmod{4}$, show that there are three consecutive integers which are all primitive roots of q, please give an example.

Proof. Note that $q - 1 = 2p$, which has prime factors $2, p = \frac{q-1}{2}$.

When $p \equiv 1 \pmod 4$, $q \equiv 3 \pmod 8$, then $\left(\frac{2}{q}\right) = -1$. By Euler's criterion, we have $2^{\frac{q-1}{2}} \equiv -1 \pmod q$. Hence $2^{q-1} \equiv 1 \pmod q$. Note that $2^2 = 4 \not\equiv 1 \pmod q$, so 2 is a primitive root of q.

When $p \equiv 3 \pmod 4$, $q \equiv 7 \pmod 8$. Then $\left(\frac{-2}{q}\right) = \left(\frac{-4}{q}\right) = -1$, which means $(-2)^{\frac{q-1}{2}} \equiv (-4)^{\frac{q-1}{2}} \equiv -1 \pmod q$. Hence $(-2)^{q-1} \equiv (-4)^{q-1} \equiv 1 \pmod q$.

Let $p = 4n+3, q = 8n+7$, only if $n \equiv 2 \pmod 3$, p, q can be both primes. In this case, $\left(\frac{-3}{q}\right) = (-1)^{\frac{q-1}{2}}(-1)^{\frac{q-1}{2} \cdot \frac{3-1}{2}}\left(\frac{q}{3}\right) = \left(\frac{q}{3}\right) = \left(\frac{2}{3}\right) = -1$, i.e., $(-3)^{\frac{q-1}{2}} \equiv -1 \pmod q$. Hence $(-3)^{q-1} \equiv 1 \pmod q$. From $p > 3, q \geq 23$, we have

$$(-2)^2 = 4 \not\equiv 1 \pmod q, (-3)^2 = 9 \not\equiv 1 \pmod q,$$

$$(-4)^2 = 16 \not\equiv 1 \pmod q.$$

Therefore $-2, -3, -4$ are primitive roots of q.

For example, when $p = 11, q = 23$, $19(-4), 20(-3), 21(-2)$ are three consecutive primitive roots of 23. $\qquad\square$

2. Show that a primitive root of m must be non-quadratic residue. It is not true in the converse. If p is a prime, show that only if p is Fermat prime then the primitive root is the same as non-quadratic residue.

Proof. It is known that there exist primitive roots if and only if the module $m = 2, 4, p^\alpha, 2p^\alpha$. Here p is odd prime.

- When $m = 4$, 3 is the unique primitive root of 4, and 3 is a non-quadratic residue of 4.
- When $m = p^\alpha, 2p^\alpha$, if g is a quadratic residue of m, then $\left(\frac{g}{p}\right) = 1$, i.e., $g^{\frac{p-1}{2}} \equiv 1 \pmod p$. Further, we have $g^{\frac{\phi(m)}{2}} \equiv 1 \pmod m$, hence g is not a primitive root of m.

Now we need only to show that when a non-quadratic residue of p is a primitive root of p, p must be Fermat prime.

It is known that a primitive root of p must be a non-quadratic residue of p, so we need only to show that the number of primitive roots of p is equal to the number of non-quadratic residues of p, i.e., $\phi(\phi(p)) = \frac{p-1}{2}$, or $2\phi(p-1) = p-1$.

From Example 2.8, we know that $\phi(n) \mid n$ if and only if $n = 1$ or $n = 2^k 3^t (k \geq 1, t \geq 0)$. It's easy to check that, only if $n = 2^k, k \geq 1$, we have $2\phi(n) = n$. Hence $p = 2^k + 1, k \geq 1$. We will show that k must be a power of 2 as follows.

When $k = 1, 2$, it is obvious; when $k > 2$, suppose k has odd prime factor, then the factorization shows that p is a composite number. Hence k must be a power of 2, i.e., p is Fermat prime. □

3. Show that each positive rational number can be a sum of distinct unit fractions, i.e., sum of finite distinct terms from

$$1 + \frac{1}{2} + \cdots + \frac{1}{n} + \cdots .$$

Proof. Let $\frac{a}{b} \in Q^+ (a, b \in N)$, since the harmonic series is divergent, there exist a unique positive integer n_0 such that

$$\sum_{i=0}^{n_0} \frac{1}{i} < \frac{a}{b} < \sum_{i=0}^{n_0+1} \frac{1}{i} \left(\text{define } \frac{1}{0} = 0 \right).$$

Let $\frac{c}{d} = \frac{a}{b} - \sum_{i=0}^{n_0} \frac{1}{i} < \frac{1}{n_0+1}$ (c, d coprime). Then there exists a unique positive integer $n_1 > n_0$ such that

$$\frac{1}{n_1 + 1} < \frac{c}{d} < \frac{1}{n_1} \leq \frac{1}{n_0 + 1}.$$

Let $\frac{e}{f} = \frac{c}{d} - \frac{1}{n_1+1} = \frac{c(n_1+1)-d}{d(n_1+1)} > 0$ (e, f coprime). Since $\frac{e}{f} < \frac{1}{n_1(n_1+1)}$, then there exists a unique positive integer $n_2 > n_1$, $\frac{1}{n_2+1} \leq \frac{e}{f} < \frac{1}{n_2}$. Continue going on, note that $e \leq c(n_1 + 1) - d < c$, in the above process the numerator must turn to be 1 in finite steps. This completes the proof. □

4. Find a primitive root of 5, which is not a primitive root of 25.

Proof. It's easy to check that all the primitive roots of 5 are $i + 5k$ ($i = 2, 3$, $k \in Z$), and all the primitive roots of 25 are $j + 25k$ ($j = 2, 3, 8, 12, 13, 17, 22, 23$, $k \in Z$).

Therefore, $\pm 7 + 25k$ ($k \in Z$) are primitive roots of 5, but not the primitive roots of 25. □

5. If p is prime, show that for any positive integer k,

$$1^k + 2^k + \cdots + (p-1)^k \equiv \begin{cases} 0 \pmod{p} & \text{if } p - 1 \nmid k, \\ -1 \pmod{p} & \text{if } p - 1 \mid k. \end{cases}$$

Proof. If $p - 1 \mid k$, then $i^k \equiv 1 \pmod{p}, 1 \leq i \leq p - 1$. Hence,

$$1^k + 2^k + \cdots + (p-1)^k \equiv p - 1 \equiv -1 \pmod{p}.$$

If $p - 1 \nmid k$, let g be a primitive root of p, then g, g^2, \ldots, g^{p-1} is a permutation of $1, 2, \ldots, p - 1$ modulo p. Hence

$$1^k + 2^k + \cdots + (p-1)^k \equiv g^k(1 + g^k + \cdots + (g^k)^{p-2}) \pmod{p}.$$

Note that $g^k \not\equiv 1 \pmod{p}$, we have

$$1^k + 2^k + \cdots + (p-1)^k \equiv g^k(g^k - 1)^{-1}((g^k)^{p-1} - 1) \equiv 0 \pmod{p}.$$
□

6. If p is odd prime, prove that

(1) $1^p + 2^p + \cdots + (p-1)^p \equiv 0 \pmod{p}$.
(2) If m is positive integer and $2^m \not\equiv 1 \pmod{p}$, then $1^m + 2^m + \cdots + (p-1)^m \equiv 0 \pmod{p}$.

Proof. (1) If follows from Fermat's Little Theorem and the sum formula of continuous integers.

(2) **Proof 1.** Note that $p - 1 \nmid p$, $p - 1 \nmid m$ (otherwise $2^m \equiv 1 \pmod{p}$, a contradiction!). It follows from Exercise 5.

Proof 2. Since $2, 4, \ldots, 2(p-1)$ is a permutation of $1, 2, \ldots, p-1$ modulo p, hence

$$2^m(1^m + 2^m + \cdots + (p-1)^m) \equiv 1^m + 2^m + \cdots + (p-1)^m \pmod{p},$$

which means

$$(2^m - 1)(1^m + 2^m + \cdots + (p-1)^m) \equiv 0 \ (\text{mod } p).$$

From $2^m \not\equiv 1 \ (\text{mod } p)$, we completes the proof. □

7. Choosing 5 as primitive root, construct a table of indices modulo 23.

Solution. The Indices Table:

	0	1	2	3	4	5	6	7	8	9
0		0	2	16	4	1	18	19	6	10
1	3	9	20	14	21	17	8	7	12	15
2	5	13	11							

The Inverse Indices Table:

	0	1	2	3	4	5	6	7	8	9
0	1	5	2	10	4	20	8	17	16	11
1	9	22	18	21	13	19	3	15	6	7
2	12	14								

8. Find the solution of $x^{15} \equiv 14 \ (\text{mod } 41)$ by tables of indices.

Solution. Taking indices for the congruence, we have

$$15\,\text{ind}_6 x \equiv 25 \ (\text{mod } 40).$$

Since $n = 15$, $\phi(41) = 40$, $d = (15, 40) = 5$, $\text{ind}_6(14) = 25$, $5 \mid 25$, hence the congruence has five solutions, they are the solutions of $\text{ind}_6 x \equiv 7 \ (\text{mod } 8)$. By the table of indices modulo 41, we find $\text{ind}_6 x \equiv 7, 15, 23, 31, 39 \ (\text{mod } 40)$. Therefore, by the inverse indices table modulo 41, the solutions of the original congruence are $x \equiv 29, 3, 30, 13, 7 \ (\text{mod } 41)$.

9. Construct a table of indices modulo 2^5.

Solution. The Indices Table:

a	1	3	5	7	9	11	13	15	17	19	21	23	25	27	29	31
$r_{-1}(a)$	0	1	0	1	0	1	0	1	0	1	0	1	0	1	0	1
$r_0(a)$	0	3	1	2	6	5	7	4	4	7	5	6	2	1	3	0

10. If p is prime, show that $x^8 \equiv 16 \pmod{p}$ has solutions.

Proof. When $p = 2$, it is obvious.

When p is odd prime, note that $16^{\frac{p-1}{(8,p-1)}} = 2^{\frac{p-1}{2} \cdot \frac{8}{(8,p-1)}}$.

If $8 \mid p - 1$, then 2 is a quadratic residue of p, hence $16^{\frac{p-1}{(8,p-1)}} = 2^{\frac{p-1}{2}} \equiv 1 \pmod{p}$.

If $8 \nmid p - 1$, then $(8, p - 1) = 2, 4$. Let $t = \frac{4}{(8,p-1)}$, t is an integer. Hence,

$$2^{\frac{4(p-1)}{(8,p-1)}} = 2^{(p-1)t} \equiv 1 \pmod{p}.$$

This completes the proof. $\qquad\square$

11. Determine the number of solutions of $x^4 \equiv 61 \pmod{117}$.

Solution. Since $n = 4, \phi(117) = 72, d = (4, 72) = 4, \frac{\phi(117)}{d} = 18$. Note that $61^{18} \equiv (-2)^{18} \equiv 2^{18} \equiv (-1)^6 \equiv 1 \pmod 9$, $61^{18} \equiv (-4)^{18} \equiv 4^{18} \equiv (-1)^6 \equiv 1 \pmod{13}$.

Hence $61^{18} \equiv 1 \pmod{117}$. By Consequence 5.2, the congruence has solution.

It follows from Theorem 5.10, the number of solutions is $d = 4$.

12. Evaluate the character functions $\chi(a)$ modulo 3 and modulo 5.

Solution. (1) If $m = 3, c = \phi(m) = 2, \rho = \pm 1, g = 2, \text{ind}_2 1 = 0, \text{ind}_2 2 = 1$. Hence the two characteristic functions of 3 are

$$\chi_3^1(a) = \begin{cases} 1, 3 \nmid a, \\ 0, 3 \mid a, \end{cases}, \quad \chi_3^2(a) = \begin{cases} 1, & a \equiv 1 \pmod 3, \\ -1, & a \equiv 2 \pmod 3, \\ 0, & a \equiv 0 \pmod 3. \end{cases}$$

(2) If $m = 5, c = \phi(m) = 4, \rho = \pm 1, \pm i, g = 2, \operatorname{ind}_2 1 = 0, \operatorname{ind}_2 2 = 1, \operatorname{ind}_2 3 = 3, \operatorname{ind}_2 4 = 2$.

Hence the four characteristic functions of 5 are

$$\chi_5^1(a) = \begin{cases} 1, 5 \nmid a, \\ 0, 5 \mid a, \end{cases} \qquad \chi_5^2(a) = \begin{cases} 1, & a = 1 \pmod 5, \\ -1, & a \equiv 2 \pmod 5, \\ -1, & a \equiv 3 \pmod 5, \\ 1, & a \equiv 4 \pmod 5, \\ 0, & a \equiv 0 \pmod 5. \end{cases}$$

$$\chi_5^3(a) = \begin{cases} 1, & a \equiv 1 \pmod 5, \\ i, & a \equiv 2 \pmod 5, \\ -i, & a \equiv 3 \pmod 5, \\ -1, & a \equiv 4 \pmod 5, \\ 0, & a \equiv 0 \pmod 5, \end{cases}$$

$$\chi_5^4(a) = \begin{cases} 1, & a \equiv 1 \pmod 5, \\ -i, & a \equiv 2 \pmod 5, \\ i, & a \equiv 3 \pmod 5, \\ -1, & a \equiv 4 \pmod 5, \\ 0, & a \equiv 0 \pmod 5. \end{cases}$$

Appendix

The List of Prime Numbers Less Than 10000

2 3 5 7 11 13 17 19 23 29 31 37 41 43 47 53 59 61 67 71 73 79 83 89
97 101 103 107 109 113 127 131 137 139 149 151 157 163 167 173 179
181 191 193 197 199 211 223 227 229 233 239 241 251 257 263 269
271 277 281 283 293 307 311 313 317 331 337 347 349 353 359 367
373 379 383 389 397 401 409 419 421 431 433 439 443 449 457 461
463 467 479 487 491 499 503 509 521 523 541 547 557 563 569 571
577 587 593 599 601 607 613 617 619 631 641 643 647 653 659 661
673 677 683 691 701 709 719 727 733 739 743 751 757 761 769 773
787 797 809 811 821 823 827 829 839 853 857 859 863 877 881 883
887 907 911 919 929 937 941 947 953 967 971 977 983 991 997 1009
1013 1019 1021 1031 1033 1039 1049 1051 1061 1063 1069 1087 1091
1093 1097 1103 1109 1117 1123 1129 1151 1153 1163 1171 1181 1187
1193 1201 1213 1217 1223 1229 1231 1237 1249 1259 1277 1279 1283
1289 1291 1297 1301 1303 1307 1319 1321 1327 1361 1367 1373 1381
1399 1409 1423 1427 1429 1433 1439 1447 1451 1453 1459 1471 1481
1483 1487 1489 1493 1499 1511 1523 1531 1543 1549 1553 1559 1567
1571 1579 1583 1597 1601 1607 1609 1613 1619 1621 1627 1637 1657
1663 1667 1669 1693 1697 1699 1709 1721 1723 1733 1741 1747 1753
1759 1777 1783 1787 1789 1801 1811 1823 1831 1847 1861 1867 1871
1873 1877 1879 1889 1901 1907 1913 1931 1933 1949 1951 1973 1979
1987 1993 1997 1999 2003 2011 2017 2027 2029 2039 2053 2063 2069
2081 2083 2087 2089 2099 2111 2113 2129 2131 2137 2141 2143 2153
2161 2179 2203 2207 2213 2221 2237 2239 2243 2251 2267 2269 2273

2281 2287 2293 2297 2309 2311 2333 2339 2341 2347 2351 2357 2371
2377 2381 2383 2389 2393 2399 2411 2417 2423 2437 2441 2447 2459
2467 2473 2477 2503 2521 2531 2539 2543 2549 2551 2557 2579 2591
2593 2609 2617 2621 2633 2647 2657 2659 2663 2671 2677 2683 2687
2689 2693 2699 2707 2711 2713 2719 2729 2731 2741 2749 2753 2767
2777 2789 2791 2797 2801 2803 2819 2833 2837 2843 2851 2857 2861
2879 2887 2897 2903 2909 2917 2927 2939 2953 2957 2963 2969 2971
2999 3001 3011 3019 3023 3037 3041 3049 3061 3067 3079 3083 3089
3109 3119 3121 3137 3163 3167 3169 3181 3187 3191 3203 3209 3217
3221 3229 3251 3253 3257 3259 3271 3299 3301 3307 3313 3319 3323
3329 3331 3343 3347 3359 3361 3371 3373 3389 3391 3407 3413 3433
3449 3457 3461 3463 3467 3469 3491 3499 3511 3517 3527 3529 3533
3539 3541 3547 3557 3559 3571 3581 3583 3593 3607 3613 3617 3623
3631 3637 3643 3659 3671 3673 3677 3691 3697 3701 3709 3719 3727
3733 3739 3761 3767 3769 3779 3793 3797 3803 3821 3823 3833 3847
3851 3853 3863 3877 3881 3889 3907 3911 3917 3919 3923 3929 3931
3943 3947 3967 3989 4001 4003 4007 4013 4019 4021 4027 4049 4051
4057 4073 4079 4091 4093 4099 4111 4127 4129 4133 4139 4153 4157
4159 4177 4201 4211 4217 4219 4229 4231 4241 4243 4253 4259 4261
4271 4273 4283 4289 4297 4327 4337 4339 4349 4357 4363 4373 4391
4397 4409 4421 4423 4441 4447 4451 4457 4463 4481 4483 4493 4507
4513 4517 4519 4523 4547 4549 4561 4567 4583 4591 4597 4603 4621
4637 4639 4643 4649 4651 4657 4663 4673 4679 4691 4703 4721 4723
4729 4733 4751 4759 4783 4787 4789 4793 4799 4801 4813 4817 4831
4861 4871 4877 4889 4903 4909 4919 4931 4933 4937 4943 4951 4957
4967 4969 4973 4987 4993 4999 5003 5009 5011 5021 5023 5039 5051
5059 5077 5081 5087 5099 5101 5107 5113 5119 5147 5153 5167 5171
5179 5189 5197 5209 5227 5231 5233 5237 5261 5273 5279 5281 5297
5303 5309 5323 5333 5347 5351 5381 5387 5393 5399 5407 5413 5417
5419 5431 5437 5441 5443 5449 5471 5477 5479 5483 5501 5503 5507
5519 5521 5527 5531 5557 5563 5569 5573 5581 5591 5623 5639 5641
5647 5651 5653 5657 5659 5669 5683 5689 5693 5701 5711 5717 5737
5741 5743 5749 5779 5783 5791 5801 5807 5813 5821 5827 5839 5843
5849 5851 5857 5861 5867 5869 5879 5881 5897 5903 5923 5927 5939
5953 5981 5987 6007 6011 6029 6037 6043 6047 6053 6067 6073 6079
6089 6091 6101 6113 6121 6131 6133 6143 6151 6163 6173 6197 6199

6203 6211 6217 6221 6229 6247 6257 6263 6269 6271 6277 6287 6299
6301 6311 6317 6323 6329 6337 6343 6353 6359 6361 6367 6373 6379
6389 6397 6421 6427 6449 6451 6469 6473 6481 6491 6521 6529 6547
6551 6553 6563 6569 6571 6577 6581 6599 6607 6619 6637 6653 6659
6661 6673 6679 6689 6691 6701 6703 6709 6719 6733 6737 6761 6763
6779 6781 6791 6793 6803 6823 6827 6829 6833 6841 6857 6863 6869
6871 6883 6899 6907 6911 6917 6947 6949 6959 6961 6967 6971 6977
6983 6991 6997 7001 7013 7019 7027 7039 7043 7057 7069 7079 7103
7109 7121 7127 7129 7151 7159 7177 7187 7193 7207 7211 7213 7219
7229 7237 7243 7247 7253 7283 7297 7307 7309 7321 7331 7333 7349
7351 7369 7393 7411 7417 7433 7451 7457 7459 7477 7481 7487 7489
7499 7507 7517 7523 7529 7537 7541 7547 7549 7559 7561 7573 7577
7583 7589 7591 7603 7607 7621 7639 7643 7649 7669 7673 7681 7687
7691 7699 7703 7717 7723 7727 7741 7753 7757 7759 7789 7793 7817
7823 7829 7841 7853 7867 7873 7877 7879 7883 7901 7907 7919 7927
7933 7937 7949 7951 7963 7993 8009 8011 8017 8039 8053 8059 8069
8081 8087 8089 8093 8101 8111 8117 8123 8147 8161 8167 8171 8179
8191 8209 8219 8221 8231 8233 8237 8243 8263 8269 8273 8287 8291
8293 8297 8311 8317 8329 8353 8363 8369 8377 8387 8389 8419 8423
8429 8431 8443 8447 8461 8467 8501 8513 8521 8527 8537 8539 8543
8563 8573 8581 8597 8599 8609 8623 8627 8629 8641 8647 8663 8669
8677 8681 8689 8693 8699 8707 8713 8719 8731 8737 8741 8747 8753
8761 8779 8783 8803 8807 8819 8821 8831 8837 8839 8849 8861 8863
8867 8887 8893 8923 8929 8933 8941 8951 8963 8969 8971 8999 9001
9007 9011 9013 9029 9041 9043 9049 9059 9067 9091 9103 9109 9127
9133 9137 9151 9157 9161 9173 9181 9187 9199 9203 9209 9221 9227
9239 9241 9257 9277 9281 9283 9293 9311 9319 9323 9337 9341 9343
9349 9371 9377 9391 9397 9403 9413 9419 9421 9431 9433 9437 9439
9461 9463 9467 9473 9479 9491 9497 9511 9521 9533 9539 9547 9551
9587 9601 9613 9619 9623 9629 9631 9643 9649 9661 9677 9679 9689
9697 9719 9721 9733 9739 9743 9749 9767 9769 9781 9787 9791 9803
9811 9817 9829 9833 9839 9851 9857 9859 9871 9883 9887 9901 9907
9923 9929 9931 9941 9949 9967 9973

Bibliography

1. M. Aigner and G. M. Ziegler, *Proofs from The Book*, 4th edition, Springer-Verlag, 2004.
2. T. M. Apostol, *Introduction to Analytic Number Theory*, Springer-Verlag, 1984.
3. A. Baker, *A Concise Introduction to the Theory of Numbers*, Cambridge University Press, 1986.
4. H. Davenport, *The Higher Arithmetic*, 5th edition, Cambridge University Press, 1982.
5. F. Diamond and J. Shurman, *A First Course in Modular Forms*, Springer-Verlag, 2004.
6. L. E. Dickson, *History of the Theory of Numbers I-III*, AMS Chelsea Publishing, 1966.
7. M. Erickson and A. Vazzana, *Introduction to Number Theory*, Chapman & Hall, 2008.
8. D. E. Flath, *Introduction to Number Theory*, JohnWiley & Sons, 1989.
9. C. F. Gauss, *Disquisitiones Arithmeticae*, Springer-Verlag, 1986.
10. R. Guy, *Unsolved Problems in Number Theory*, 3rd edition, Springer, 2007.
11. H. Halberstam and H.-E. Richert, *Sieve Methods*, Academic Press, 1976.
12. T. Hall, *Biography of Carl Gauss* (in Chinese), Fanyi Press, 1977.
13. G. H. Hardy and E. M. Wright, *An Introduction to the Theory of Numbers*, 5th edition, Oxford University Press, 1979.
14. E. Hecke, *The Lectures on the Theory of Algebraic Numbers*, Springer-Verlag, 1981, in Chinese by Science Press, 2005.
15. L. G. Hua, *Additive Theory of Prime Numbers*, Amer. Math. Soc., 1966, in Chinese by Science Press, 1979.

16. L. G. Hua, *An Introduction to Number Theory*, Springer-Verlag, 1987, in Chinese by Science Press, 1979.

17. K. Ireland and M. Rosen, *A Classic Introduction to Number Theory*, Springer-Verlag, 1982.

18. K. Kato, N. Kurokawa, and T. Saito, *Suron, I: Fermat no yume to ruitairon, N. Kurokawa, M. Kurihara, and T. Saito, Suron II: Iwasawa riron to hokei keisiki*, in Chinese by Higher Education Press, 2009.

19. W. J. LeVeque, *Elementary Theory of Numbers*, Dover, 1990.

20. M. R. Murty, *Introduction to p-adic Analytic Number Theory*, International Press, 2002.

21. O. Ore, *Number Theory and Its History*, Dover, 1976.

22. C. D. Pan and C. B. Pan, *Goldbach Conjecture*, Science Press, 1992, in Chinese by Science Press, 1981, 2nd edition, 2011.

23. P. Ribenboim, 13 *Lectures on Fermats Last Theorem*, Springer-Verlag, 1979.

24. W. Scharlau and H. Opolka, *From Fermat to Minkowski*, translated by W. Buhler and G. Cornell, Springer-Verlag, 1984.

25. M. R. Schroeder, *Number Theory in Science and Communication*, Springer-Verlag, 1984.

26. I. M. Vinogradov, *Basic Number Theory* (in Chinese), Higher Education Press, 1979.

27. Y. Wang, *Goldbach Conjecture*, World Scientific, 1984.

Author Index

Adleman, 94
Ahlgren, 238
Al-Karaji, 59, 342
Al-Khwarezmi, 58
al-Tūsī, Nasīr al-Dīn, 119
Alhazen, 118, 119
Apéry, Roger, 247
Archimedes, 13–15
Archytas, 27
Aristotle, 3
Artin, Emil, 208, 209
Aryabhata, 23
Atkin, O. L., 238
Augustin, Saint, 8
Avicenna, 119

Babbage, 147
Bachet, 303
Bacon, Roger, 119
Baker, Alan, xxiv
Balasubramanian, 38
Beiler, 175
Bernoulli, Jacob, 244
Bernoulli, Nicolas, 112
Bertrand, Joseph Louis, 44
Bhaskara II, 220
Bill, xv
Birch, 263
Boylan, 238
Brahmagupta, 220, 221

Brudno, 326
Burgess, 210

Cai, Tianxin, 23, 175, 177, 251, 285, 300, 331, 347, 352
Cantor, Georg, 32
Cantor, Moritz, ix
Cao, Hui-Qin, 272
Carlyle, Thomas, 165
Carmichael, 100
Cassini, 336, 354
Catalan, xv, xix, xxiv
Cauchy, Augustin Louis, 182
Chebyshev, Pafnuty, 45
Chen, Deyi, 175, 254, 315, 331, 352
Chen, Jingrun, xviii
Chen, Xigeng, 234
Chen, Yonggao, 22
Chen, Zi, 5, 6
Chern, Shane, 23, 53
Chowla, 143, 144
Chung, Fan, 38
Clausen, x
Coates, 263
Cole, Nelson, 26
Collatz, Lothar, 71
Columbus, Christopher, 44
Confucius, 148
Cook, J. P., 176
Cosgrave, 269

395

Darmon, Henri, 275
Davenport, Harold, 228
de Bessy, Bernard Frénicle, 78
de Fermat, Pierre, 9
De Koninck, 91
de Montaigne, Michel, 46
De Morvre, Abraham, 112
de Polignac, Alphonse, xxiv
Deligne, 292
Descartes, René, 9
Diffie, 94
Dilcher, 269
Diophantine, x, xiii–xvi, xviii, xix,
 xxiii, xxiv, xxviii, xxix
Diophantus, xviii, xxiv
Dirichlet, xxiv, xxix
Du, Bing, 178
Dujella, Andrej, 228

Einstein, Albert, vii
Eisenstein, Ferdinand Gotthold, xxii,
 xxv
Elkies, 325
Eratosthenes, 42–44, 49
Erdős, Paul, 10
Escher, M. C., 17, 18
Ettingshausen, 135
Euclid, 19, 21, 27, 44
Euclidean, 23, 33–35
Euler, J. A., 305
Euler, Leonhard, 9
Evra, Shai, 324

Faltings, Gerd, 259, 276
Fibonacci, ix, xxi, xxii, xxiv, xxviii
Fields, 259, 275, 292
Fourier, 280, 289
Frey, Gerhard, 325
Fridlander, 210
Frobenius, 131
Frye, Jim, 325
Frye, Roger, 325
Fu, Xudan, 272

Galilei, Galileo, 14
Galois, 276
Gao, Weidong, 81

Gauss, Karl Friedrich, ix, x, xii,
 xv–xvii, xxiv, xxv
Genocchi, 296
Germain, Sophie, ix, xxiv
Gersonides, 276
Gessel, I., 337
Ginzburg, 81
Glaisher, J. W. L., 255
Godel, 108
Goldbach, Christian, 18
Goldbach, David, 30
Goldfeld, 344
Goldston, Daniel, 51
Graham, R. L., 38–40
Granville, Andrew, 38
Green, Ben, 50
Gross, 263
Grothendieck, 292
Gupta, 209
Gyory, 300

Hadamard, Jacques, xix, xxviii
Halberstadt, E., 318
Hardy, xxiv
Han, Xin, 107
Hasse, Helmut, 208
Heath-Brown, Roger, 209
Heegner, Kurt, 344
Helfgott, Harald, 52
Hellman, 94
Hensel, Kurt, 4
Hooley, Christopher, 209
Hua, L. K., 303, 305
Hua, Hengfang, 148

Ivić, 91
Iwaniec, 178

Jacobi, Lee W., 326
Jacobstahl, 278, 285, 286
Ji, Chungang, 293
Jia, Chaohua, 91
Jia, Lirui, 294
Jia, Xian, 14
Jiao, Rongzheng, 234
Jones, 322, 345
Jordan, Camille, 97

Kakutani, Shizuo, 72
Kandinsky, Wassily, viii
Kant, Immanuel, 205, 206
Kaplan, P., 336
Khayyám, Omar, 119
Kihel, 141
King Charlie X, 50
Kloosterman, 237
Ko, 250
Kolmogorov, Andrey, 11
Korselt, 150
Kraus, A., 318
Kronecker, 63
Kummer, x, xxiv, xxix

Lagrange, x, xii, xxiv
Lambert, Johann Heinrich, 205
Landau, 91
Lander, 325
Laplace, Pierre-Simon, 14
Le, Maohua, 234
Lebesgue, H.-L., 318
Lebesgue, V. A., 318
Legendre, Adrien-Marie, xi, xxiv
Lehmer, 17
Lehmer, D. H., 100
Lehmer, D. N., 264
Lehmer, E., 124, 263, 264
Leibniz, Gottfried Wilhelm, 4
Leudesdorf, 256
Li, Shanlan, 148
Liouville, 89
Littlewood, 98
Ljunggren, 255
Luca, 84
Lucas, Edouard, x, xvi, xxii, xxiv
Lutz, 261, 262

Madden, Daniel J., 326
Mans, 147
Marques, Gabriel Garcia, xix
Masser, David, 273
Mathews, G. B., 278
Matsumoto, Kohji, 334
McDaniel, 150
Mersenne, Marin, 26
Mihailescu, Preda, xxiv

Möbius, xii
Mochizuki, Schinichi, 278
Mordell, ix, xxiv
Morley, xii, xxix

Nagell, 261, 262
Nagura, 45
Nair, M., 37
Newton, Issac, 14
Niven, 22
Noe, 83, 84
Norrie, 324

Odifredi, xx
Oesterlé, Joseph, 273
Ono, Ken, 238

Paganini, Nicolò I., 9
Paley, Raymond, 190
Pan, Chengbiao, 64
Pan, Chengdong, 108
Pan, Hao, 272
Parkin, 325
Pascal, 14
Peano, Giuseppe, 12
Pell, John, xix
Pfaff, 64
Pillai, 304, 305
Pintz, János, 51
Plato, 19
Poincaré, 261, 262, 329
Pollock, xxiv
Pound, Azra, vii
Poussin, Nicolas, 46
Proclus, 132
Pythagoras, 6–9, 12, 15

Qin, Hourong, 178, 345
Qin, Jiushao, x, xxiv, xxviii

Rado, 250, 251
Ramanujan, xxiv
Ramsay, 50
Recorde, Robert, 332
Ribet, Kenneth, 259
Ricci, Matteo, 41
Riemann, xviii

Rivest, 94, 97
Roesler, 38
Rong, Fang, 5, 6
Rosen, K. H., 338
Roth, 275
Rotkiewicz, 150
Rouse, 322, 345
Russell, Bertrand, 2

Salie, 210
Sankaranarayanan, 91
Saradha, 300
Sarton, George, 109
Schinzel, xviii
Selberg, Atle, 46
Shamir, 94
Shang, Gao, 5
Shanks, 147
Shapiro, 84
Shen, Lixing, 72
Shen, Zhongyan, 175
Shimura, Goro, 259
Shoup, 210
Siegel, 258
Sierpiński, Waclaw, 204
Silverman, 278
Smith, A., 344
Soundararajan, 38
Stewart, Bonnie M., 204
Straus, Ernst Gabor, 203
Sun, Wu, 104
Sun, Zi, x
Sun, Zhihong, 296
Sun, Zhiwei, 104, 105, 108–110
Suryanarayana, 333
Swinnerton-Dyer, 263
Sylvester, J. J., 45
Szmidt, 270

Takakazu, Seki, 244, 245
Tang, Min, 22
Taniyama, Yutaka, 259
Tao, Terence, 50
Tate, J., 360
Thābit, Sinan Ibn, 9
Theaetetus, 19
Tian, Ye, 344, 362

Tornheim, 247
Tunnell, 345

Urbanowicz, 270

Vinogradov, I. M., 49
von Neumann, John, 4
von Staudt, x

Wade, Sir, 148
Wang, Feifeng, 150
Wang, Liuquan, 40
Wang, Xiaoyun, 97
Waring, xiii, xxii, xxiv, xxix
Weil, Andre, 263
Weil, Simone, 263
Wieferich, Arthur, 32
Wiles, Andrew, xii, xxiv
Williams, K. S., 336
Wilson, 12
Witten, 247
Wolf, 46
Wolstenholme, 252–255, 265, 281, 285, 294, 296
Woolf, Virginia, 254
Wróblewski, Jarosław, 50
Wylie, Alexander, 109

Xia, Binzhou, 294
Xu, Guangqi, 41
Xu, Shengli, 72

Yang, Hui, 14
Yang, Peng, 251, 300
Yang, Quanhui, 135, 136
Yang, Xunqiang, 204
Yao, Qizhi, 97
Yazdi, Muhammad Baqir, 9
Yildirim, Cem, 51
Yixing, 105
Yu, Chengyan, 125
Yu, Jianchun, 150
Yuan, Xinyi, 344

Zagier, 247, 248, 263, 270, 271
Zhang, Hao, 23
Zhang, Mingyao, 64

Author Index

Zhang, Mingzhi, 69
Zhang, Qiujian, 110, 126
Zhang, Shouwu, 344
Zhang, Yitang, xviii
Zhang, Yong, 260, 315, 326, 334, 340, 347
Zhao, Jianqiang, 292, 293, 296

Zhao, Xiaodong, 38
Zhou, 5
Zhou, Xia, 125, 151
Zhu, Shijie, 109
Ziegler, 299
Ziv, 81
Zudilin, Wadim, 247

Printed in the United States
by Baker & Taylor Publisher Services